# THE SECRET
# HISTORY OF
# SHARKS

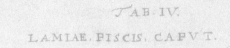

TAB. IV.

LAMIAE. PISCIS. CAPVT.

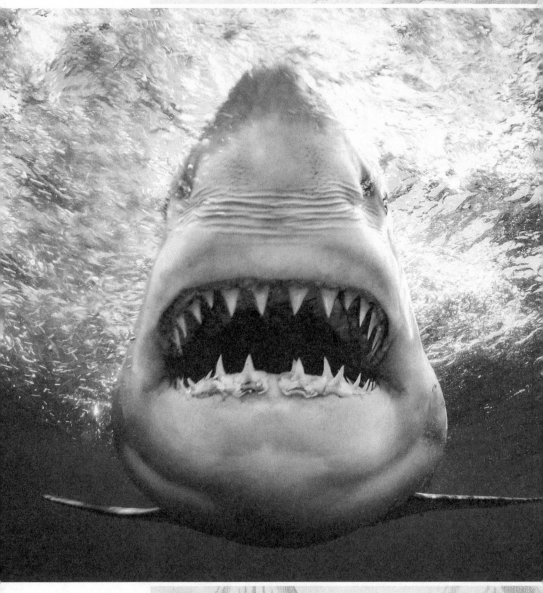

# THE SECRET HISTORY OF SHARKS

## The Rise of the Ocean's Most Fearsome Predators

# JOHN LONG

BALLANTINE BOOKS
*New York*

Published in the United States by Ballantine Books, an imprint of
Random House, a division of Penguin Random House LLC, New York.

BALLANTINE BOOKS & colophon are registered trademarks of
Penguin Random House LLC.

LIBRARY OF CONGRESS CATALOGING-IN-PUBLICATION DATA
Names: Long, John A., 1957- author.
Title: The secret history of sharks: the rise of the ocean's most
fearsome predators / John Long.
Description: First edition. | New York: Ballantine Books, [2024] |
Includes bibliographical references and index. |
Identifiers: LCCN 2023052843 (print) | LCCN 2023052844 (ebook) |
ISBN 9780593598078 (hardcover) | ISBN 9780593598085 (ebook)
Subjects: LCSH: Sharks. | Sharks—Evolution. | Sharks, Fossil.
Classification: LCC QL638.9 .L627 2024 (print) | LCC QL638.9 (ebook) |
DDC 567/.3—dc23/eng/20231221
LC record available at https://lccn.loc.gov/2023052843
LC ebook record available at https://lccn.loc.gov/2023052844

Printed in the United States of America on acid-free paper

randomhousebooks.com

2 4 6 8 9 7 5 3 1

FIRST EDITION

*Book design by Simon M. Sullivan*

*Frontispiece photo by Andrew Fox*

Dedicated to the memory of Josephine and Keith Long,
who encouraged me as a young boy to follow
my fossil-hunting dreams

# CONTENTS

## PART I—THE FIRST SHARKS

## PART 2—SHARKS RULE

# PART 3—SHARKS UNDER PRESSURE

# PART 4—THE AGE OF THE MEGASHARKS

# PART 5—SHARKS TODAY

# PART 1
# THE FIRST SHARKS

*Shenacanthus* and *Xiushanosteus*

## CHAPTER I

# THE HUNT FOR
# THE SECRETS OF SHARKS

*In Search of Shark Fossils*

---

**JANUARY 1, 1992, ANTARCTICA:** Paleontologists will do anything to find spectacular new fossils. I was no exception. I walked alone, pushing my way forward in thigh-deep snow toward the brooding mountain of ice and rock directly ahead. In the light snow of that New Year's Day I was rugged up in my thickest survival feather-down jacket with several layers underneath my outer woollen pants. Heavy double mountaineering boots protected my feet and a thick woollen balaclava covered nearly all my face, save the eyes, which were protected by sunglasses. It was around minus 4°F and uncomfortably windy. The other three members of our expedition had decided to stay in their tents and rest after the previous long, hard day of sledging, pulling our four Nansen sleds with our two Alpine Ski-Doos about forty miles uphill from our last campsite, sleds loaded to the hilt with our gear and precious fossil samples.

As the only one on our team who specialized in the study of fish fossils, I had waited eagerly for this day to come, losing sleep over the prospect of finding some highly significant fossils. I felt a little frightened going out this day alone, but the past two months working in remote mountainous regions of Antarctica, navigating our way through many perils, had given me an uncanny confidence that I could safely manage this short trek to the mountain without any hitches. The other team members decided it was too claggy a day for going outside, opting to catch up on daily chores and letters, quietly resting. I told them I would not be long, just a few hours, and if the snow got any heavier I'd head home immediately. We also had a plan in place that if I didn't make it back by a certain time, they would come searching for me.

I was excited to be finally heading to the famous Portal Mountain, a fossil site high in the remote Transantarctic Mountains, where strange fossil sharks' teeth had been found by a previous geological party some twenty years earlier. I could possibly find a new species of ancient shark that could shed light on what sharks had been doing in this great southern land of Gondwana nearly 400 million years ago.

I just had to get onto the rocky ledges a couple hundred feet away. Each wind gust felt like pinpricks sucking the warmth from my whiskery, ice-covered face. The snow was waist deep and getting harder to plow through. In places it felt light, kind of hollow underfoot. Then it happened—I broke through the ice—and for a fleeting second, I felt my feet dangling in midair as I began to drop. Immediately thrashing my body to one side, I landed facedown, sideways to the ice crack, buried deep in snow. I was panting vigorously and my heart was pounding as I slowly pulled myself away from the death trap, stood up, shook the snow off, and looked back at where I had fallen. There, just behind me, loomed a deep, dark, ominous crevasse about two feet wide. I couldn't see the bottom.

I struggled on toward the Portal, placing each foot carefully in the deep snow, and after half an hour managed to make it safely to the base of the mountain. Once secure on the solid ledge of pale sandstone, I let it all out. I cried and lamented the near loss of my life, thinking of my family at home, especially my three young children. Before long I snapped back to reality, as it dawned on me that I was at the foot of the Holy Grail of fish paleontology. I was finally, after six years of planning and training for this trip, on the remote mountain where big breakthroughs could be made. I needed to pull myself together and continue, so up the mountain I went.

It had been snowing heavily the past week, leaving the mountain slopes covered in a precariously thick buildup of snow. I pushed my way across a platform of rock that ended in a wide, snow-laden ledge, aiming to get to a better rocky outcrop about a hundred feet away. I had not made two steps when I suddenly noticed pieces of ice whizzing past me from above. Looking up, I clocked a wall of snow rapidly descending upon me. The impact knocked me off my feet; I tumbled down the slope, caught up in a small avalanche.

I swore profusely, thinking it was really not my lucky day. I decided not to press my fate any further, instead focusing on one task: making my way to camp as carefully as possible. Step by step, I traced my path back, steering well wide of the dreadful spot where I had broken through the ice. Once back at camp safely, in another emotional outpour, I retold my story to the group. I was administered a couple shots of medicinal whiskey before crawling inside my Scott Polar tent to rest.

**Sledging toward Portal Mountain, Antarctica, December 1991 (the author on the right)** AUTHOR PHOTOS

The next day, Brian, our team safety expert, led us back to Portal Mountain. He walked carefully over my tracks. Using a six-foot crevasse pole, he tested the ground each step of the way, forging a safe access route. He discovered that I had walked over and back across a field of seven crevasses but had broken through only one of them. To this day I feel incredibly lucky to have survived.

That day we found the fossil site I had been looking for and collected some excellent fossil shark remains. For me these fossils made the whole expedition worthwhile. One of them was a giant shark's tooth, almost an inch in height, sporting two distinct prongs jutting out of a rounded, flat root, which I would later name *Portalodus,* after the near-fatal mountain. Over the next few days, I found other new species of unusual fossil sharks' teeth in the nearby Lashly Range, some thirty miles farther north. At the time I had no idea how important these finds would prove to be. Later, back in my lab, they revealed new evidence of a major shift in shark evolution defining the first time, around 390 million years ago, when sharks suddenly grew to much larger sizes and began invading freshwater habitats.

This is one of many stories you'll read about how humans hunt for and find shark fossils. Sometimes it's a dangerous, tiring job involving lengthy exploration of remote sites, hard manual work, and repeated failures where one small clue to the shark puzzle is found after days or weeks of searching. Other fossil sites are easy to access and excavate, with a reliable rate of expected finds, so it's just a matter of diligence and time (a magic number of hours/days/weeks/months/years of labor) required to find a very special fossil in the right layers of rock.

One of my childhood and student mentors, Dr. Tom Rich, a stalwart American who landed his dream job as a museum curator in Australia back in the 1970s, told me, when I was a teenager on one of his digs, that to succeed in paleontology you need "the will to fail." He searched the southern cliffs along the Victorian coastline for more than twenty years, hoping to find Australia's oldest mammal fossils, without success. He never gave up. Then one year his team found a tiny mammal jaw about the size of your little fingernail, the oldest fossil mammal ever found in Australia! Since that day he has found many new species, each helping to dramatically rewrite the story of how Australia's mammals evolved. Dogged, sheer persistence always prevails.

The human side of the shark story is vital to understanding how our theories concerning the evolution of sharks have changed over time and why. Many of the best fossil sharks featured in this book were found by paleontologists on regular fossil-hunting expeditions. Some significant discoveries were made by young paleontologists, new to their field, others by veterans having fifty years of experience under their belts. Some of the most important finds were made by amateur fossil collectors, who then worked closely with scientists so that their finds could be studied and published. Each fossil hunter has their own story to tell about why they dedicate their lives to this odd pursuit. Throughout this book I feature the stories of a number of extraordinary people from the past and present who have studied ancient sharks and made significant scientific discoveries. I've been lucky to work with several of them and call them my friends.

What got me into this crazy fossil shark business? Sharks have fascinated me from a very young age. I grew up in Melbourne, Australia, a city blessed with many good fossil sites within its urban areas. At age seven I collected my first fossils on a dig with my friend and his dad, then quickly developed a passion for collecting all kinds of fossils, particularly sharks' teeth. I would regularly head down to Beaumaris Beach in Port Phillip Bay, Australia, about ten miles from where I lived, to go snorkeling in the shallows and find fossils of ancient marine life that inhabited the seas 6 million years ago. My most prized finds were the teeth of giant extinct sharks, including big megalodon teeth (*Otodus megalodon*). At age thirteen I was one of those overly enthusiastic kids who could identify each shark species by their teeth and tell you what part of the mouth they came from—upper or lower jaw, front or back. I documented my entire fossil collection in two large exercise books, each a hundred pages long, filled with my color drawings and lengthy descriptions of each fossil. It won me the state's top science prize in 1972, which came with the hefty sum of sixty bucks, a fortune back then.

By 1975, when I completed high school, my fossil collection included more than five hundred sharks' teeth collected from many sites around my state. I had classified them into more than twenty different species. Most represented the common sharks alive today, such as great whites, hammerheads, tiger sharks, whaler (or requiem) sharks, and the little bullhead sharks, as well as flat teeth from rays and tooth

**Some fossil sharks' teeth I collected around age nine from the Beaumaris site in Australia**   AUTHOR PHOTO

plates from a chimaerid. My childhood inner voice urged me to study paleontology in college, and in 1984 I earned a doctoral degree specializing in the evolutionary history of fishes.

Why fishes, you ask? At the time, Professor Jim Warren, an American who had moved to Australia in 1962, was chair of the Zoology Department at Monash University. He had just finished excavating a new fossil site full of exquisite, complete fishes of Devonian age, roughly 385 million years old. All the other paleontology students at the time were queuing up to work on Australia's giant fossil mammals and birds, turning their noses up at these spectacular fish fossils. My childhood interest in fossil sharks gave me the confidence to take on this study. After completing my doctorate, I held a series of curatorial and senior management positions at various museums across Australia and in the United States, where I had the good fortune to work on all kinds of ancient fish species, including sharks. For the past decade I've been working as a professor at Flinders University in South Australia, where I continue my research into the deep-time history of fishes of all kinds, including sharks.

This lifelong pursuit has instilled in me a deep respect for the role sharks have played in regulating their ecosystems, and how they have adapted over hundreds of millions of years to Earth's constant engine of change. As tectonic plates drifted slowly—at the speed your hair grows—and formed new continents, the ocean currents and sea levels were in constant flux, and global climates shifted. Supervolcanoes erupting, oceans choking for oxygen, and massive asteroids striking Earth, all caused global mass extinction events. These negative environmental factors proved to be positive drivers of evolution, resulting in the sudden appearance of many new species—called a species "radiation" when they appear rapidly and spread out into new areas. Sharks can do this, as they have been perfecting their intimate way of reproduction for such a long time. They are the poster children for evolutionary success, the prism through which we can see the struggles of all life—maybe even our own.

Today, sharks include some of the weirdest, wildest, and most spectacular creatures on our planet—the ocean's most feared predator, the white shark; the oceans' weirdest-shaped fish, the hammerhead; the magnificent giant filter-feeding whale sharks and manta rays; the voracious garbage eater of the seas, the tiger shark; and the extraordinary bullhead sharks, docile little fishes that are true living fossils, virtually unchanged from when dinosaurs ruled the land 150 million years ago. Each of these species tells us a story not only about how sharks evolved but also about the very nature of survival of life on this planet.

We need sharks today to keep our oceans vibrant and healthy. Sharks play a vital role in regulating our ocean's food chains, transporting or recycling nutrients from one zone of the ocean to another, marshaling the correct balance in nature required for all life to thrive, from microscopic plankton to gigantic blue whales. Without sharks, our marine resources would diminish; our oceans would die.

Sharks are incredible organisms with special abilities that set them apart from all other fishes. These capabilities are the fruits of their long evolution, spanning some 465 million years—they are the superpowers that enabled sharks to adapt and survive a series of devastating/mass extinction events that nearly wiped out life in the seas on several occasions. Whatever nature has thrown at them, sharks, like the Terminator, just kept coming back. Some of these characteristics are well established, while others are new to science.

Their gifts include a superb sense of smell (olfaction) capable of detecting minute amounts of blood or other organic compounds in the water from hundreds of yards away. Unlike other marine and land animals, they sway their heads to home in on the direction of the scent, detecting minute differences at each sway between the left and right sides of their nasal capsules. White sharks raise their snouts out of the water periodically to sniff the land to locate themselves when navigating across the globe.

Sharks also have a special power that detects the faint electrical fields of other living creatures—handy when your prey might be buried under sand, or when used as powerful electrical discharge to stun or kill your prey (electric rays can do this as well, but as we will see, all rays are just a flattened kind of shark). While a few other fishes and even some aquatic land animals also have electrosensory organs, none have developed this power. It has been claimed that sharks can detect electric currents as weak as one-billionth of a volt, and that if two AA batteries were connected under the sea, a shark could sense the charge from a thousand miles away.

We all know about sharks' teeth, the gaping maw of death that we fear in our worst nightmares, yet there are many who admire them, collect them, and study them. They are part of our lore, our history, and in some cases our religion. Sharks have evolved an incredible ability to develop new teeth of all shapes and sizes. Every living species of shark can be identified from its teeth alone, they are that distinctive. Over millions of years, sharks have evolved a myriad of bizarre tooth shapes, from sharp killing and sawing types to bulbous crushing tooth plates and saw-like wheels of death specially designed for killing one kind of prey. Even more outstanding is that they have evolved entirely new kinds of tooth tissues to make these weird teeth super-resilient (which also makes them great fossils).

Each of these powers will be examined in detail as we explore sharks' evolutionary journey unfolding through time. Our story will follow the trail left by millions of years of fossil evidence and will be built upon meticulous, new scientific techniques that enable us to discover detailed new information about their past lives.

## The mystery of sharks

Let's begin our grand tour of the deep-time history of sharks with a bold statement: The origin of sharks is one of the last great unsolved mysteries in the five-hundred-million-year-old evolution of the backboned animals we call "vertebrates" (fishes, amphibians, birds, reptiles, and mammals). We have learned a great deal about the early origins of bony fishes, amphibians, reptiles, and mammals in recent years through stunning new "transitional" fossil finds. While there have not been any significant advances in our knowledge of shark origins, we still know them mostly by fragments for their first 56 million years. The enigma heightens around the age-old question of whether sharks evolved the first jaws and teeth, or some other ancestral archaic fish group before them.

The story of how backboned animals evolved jaws is *our story*, how we humans evolved from simple jawless lamprey-like things through to jawed fishes such as sharks, then ultimately to land animals that can walk upright and chew gum at the same time. Evolution has tattooed its trademark on all creatures. We need to remember this fact when we start exploring the origin of sharks' anatomical features to discover how they carried on through hundreds of millions of years of evolution to end up as parts of our own human bodies.

The evolutionary story of sharks helps us better understand where we humans have come from as a species and to position ourselves within the deep-time context of nature's complex web of life. For example, the way our arms and legs, our jaws, teeth, and even our brains develop is revealed from the study of shark **homeobox** genes—the blueprint genes present in all creatures that direct the sequence of events in our developing embryonic bodies. The study of fossil sharks can reveal when some of these genes first evolved and began crafting the modern shark skeleton, eventually leading to the human skeleton.

The big missing piece of the shark story is where they came from, and how they survived for so long—close to half a billion years—escaping mass extinctions and major environmental crises with very little physical change. The fossil record of sharks is important for us to explore for other reasons. First, it shows the diversity of extinct shark species was far, far greater than our living shark diversity today (by a factor of at least three or four). Second, it shows how sharks have adapted to survive after many challenging situations—catastrophic meteorite impacts, globally destructive gigantic volcanic eruptions, rapidly changing sea levels and temperatures, deadly fluctuations in global oxygen and carbon dioxide levels, and changing continental configurations rerouting vital ocean currents and redirecting nutrient supplies. Coupled with the repeated rise of new superpredators in their oceans, like giant marine reptiles, sharks periodically lost their crown as the ocean's apex predators. Sometimes they had to radically reinvent their body plans in order to reclaim it.

The patterns of adaptation, survival, and extinction that we see in the long evolutionary history of sharks mirror trends in the evolution of all life; they can even teach us about our own evolutionary journey. For example, sharks survived five global mass extinction events, several of which involve harsh conditions similar to what climate change models are predicting will occur on our planet in the not too distant future. I think we can learn a valuable lesson from these consummate survivors in how to cope with rapid changes to our environment. What worked for them might well work for us.

# A life story in the rocks

The French filmmaker Jean-Luc Godard mused that "sometimes reality is too complex. Stories give it form." Before we go diving headfirst into the lost world of ancient sharks, you should first understand the background story—how and what paleontology is, and how paleontologists work. The history of sharks through time can be told only through the voice of paleontology spoken over the music of geology.

Paleontology is the study of ancient life and its environments, as told through remains or traces of past life called fossils. Fossils can be bones, petrified wood, shells, impressions of trackways where creature once walked, eggs, or even fossilized poop (coprolites). Fossils have extra significance when the rocks they are found in can be accurately dated and studied to show what kind of environment the creature lived in. I'm going to give you some tools to help you understand how we do this. This understanding is important, as we need to be confident in knowing the timing of when all the main evolutionary events took place.

Our main storyline here concerns how sharks evolved and diversified over a very long period of time. Geologists refer to periods of time that are measured in tens or hundreds of millions of years as **deep time** (and hereafter throughout this book I use MYA as shorthand for "million years ago"). The geological time scale shows the sequence of ages through time, each with its own distinct life forms, so for convenience we can refer to each package of time by a name, like the Devonian period. These periods are the playgrounds of evolution where big transformations in species occur, where whole new lines of organisms emerge—the playgrounds where species periodically seesaw on the edge of extinction. If they survive, they can keep playing in the next age.

Dating a fossil is a job done by geologists using a range of reliable methods that measure the decay of radioactive elements. Once rocks form from the cooling of molten lava, erupted from volcanoes or deep-sea volcanic vents, the decay of their radioactive minerals starts—the geological hourglass starts running down. Certain elements, like uranium, break down at a known decay rate, so they can be measured to determine the age when the rock first crystallized, usually within ±0.1 percent accuracy. Cross-checking the dates using other minerals with known decay rates helps to verify the dates for certain key layers. We

| TIME (million years ago) | | | LIFE EVENTS | SHARKS & KIN | FOSSIL SITES |
|---|---|---|---|---|---|
| CENOZOIC ERA | Pleistocene / Pliocene | 2.6 | First hominids (human family) | First modern white shark | Pisco Fm, Chile |
| | Miocene | 5.3 | | First *Otodus megalodon* | |
| | Oligocene | 23 | | | |
| | Eocene | 34 | First whales | First requiem and tiger sharks | Kazakhstan sites |
| | Paleocene | 56 | *Extinction* | First *Otodus* | Monte bolca sharks & rays |
| MESOZOIC ERA | Cretaceous | 66 | First primates | *Aquilolamna* | Lebanese sharks |
| | | | First dinosaurs | *Ptychodus* | Western Interior Seaway, USA |
| | | | | *Cretoxyrhina* | Lastame, Italy |
| | | | | *Cardabiodon* | Cardabia, Australia |
| | | | | First saw sharks | Las Hoyas |
| | | 145 | First flowering plants | | |
| | Jurassic | | First birds | *Palaeocarcharias* | Solnhofen sharks |
| | | | First large pliosaurs | *Ostenoselache* | Dorset, UK |
| | | 201 | *Extinction* | First modern shark families | Lyme Regis, UK |
| | Triassic | | First dinosaurs, mammals, pterosaurs, crocodiles | First rays | |
| | | | First ichthyosaurs | Hybodont sharks abundant | |
| | | 252 | *Extinction* | First neoselachian sharks | |
| | | | | *Hybodus* appears | Wapiti Lake |
| PALEOZOIC ERA | Permian | | | *Helicoprion* | Perm Mountains |
| | | | First mammal-like reptiles | | Idaho sites |
| | | 299 | | *Dracopristis* | Mazon Creek sites |
| | Carboniferous | | First reptiles | *Bandringa* | Bear Gulch sites |
| | | | | *Akmonistion* | Bearsden & Wardie |
| | | 359 | *Extinction* | Holocephalan peak diversity | Cleveland Shale/Morocco |
| | Devonian | | First trees forming forests | First holocephalan | Gogo, Australia |
| | | | First amphibians (tetrapods) | First selachian sharks | Antarctica |
| | | | | | Atholville, Canada |
| | | | First insects | First shark body fossil | Leon, Spain |
| | | 419 | First bony fishes | | |
| | Silurian | | First placoderms | *Shenacanthus* | Chongqing, China |
| | | 444 | *Extinction* | First 'stem-chondrichthyans' | Colorado, USA |
| | Ordovician | | First land plants (mosses) | First 'chondrichthyan' scales | Central Australia |
| | Cambrian | 485 | First fishes (jawless forms) | | |

**Geological time scale showing selected major events for chondrichthyans and other life**

can also date fossils by the associated faunas of microfossils in the rocks. Throughout the world, marine rocks contain fossils of microscopic single-celled organisms with hard shells around them, like foraminifera or radiolarians. These have been dated by radiometric methods using layers of volcanic ash or rock above and below the sediment layers containing the microscopic fossils. The use of fossils to date rocks within a narrow time range is called **biostratigraphy**. The presence of certain key species of these tiny fossils can indicate narrow time ranges, usually within one or two million years. Fossil sharks' teeth are often found as microfossils, so they can be useful for this purpose.

Reconstructing the environment in which a fossil shark lived is often the job of specialist geologists who study the sedimentary rocks in which the fossil was found. Studying the rocks surrounding a fossil is our equivalent of a forensic examination of the remains. The clues in the rocks tell us in what environment the creature had lived, and sometimes its cause of death. Sedimentary rocks are formed when sands, silts, or muds are deposited in rivers, lakes, or the sea, then harden from burial in the basin or channel where they were laid down. Chemical agents then cement the sediment grains together, and, combined with pressure from the Earth's tectonic movements, or from the weight of thicker piles of sediments above pushing down, compact the sediment into hard rocks. Sand can form sandstone, mud hardens to make mudstone, and shales are formed when muds and silts with many flat mineral grains in them harden into layered rocks. Limestones are mostly made up of particles of ancient sea creatures, like shells or bits of coral, so we know these were deposited underwater in shallow marine conditions. Deep-sea sediments are mostly made up of fine mudstones formed by minute clay and silt particles that are so tiny they take a long time to settle to the ocean floor.

Throughout the story of life on Earth we see groups of animals and plants that become successful, then either survive to today or go extinct. While most of us might think of dinosaurs as an unsuccessful group, simply because the giant dinosaurs went extinct 66 MYA, we forget that they ruled the planet unchallenged for some 160 million years. And not all dinosaur groups went extinct, as all birds alive today are descendants of the **theropod** ("beast foot") dinosaurs (the group containing big predators like *T. rex*), and thus are living dinosaurs. Similarly, insects have been around on Earth for about 400 million years and have evolved into millions of different species. Yet many insect groups evolved, were successful for a long time, then went extinct. Take, for example, the Odonatoptera—a superorder of giant dragonfly-like insects, some with wingspans up to 28 inches. Although they were abundant for around 50 million years, they eventually went extinct well before the dinosaurs appeared. While sharks have been around for almost half a billion years, we humans (genus *Homo*) have been here just over 2 million years, and our modern human species is only around three hundred thousand or so years old.

When we think of any living group of animals or plants, we only ever see a small part of their evolutionary story from observing their living diversity. It pays to consider their total diversity (living plus extinct species), as only then can we comprehend the full story of what living forms survived. The ones that didn't make it to today show us the evolutionary experiments that failed, yet some still invented and passed on new anatomical features that led to the success of modern species. Sharks are no different. We will see throughout our story that some of the most bizarre extinct forms were crucial steps on the evolutionary stairway leading to modern sharks.

The history of sharks stars an incredible range of extinct organisms, including a great diversity of fascinating, bizarre-looking sharks, many of which have no counterparts today. Imagine a thirty-foot shark sporting one deadly row of enlarged jagged teeth jutting out of its lower jaw, resembling a buzz saw from Hades. Imagine another monster about the same size, capable of scooping up giant clams three feet wide so it could grind them to powder with its mouth full of crenulated pavement teeth. Not to mention the greatest killer the world has ever known, the mighty megalodon (short for *Otodus megalodon*), whose sixty-six-foot body carried massively powerful jaws sporting seven-inch serrated teeth—the perfect weapons to hunt and dismember huge baleen whales. I will delve into why each of these species, along with many other bizarre ancient sharks, was so successful in its time and discuss the underlying reasons for its success—changes in past climates, sea levels, ocean chemistry, food resources, and other biological factors that enabled each of them to thrive for millions of years. First, we need to understand the evidence we have for knowing about ancient sharks: their fossil remains.

## Sharks preserved as fossils

Sharks present a fascinating record of fossil remains, yet they are still relatively poorly known by us paleontologists. Most fossil sharks are known primarily from their isolated teeth, scales, or fin spines. Only very occasionally, when chemical conditions in the water were just right and burial was rapid, some ancient sharks became preserved as whole organisms. Still, we specialists who study fossil sharks know how to see beyond the diaphanous shape of a fossil shark's outline impressed in the rock. Teeth stuck in the bones of their prey can tell us what they hunted, or even how they made their kill. We can study the physical features and mechanical constraints of their scales, fins, and body shapes to reconstruct what these sharks of the past looked like or how they might have lived, or even to calculate how fast they could swim or how hard they could bite.

The main problem with studying the fossil remains of sharks is that they all have cartilage skeletons that decay more quickly than bone once the animal is dead. I would estimate that more than 99 percent of every known fossil of a shark ever studied is either a tooth, a scale, or a fin spine. Probably less than one percent of all fossil sharks are known from actual cartilage skeletons that have somehow miraculously been preserved under exacting conditions. Most of these fossil skeletons would be partially articulated skeletons, meaning they are missing some parts, like the tail, head, or a section of the body. An "articulated" skeleton is one with the bones preserved still joined in their natural life position, like the superb fossil shark shown on the opposite page.

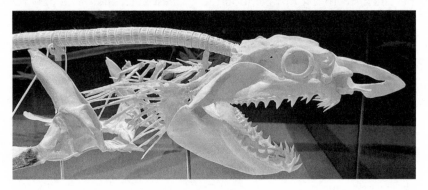

**Mako shark skeleton showing cartilaginous tissues**
AUSTRALIAN MUSEUM, REGISTRATION NUMBER AMS I.47391-001    AUTHOR PHOTO

Complete preservation of a fossil shark occurs only when the dead shark carcass undergoes rapid burial in a quiet marine setting, or a lake basin, depleted in oxygen, preventing scavengers, or even bacteria, from destroying the delicate cartilage. In very rare conditions, where the physical conditions are perfect and the underwater chemistry is just right, we find exceptionally good fossils of complete sharks. Sometimes we find them with soft organs like livers, or even brains, preserved, sometimes with guts containing their last meals intact, and sometimes displaying the pigments of their skin, reflecting the faint ghostly patterns seen on the living creature. These "hero fossils" can tell us a lot about the lives of ancient sharks and inform the bigger picture of what drove their extraordinary evolutionary success.

*Cladoselache*, **a well-preserved Devonian shark fossil from Ohio. Here the cartilages and some muscle tissues are preserved intact** AUTHOR PHOTO

The overarching narrative of shark evolution can be assembled from these superb "snapshots" of complete ancient sharks, and from information gleaned from the studies of a vast number of isolated remains. It's a lot like compiling a biography of a long-dead historical figure using birth and death certificates, some random shire records showing trades or taxes paid, and a couple of private letters. The big gaps in such stories can be filled in with your imagination to reconstruct what might have happened, guided by a thorough knowledge of the dates, places, and contextual events in history. If you know your background history well, it is possible to construct a gripping five-hundred-million-year-old tale.

## The right time to tell this story

Now is the best time ever to study fossil sharks and tell their story. Most of the breakthroughs in understanding shark evolution revealed in this book were announced within the past decade, in this exciting new era of high-tech paleontology. For the past few hundred years, scientists had only simple hand tools to excavate and prepare their fossils, with electric drills and other more sophisticated preparation instruments appearing in the mid-twentieth century. The parts of the fossil they could see were all they could study, so the nineteenth-century images of fossil sharks are often nebulous wood-block prints showing vague shark-like outlines, with parts of the skeletons still obscured by layers of rock. From the mid-twentieth century on, we see the use of chemical preparation techniques to extract fossils out of certain rocks, like limestone, as they dissolve easily and do not damage the bones inside them. In such cases we can use certain types of acid solutions to extract sharks' teeth, spines, scales, and sometimes their cartilage skeletons in 3-D from the rock to study them.

However, without doubt the biggest advances in paleontological methods have been the advent of powerful new high-tech imaging tools like CT scanners (CT stands for "computer tomography," or imagery). These instruments use X-rays to image fossils as multiple slices through the specimen. Synchrotrons are large-particle accelerators that shoot intensified beams of electrons through objects to image the ancient remains of sharks inside the rock in great detail, right down to the cellular level. Then, using the scan data, we can make 3-D models as printouts of our fossil specimens at any size we like.

Recently, I've been fortunate to access the latest high-powered imaging device that shoots beams of neutrons through the fossil to create stunning 3-D images of its anatomical features. While the synchrotron can shoot electrons in beams of light a million times brighter than the sun, the neutron beams can pass through much denser specimens in thicker slabs of rock and give even higher resolution of the fossil, revealing its innermost secrets. Using such instruments has even revealed structures preserved in extinct organisms we never expected to find, like the remains of muscles, nerves, soft organs like hearts and livers, embryos—even the 3-D brain in one ancient fossil shark from Kansas. All these incredible discoveries will be discussed later in the book.

Other powerful instruments and techniques are revealing the biological secrets of ancient sharks. For example, by analyzing the isotopes of certain elements like nitrogen, carbon, and oxygen in their teeth and hard tissue, we can determine the ancient food chains of extinct marine ecosystems, drilling down on what sharks of the past were eating. An even more startling new technique, which I'll explain in more detail later, has allowed scientists in late 2023 to calculate the body tempera-

**The author imaging a fossil at the Dingo neutron beam imaging facility at ANSTO in Sydney, Australia**  PHOTO BY JOSEPH BEVITT

tures of gigantic extinct predatory sharks, which helps us estimate their daily food needs, their capability for transoceanic migration, and how fluctuating sea temperatures might have brought about their decline.

Finally, today we also have advanced computer programs that enable us to analyze the evolutionary relationships of fossil sharks with a high degree of confidence. Using these phylogenetic software programs provides a great deal more certainty in defining the role each fossil shark species played in the bigger story by telling us on which branch they sit in the big evolutionary tree of sharks. All these powerful tools and new techniques have changed the game for paleontologists, as we can now reconstruct many aspects of ancient shark physiology, their behaviors, and their evolutionary story.

In this book I will reveal how and when the major groups of sharks evolved, present the scientific evidence, and discuss the key anatomical innovations that enabled them to diversify into the living species. The mission we are on is also centered on one big question about how sharks survived the great mass extinction events, and what we as humans can learn from their experience. Can they teach us a lesson to help save our own species from extinction? Think of this book as the ancestry.com of sharks, showing how over the course of 465 million years they were shaped and honed by a constantly changing world—a never-before-told tale, the now-revealed *secret history of sharks*.

Are we ready to go back in time to begin our story of the mysterious origins of sharks? Let's now visit the desert heart of Australia, as that is where we have found the oldest evidence of sharks anywhere.

CHAPTER 2

# THE ENIGMATIC OLDEST SHARK FOSSILS

*How Tiny Scales Teach Us Big Things*

## Fossil hunting in the dead heart of Australia

NORTHERN TERRITORY, AUGUST 1993. It was a scorching hot afternoon in a dry desert valley in central Australia, part of the vast Amadeus Basin in the southwest corner of the Northern Territory, a place where glistening venomous snakes and long-necked goanna lizards were sliding and scuttling about between the rocky outcrops. A small rocky mesa lay ahead, named Mount Watt, which was drawn on my geological map as a tiny colored patch representing an outcrop of fine-grained sandy rock that geologists call the Stairway Sandstone. This rock was deposited as sand in a clear, shallow seaway in the middle of the Ordovician period, about 465 MYA. I grabbed a large sledgehammer and my biggest metal chisel, then went to work. With each swing of the hammer, the chisel thrust deeper between the buff-colored rock layers, sending dust and chunks of stone splintering off in all directions.

Peering at the shattered pieces up close through my trusty 10x hand lens enabled me to examine each tiny mark on the rock. I was rewarded by finding the telltale patterns made by ancient bone entombed in the rock—small depressions on the surface, distinct shapes with complex dimpled indentations. Such dimples were the hidden treasures of this harsh land. Deep weathering of the porous rock over millions of years

The desolate outcrop called Mount Watt in central Australia. The oldest known shark scales came from these layers   AUTHOR PHOTO

had long since destroyed any trace of the original bone from these layers. These patterns of dimples were scientifically significant; the geometry of the pattern, its regularity and juxtaposition of larger depressions with smaller ones around it, told me I was looking at an impression of the bony plate belonging to a fish named *Arandaspis*, a name meaning "shield of the Aranda." It honors the First Nations people who have inhabited this land for tens of thousands of years, the Arrente people, also referred to as the Aranda or Arunta people. *Arandaspis* looked like a tube of bony plates about a foot long, with headlamp-style forward-facing eyes and a circular mouth. It had no fins and likely lived on the sandy bottom of the sea eating small critters it sucked out of the sediment.

To get to this remote site, my two mates and I had driven about sixteen hundred miles northeastward from Perth, along the Great Central Road and the infamous Gunbarrel Highway, one of the longest stretches of barren desert road in Australia. There were expanses of angry corrugated gravel road extending through the flat sandy desert for more than six hundred miles, with no towns in between. The two men accompanying me on this venture were young geologists studying other rocks of the region, while my mission was to collect fish fossils. I worked hard at the outcrop that afternoon, chipping away for some hours as dark turbulent clouds began closing in.

My two buddies had just made it back to camp when I returned with my fossil booty. There was no time to chat about the day's work over a cold beer, as we had to quickly take shelter in our tents because a hard rain began pelting down. As I huddled in my tent, the winds grew stronger and more menacing until eventually the demonic gale ripped my tent apart, forcing me to flee for cover into the back of our truck. Watching, horrified, out the window, I saw my now destroyed tent blow away into the night. I spent the rest of that field trip (another two weeks) sleeping in the back of the old truck, stretched out on top of our rather lumpy supplies.

Yet despite my discomfort, my efforts that day were rewarded with some excellent specimens of *Arandaspis* bones. These would later be showcased in a new gallery of evolution at the Western Australian Museum, where I then worked as the curator of vertebrate paleontology. On reflection, though, I still felt a little disappointed that I had not found anything new that day to add to our knowledge of what else might have lived alongside *Arandaspis*. We aim high at the field site, hoping that something truly remarkable will be revealed with each swing of our hammers, addicted to that adrenaline rush as the rock splits to reveal the remains of an ancient life once lived, showcasing some new creature that is totally unknown to science. Typically, such discoveries are few and far between. About twenty years later it would be revealed that the oldest known fossil shark scales were found in the same layers of rock I was excavating.

Before we talk about this first mysterious shark, let's orient ourselves

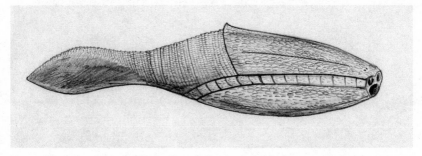

**Reconstruction of an early jawless fish,** *Arandaspis,* **from central Australia**  PENCIL DRAWING BY AUTHOR

in this deep-time space to see what Earth was like in the Ordovician period and what kinds of life inhabited these oceans.

## The Ordovician world

If you had stood 465 MYA in the exact spot where I swung my hammer that day, you would have been about 60 feet or so below the surface of a warm tropical sea that cut right through the guts of the Australian continent. Australia was then encapsulated within the gigantic supercontinent of Gondwana, a landmass comprising South America, Africa, India, Australia, and Antarctica joined together. Laurentia (North America) was around three thousand miles away, sitting close to the ancient continental blocks of Siberia and Baltica. The northern boundaries of Gondwana straddled the equator and nestled against a large continental block of land now known to be a part of northern China; in the south it stretched out over the pole. Global sea temperatures in the Ordovician period were sweltering, estimated at around 100°F at the equator, with high seas, between 380 and 680 feet above today's levels.

The shallow marine seaway that traversed Australia was named the Larapintine Sea, a vast watery home to a great diversity of primeval marine life. Just imagine *Endoceras*, a 12-foot-long squid-like creature

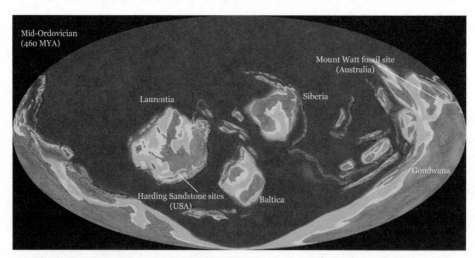

**Ordovician world map, 460 MYA, showing sites mentioned in this book**  RON BLAKEY, DEEP TIME MAPS (DEEPTIMEMAPS.COM)

living inside a gigantic tusk-like shell, an orthoconic **nautiloid**. They hovered like vampires above the seafloor teeming with a myriad of small, multilegged crustacean-like critters called **trilobites**. Here and there a few clams and worms burrowed their way down to safety as the nautiloids plucked fat trilobites off the seafloor with their tentacles, then fed upon them by ripping them apart with their horny beaks. Up to this point in time we had very few creatures we could call fishes in our oceans.

The oldest fishes were jawless forms that looked like finless compressed worms with gill openings and rounded mouths at the end of their bodies. *Metaspriggina* is one such form; it was found at the famous Burgess Shale in British Columbia, Canada, and lived around 508 MYA. Its weak mouth and well-developed gills suggested it either filter-fed on plankton or hunted small swimming worms and other tiny creatures. Some other similar worm-like swimming jawless fishes are known from slightly older layers in China as well.

The early part of the Ordovician had a diverse number of swimming protovertebrates called **conodonts**. These elongated eel-like creatures had jaw-like structures in their mouths, with a cartilage notochord supporting their body instead of a backbone, so no hard skeletal features that would tag them as a vertebrate. They are mostly known from their phosphatic jaw-like mouthparts made of tissues constructed like in vertebrates. We know almost nothing about their lifestyle. Although some were thought to be gentle filter feeders, others might have actively hunted live prey. The biggest conodonts lived at the same time as *Arandaspis*, between 467 and 458 MYA, when forms like *Iowagnathus* might have reached up to 4 feet long. It's likely that some of the conodont animals hunted the first jawless fishes. Jawed vertebrates were not known at this time—that is, until the first sharks appeared on the scene.

The warm inland Larapintine Sea was home to the world's first shark, aptly named *Tantalepis*. I'll tell you more about this creature's tiny fossil scales soon, after a brief introduction to sharks. Then we can dig down into the finer detail of what its scales can tell us.

# Sharks for beginners

Sharks are very highly evolved supercharged jawed fishes, and like their competitors, the bony fishes (the majority of fishes in our oceans and rivers, like catfishes, salmon, or trout), sharks share the same positions of all their fins. Up front they have paired pectoral fins coming out from behind their heads and paired pelvic fins coming out from the rear of their streamlined bodies. These two pairs of movable fins help the shark maneuver in water, for efficient turning or braking.

They also have a powerful stiff tail fin that pushes the shark through the water; think of it as the shark's main thruster jet. Some sharks have a small triangular anal fin behind the cloaca (the opening for excretion of waste and for reproductive purposes) and a prominent triangular dorsal fin, or two in some species. These are all median fins that do not move but project from the midline of the shark's body and help to stabilize the shark when swimming, like shafts on an arrow.

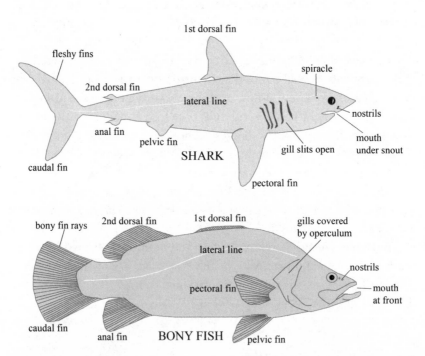

**Main external differences between a shark (chondrichthyan) and a bony fish (osteichthyan)** DRAWING BY AUTHOR

Some species have sharp fin spines in front of their dorsal fins for protection against large predators, as in the bullhead and dogfish sharks. When larger sharks take them whole as prey, the dorsal fin spines pierce the roof of the mouth and the predator will often spit them out immediately. Sharks also have thick, fleshy, stout fins, whereas bony fishes like trout have thin, flexible fins. These thicker fins give sharks greater stability at high speeds, like the pulled-in wings on a fighter jet, but the trade-off is they are less maneuverable and can't pull up as quickly compared to most ray-finned fishes, which can move with greater agility, like a fixed-winged propeller airplane doing tight aerobatics.

Sharks are distinguished from other fishes by several features, including their minute pointed placoid scales, smaller than pinheads, covering their bodies, unlike the larger rounded or rectangular scales, about the size of your thumbnail, seen on the bony fishes. The tiny size of a shark's scales improves its swimming efficiency and helps to protect the shark from minute parasites. Sharks also have exposed gill slits rather than covered gills. Gill covers offer more protection against damage or infection by parasites.

Like all fishes, sharks have a visible line of little pores that run the length of their bodies and all around their heads. These are sensory cells, called **neuromasts**, which form a sensory system (the lateral lines) that enables them to detect subtle changes in water pressure indicating that something is moving nearby. It can tell the shark that a struggling, wounded fish is in the water ahead of it 300 feet away—a very handy tool in dark, murky waters. Singular among fish, they possess a well-developed electrosensory system for detecting the weak electric fields of other living creatures, including microscopic plankton in some cases. If the prey isn't moving and is buried just below the seafloor, so that the shark can't detect it with its sight or smell, then it might use its electroreception cells to locate the prey, then dig down with its snout to grab its meal.

The main features defining a shark that distinguish them from bony fishes are shown in the figure on the previous page.

Sharks have closely related living kin—the rays and **chimaerids** (also known as ratfishes). These three are grouped together as the **chondrichthyans**, and they are all "sharks" in the general sense because they all evolved from an ancestral shark body plan at different times.

Today each of these groups has quite different yet essentially shark-like body features and characteristic teeth.

The big picture of the evolution of sharks includes many offshoots from the main branch leading to what I call true "sharky sharks," like the white sharks or hammerheads, which scientists call the **selachians** (class Selachii). The group containing both sharks and rays is called **elasmobranchs**. This distinction highlights that there were many experimental offshoots from the main branch leading to modern sharks, including many weird extinct groups that all had their days of glory, reigning over the oceans at various times. In the end they went extinct when competition became too fierce, or Earth suffered a widespread extinction event.

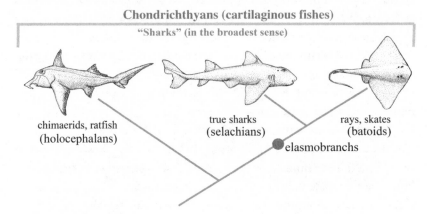

Chondrichthyans (cartilaginous fishes)

"Sharks" (in the broadest sense)

chimaerids, ratfish
(holocephalans)

true sharks
(selachians)

rays, skates
(batoids)

elasmobranchs

**Simplified tree of chondrichthyans showing relationships between sharks, rays, and holocephalans** DRAWING BY AUTHOR

There are at least 1,225 living species of sharks and their kin alive today. This book tells the story of the entire group, the **chondrichthyans**, with the main focus squarely on sharks. When discussing living sharks, I will include their scientific names in a list at the end of the book. Around the world they all have different common names, so the scientific name avoids any confusion as to the species being discussed.

Sharks and chondrichthyans all possess a skeleton made of cartilage. Cartilage is the rubbery tissue lining our joints and supporting our nose and ears. It is also the framework that bone develops around in all vertebrates. Cartilage is the reason why babies have rubbery arms and legs—their limb bones are mostly cartilage right after they are born.

Once the bone has ossified enough around the cartilage core of the leg bones, the baby is able to support its own weight and venture to take its first cute baby steps.

Shark cartilage is not the same cartilage as in our bodies. It is a very specialized type of strong yet flexible cartilage, built with tesserae (little "tiles"). It is found only in sharks, rays, and chimaerids, so it essentially defines them as a unique group of fishes. This is important to remember when we start exploring the origins of sharks way back in time. The tesserae and other markers of shark cartilage help us identify true sharks, distinguishing them from other kinds of archaic fishes. As the shark grows, the tesserae become wider and thicker and the skeleton becomes stronger by adding denser mineralization to the tesserae where it has the highest stresses. Sharks vary in the degree of calcification of their skeletons. Some sharks have very little calcification in their skeleton, so their cartilage is soft and rubbery like your ear, while others, like the powerful white shark, have heavily calcified cartilage, giving them a more rigid skeleton for their strong muscles to attach to.

While most bony fishes reproduce using external fertilization (spawning in water), all chondrichthyans have external sexual organs called "claspers" attached to the pelvic fins of males. They reproduce using internal fertilization like us, depositing sperm inside the female, and both our reproductive organs developed from the same genes found in sharks and us. Inside their bodies, sharks have enormous livers containing large amounts of squalene oil, a special reserve of energy that helps them survive for long periods when females are pregnant, or when large sharks are crossing vast oceans.

Sharks are also defined by perhaps their most identifiable feature, their special kind of teeth formed in rows. Shark teeth erupt from a soft organ inside the mouth behind the jaws called the **dental lamina**. It consists of special cells that continuously produce new teeth as if they were growing from a conveyor belt, so more keep coming out one after another. Teeth appear on the jaws as rows stacked behind each other from front to back. These teeth are generated and shed throughout life, ensuring that new teeth can pop up at the biting edge of the mouth as required. Many bony fishes can resorb and remodel tooth tissues, but sharks cannot do this. Their strategy is to replace dull or damaged teeth with new sharp ones. The peculiar way that sharks and rays make teeth and shed them is great for us paleontologists, because each shark leaves behind many thousands of teeth in its lifetime. These end up on the seafloor to eventually become fossils.

Other chondrichthyans, like rays and chimaerids, have teeth that follow similar growth processes but differ in how they develop and are used and shed (or not). Rays grow and shed their flat teeth regularly, but not as quickly as sharks do. The chimaerids are a little different: They have sets of large, continuously growing tooth plates that they do not shed.

The study of living sharks can tell us a lot about their anatomy and what special superpowers they use to detect and capture their prey, or to find a mate. Today all sharks live by hunting prey, although certain large forms, like whale sharks, take this to a refined level by extracting their tiny living prey from seawater by filtering the water through their gills. While this is also true of some bony fishes, many more are omnivores or specialized herbivores that eat a much wider range of foods including plants, algae, and plankton. (Though it's been recently discovered that some so-called "predatory" sharks consume a lot of vegetable matter, too—for instance, seagrasses can make up to 62 percent of a bonnethead shark's diet.)

Sharks are also defined by their characteristic swimming style of moving their heads from side to side to home in on odors using their powerful olfactory or nasal capsules. White sharks can sense a taint of blood and tissue in the water emanating from a rotting whale carcass several miles away if the smell is carried by a current. They use their lateral line pores to sense the movement of the current, in combination with their highly sensitive noses to sniff the water, then turn toward the source and steadily move in until the decaying feast is found.

However, the mystery of the long, slow evolution of sharks was not solved by the sum of their characteristics and superpowers alone. It was also driven by many factors beyond their control, from the shifting of continental plates that created new currents or pathways to new oceans, to the rapid rise of other scary predatory fishes and monstrous reptiles in the ancient oceans. Sharks won the day in the end because of their unique evolutionary adaptations, as we will soon see from their extraordinary fossil remains.

# Shark origins: the mystery years

The first great mysteries about sharks are when they evolved and where they came from. The oldest shark scales appeared about 465 MYA in Australia. Over the next 40 million years, climaxing at the end of the Silurian period (419 MYA), shark remains mostly consist of a few isolated scales, tooth whorls, and a handful of broken fin spines.

Some 28 million years after the first shark scales appeared, we have two remarkable finds from China. These are the first nearly complete small shark fossils, called *Shenacanthus*, and tiny shark-like teeth formed as clusters or tooth whorls, named *Qianodus* (both are 438–436 MYA). We then have a roughly 20-million-year gap until the first proper sharks' teeth suddenly become abundant in sedimentary rocks (417 MYA), then about 6 million years until the first nearly complete shark fossil (409 MYA).

My quick "back of the envelope" calculations suggest that all the known fossils of sharks found throughout the world during the first 56 million years would nearly fit into that envelope. However, good things often come in small packages. The new high-tech approach to paleontology, championed in the last decade, means that, using powerful micro-CT scanners and synchrotrons, we can scan the heck out of these tiny scales, spines, and teeth to make incredibly detailed observations on how their hard tissues evolved. Understanding these biological processes, like how cartilage can change into bone, or evolve into tooth tissues like dentine, informs the bigger evolutionary story of how the first internal skeletons formed.

Until more fossils dated at the critical stage are found, the first 56 million years of shark history—a period akin to the time stretching back from today almost to the last dinosaurs—should be referred to as their "slow cooking period." It was a time when through trial and error, survival of the fittest, they eventually split from their ancestors and evolved the characteristic features that we now associate with modern sharks.

We are always hoping to discover which fishes were the ancestors of sharks. One fact that helps us is that sharks lack external bones but have a well-developed internal skeleton made of cartilage. The first contenders for shark ancestors include the ancient jawless fishes. These had

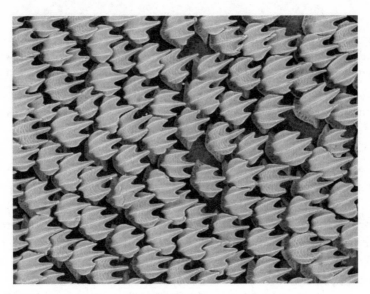

**Mako shark scales showing the crowns only, as the bases are stuck into the skin. Each scale is about 1/500 inch**
PHOTO BY GEORGE V. LAUDER, HARVARD UNIVERSITY

a full external bony skeleton and a presumed weakly developed internal cartilage skeleton. Other contenders, like the armored placoderms, early jawed fishes covered with bony plates, appeared well after the oldest sharks (more about them later). For now we have only a few fossil scales to establish the oldest shark in existence.

Scales can tell us many things about their owners. Shark scales are tiny, almost microscopic, and are called **placoid** scales, which differ from the typically flat, rounded scales of most bony fishes. They tell us something very important about the early evolution of sharks: the nature of their oldest bony tissues.

Modern sharks have tens of thousands of little placoid scales in their skin. They have many shapes and serve many functions, primarily to protect the skin of the shark against abrasion and to prevent parasites from attaching to the shark's skin. The scales are sometimes reduced in patches of skin to allow specialized bioluminescent organs to glow in some deep-sea sharks. Another functional use of placoid scales is to create microscopic layers of turbulence around the skin-water interface—this reduces overall drag for the shark swimming through the water. It works like a golf ball with an irregular surface (dimples), which can go 30 percent farther than a smooth ball. This irregular skin surface enables the shark to go much faster than if its skin was perfectly smooth.

Each shark species has many different kinds of placoid scales (as many as twenty-five different types), from simple ones with one layer and a few raised ridges to quite complex varieties with elaborately sculpted surfaces displaying many branching ridges or riblets. The sculpted part of a fish scale that sticks out of the skin into the water is called the scale's "crown," whereas the part embedded in the skin is called its "base."

The shape of a scale, and the complexity of its surface—the numbers of ribs or complex bumps on it—reveals surprising adaptations in different kinds of sharks. It's a bit like how cars have different patterns of tread on their tires depending on where they go (on pavement or on rugged terrain) and how fast they travel. The fastest-swimming sharks today, like makos, have placoid scales with just three fine ribs on the smooth crowns. These scales give the mako maximum efficiency, reducing drag in the water for swimming at high speeds. As sharks move through the water, the small scales on the head of the shark do most of the work, creating a micro layer of turbulent flow that breaks the "wall of water" as the shark's body pushes through and forms a miniature vacuum that sucks the shark forward. Mako sharks can bristle their flank scales, making them stick up for short times and giving them brief bursts of greater speed. Other sharks have placoid scales shaped to direct the flow of water around the nostrils for better smelling, or away from the specialized sensory cells along the lateral line of the body. A recent study has just revealed that even the eyes of some sharks, like the giant whale shark, have their own distinct types of minute placoid scale that provides a protective covering and also helps direct water away from the eyeballs. The scales help the shark with everything it does in its daily life.

Until we can map the shape of every scale type on every species of shark, we simply won't know how many species, living or fossil, have similar scale shapes. Given how much we still don't know about every type of scale on a living shark, how can we be sure the tiny fossil scales from central Australia really came from sharks?

## *Tantalepis,* the first shark?

Understanding the murky origins of sharks is a lot like being a detective at a murder scene where far too few clues have been left behind.

The oldest probable shark is *Tantalepis,* announced in 2012. Its scales, each roughly the size of a pinhead, closely resemble the simple placoid scales seen in some living sharks. The British team who found these scales named them after Tantalus, a son of Zeus in Greek mythology, whence the word "tantalizing." The name seductively alludes to what this new species might tell us about the murky origin of sharks.

*Tantalepis* scales are simple structures having three prominent ridges on the crown of the scale. The base of the scale is a simple, flat to slightly concave rounded plate. Detailed work on living shark scales corroborates that the three fine ribs on the *Tantalepis* scales might suggest that they were like nurse sharks, which are strong swimmers capable of fast sudden sprints when feeding at night.

0.5mm
0.02 inch

*Tantalepis* **scales, the oldest evidence of sharks on Earth at around 465 MYA** PHOTO BY IVAN SANSOM, UNIVERSITY OF BIRMINGHAM, UK

Back in 2006, Ivan Sansom from the University of Birmingham and his team scoured the dry creekbeds of central Australia about sixty miles south of Alice Springs. It was hot, dusty work, and swarms of flies were their constant companions. They bagged a sample of fine siltstone rock that was later processed in their lab. By breaking down the rock using chemicals they could pick through the insoluble residues and find the tiny fossils. All told, the sample yielded around 150 of these tantalizing shark-like placoid scales. However, only a few of the scales were complete or well preserved; most had been broken or abraded by sea currents before being buried. So, even though they resemble modern shark placoid scales, what is the hard evidence that makes us think these tiny ancient fossils once belonged to sharks?

The structure of these early scales is our best clue. The simplest and the most primitive type of early shark scale is made from just one developmental unit, called the **odontode** (meaning "tooth-like" unit). The crown of the scale is composed of a kind of hard dentine, as in our teeth, and the base is made of a bony tissue. The scales lack one feature seen in later shark scales—a canal in the neck of the scale. For this reason, we suspect *Tantalepis* is most likely an early chondrichthyan of some kind, but we can't be 100 percent sure.

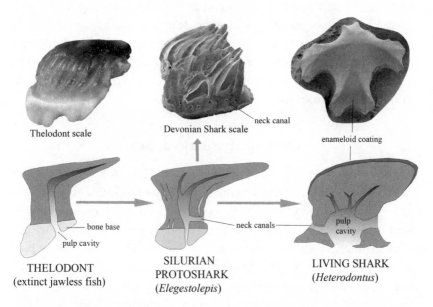

How ancient jawless fish scales (thelodont, left) might have given rise to the first shark scales (middle and right)  PHOTOS BY AUTHOR (LEFT, RIGHT) AND PLAMEN ANDREEV (MIDDLE); DRAWINGS BY AUTHOR

Similar types of simple scales are found in a group of ancient jawless fishes called **thelodonts**, which are tubular to flattened fishes. They lacked bony plates around their heads as we see in *Arandaspis*. The oldest known thelodont scales first appear in the Harding Sandstone of Colorado about 10 MYA after *Tantalepis*. These early thelodont scales have a base pierced by a canal, called the pulp canal, just like the ones we have in the base of each of our teeth. This tells us the biological processes used by the fish to make its scale were quite different to the process used by sharks. It is a subtle yet important point of difference which allows us to distinguish thelodonts from sharks by just their scales. Shark scales grow from little primordial buds in the skin, and once they reach maximum size, they stay fixed in the skin as the shark keeps growing. They fall out eventually and new scales replace them. In this way the shape of the scales can change as the shark matures.

*Tantalepis* is as close as we can get to the oldest evidence for the origins of the shark group, the chondrichthyans. The scales are remarkably similar to those of several living sharks, like the nurse shark shown on page 372. It also reflects the main problem in searching for the origins of major lineages of animals—the dearth of good fossil vertebrate material at this early time in the history of life. Luckily we don't have to wait long after *Tantalepis* before the first definite chondrichthyan scales begin to appear. We will now look at other kinds of early sharks from this enigmatic dawn of their history.

## Other early sharks appear

At the end of the Ordovician period, 443.8 MYA, we have at least five kinds of putative fossil sharks known only by their scales. They tell us two things: Simple placoid scales evolved into more complex forms with many ridges, and tissues making them began to evolve as we encounter the first scales made up of two key components, a bony base and a dentine crown. We know these sharks were likely very small (maybe a foot or less in length), and that they likely lacked teeth and fin spines. If such hard parts were present on these sharks, the large amount of sampling done by now should have turned up a few bits and pieces of them, but so far not a scrap has been found to suggest these structures existed.

We can't imagine these beasts as looking anything really like living sharks, so let's just bite the bullet and refer to them as "protosharks." The many species of living sharks that bear similar simple placoid scales to our Ordovician protosharks all have jaws, so it's safe to assume our shadowy protosharks also possessed jaws. We cannot assume that teeth were necessarily present in any of these protosharks, unless they were using rows of enlarged placoid scales around the edges of the mouth as "teeth." If so, then these early "scale-teeth" would not be easily differentiated from other body scales in our samples. I think it would be impossible to tell whether they were used like teeth unless we were to find distinctive wear patterns on the sharp edges of the scale crowns, and to date such wear patterns have not been observed.

One other kind of early "shark" was also living in the Late Ordovician seas. Understanding their anatomy is really the key to deciphering the origins of modern sharks. Their scales differ from the simple placoid scales of sharks by having a very rounded, protruding base instead of a flat base. They were at first thought to represent another extinct group of jawed shark-like fishes and were commonly called **acanthodians** (meaning "spiny ones"). They are known from isolated scales, teeth, and fin spines from across the globe, particularly common in younger rocks of the Silurian and Devonian periods, where many complete body fossils have been found. Today we know they are the ancestral stock that modern sharks came from, so we refer to them as **stem sharks**, meaning they are a small branch (a stem) coming off the main evolutionary tree of sharks. Most of the acanthodians were a dead end with no survivors, but some species at the base of the tree bridged the gap between the protosharks like *Tantalepis* and the modern line of sharks.

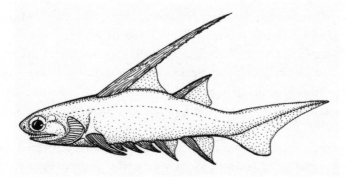

**A stem shark, or acanthodian (*Parexus*, from Scotland)**
DRAWING BY AUTHOR

You might wonder whether such detailed research on ancient shark scales has any real value to the modern world. Some living sharks, like makos and hammerheads, can swim at fast speeds aided by their scale shapes, so research on shark streamlining can have practical uses. Let's meet an incredible German scientist who took the study of fossil shark scales to an unlikely application that had huge economic benefits for his country.

## Ancient shark scale research pays off

Wolf-Ernst Reif, born in the last year of World War II, was a German child prodigy, the offspring of two brilliant scientists. Enamored by geology, he collected stones at age three, and by age seventeen he had published his first peer-reviewed scientific paper. Despite being born with spina bifida, a debilitating condition he endured throughout his life, he eventually rose to become one of Germany's most distinguished evolutionary biologists. His PhD dissertation at the University of Tübingen posed a series of major questions about shark evolution. Foremost among these were "How do shark scales form, and how are they related to teeth?" To answer this question, Wolf-Ernst spent a year in Hawaii in 1976 studying live sharks to collect data and develop ideas about the role that shark scales play in their swimming efficiency.

He was far ahead of his time in adopting a multidisciplinary approach to solving this problem. He collaborated with a wide range of professionals including biologists, physicists, and even architects to better understand the structural principles that have guided evolution to develop varying shapes of shark scales. He discovered that the scales of certain fossil sharks had many fine ribs on them, like those found on the scales of the fastest-swimming living sharks. In order to determine the effect that scale shape had on swimming dynamics, he teamed up with an equally brilliant German physicist named Dietrich Bechert, who was then working at the German aerospace organization (the DLR).

Together, the paleontologist and the physicist performed experiments measuring water flow over scales, making stunning advances to our knowledge of the hydrodynamic properties of shark skin. They began experimenting by replicating the scales' special features on different materials by creating thin foils that could be applied to the surfaces of ships and planes. The foil reduced the drag between the ship or plane surface and the water or air, thus saving loads of money on fuel. Bechert continued this research in his own lab and went on to invent a film that broke the world record for drag reduction using riblet optimization to minimize the drag in the water by up to 10 percent. This research has already paid big dividends in reducing emissions and fuel use for the aerospace industry. The German airline Lufthansa announced in 2022 that all their Boeing 777 cargo freighters are now fitted with AeroSHARK, a newly developed surface film that maximizes fuel efficiency. In their press release they specifically mentioned that the inspiration for AeroSHARK came from the work of Wolf-Ernst Reif and Dietrich Bechert.

**Wolf-Ernst Reif, 2005**
PHOTO BY HIS LATE WIFE, ROSE-
MARIE

Wolf-Ernst passed away in 2009, so he never got to see how his basic research was eventually applied to commercial use. It's incredible to think that because of his pioneering early work on fossil shark scales more than forty years ago, the Lufthansa cargo fleet will now use around 4,000 tons less fuel and prevent around 11,700 tons of carbon dioxide emissions each year. Just think of the massive environmental savings that would be achieved if *all* the world's airlines adopted the use of AeroSHARK technology.

## Sharks endure the first mass extinction

As the Ordovician period comes to a close, sharks finally seemed to be running a good race in the evolutionary sweepstakes. Then, suddenly, they almost lost it all. The last nights of the Ordovician period were like life's first New Year's Eve, anything but quiet. It went out with a devastating bang.

The end of the Ordovician marks the first of five major global mass extinction events that changed Earth's history. It began with massive tectonic movements of landmasses down south. The great southern supercontinent of Gondwana had drifted over the South Pole and expansive ice caps developed, causing sea levels to drop rapidly as water became frozen in the polar ice caps. Vast areas of widespread shallow marine habitats were suddenly lost, while a rapid decline of oxygen in the seawater caused deadly conditions on much of the seafloor. A bunch of essential trace elements necessary for life, like selenium, became depleted in the oceans, piling salt on the wound to the struggling marine life. These events took place between 445 and 443 MYA as two deadly pulses wiped out about a hundred families of marine invertebrate animals. These included many of the trilobites, the once abundant filter-feeding graptolites, and the bivalved lampshells (**brachiopods**). It is ranked as the second most devastating of Earth's five mass extinction events.

The few species of sharks around at the time suddenly faced their most dramatic challenge for survival. How did they get through? As we have seen, all Ordovician sharks lived in shallow intercontinental seaways and intertidal estuaries, so they were probably knocked sideways by the sudden loss of habitat when sea levels rapidly fell. A talented young Bulgarian paleontologist named Plamen Andreev, based at the University of Birmingham, has made a lifelong study of the oldest shark scales (see bio on p. 404). One of the new sharks he discovered from the Harding Sandstone, called *Solinalepis,* is the clue to how sharks probably survived through to today. *Solinalepis* is in the group with the most advanced scale structure, called the **mongolepid** sharks. These sharks had scales with more complex crowns made of a denser noncellular dentine with neck canals present—a feature seen in the scales of living sharks.

The abundance of mongolepids in the next period, the Silurian, from many sites around the ancient northern continent of Laurasia (China, Mongolia, and Siberia) testifies to their resilience. I suggest that other mongolepids probably lived at the same time farther from the coast in open oceans (in rocks that haven't yet been well sampled for shark scales) because some clearly survived. The deep oceans didn't suffer any dramatic changes during the extinction event, so sharks living there would have been more likely to survive the killer sea level changes.

Whatever the reason, some shark groups made it through the chaos and continued to diversify and evolve into new forms. Let's see what progress sharks made in the short but dramatic Silurian period.

# Silurian sharks

By the Early Silurian period (443–434 MYA) we find a rising diversity of sharks that are known from scales, as well as the first fossil records of isolated fin spines and one nearly complete fossil shark.

The fin spines of living sharks are made of a dense dentine tissue, similar to that forming the base of the teeth, so they are very hard structures. They work well in protecting the sharks from being eaten by larger predators. Living dogfishes and horn sharks are examples of living species bearing stout, sharp fin spines preceding the two dorsal fins on their backs. The appearance of shark fin spines in the fossil record is therefore of interest to our story, as it signals that large predators must have been lurking in the same seas.

One group of early fossil sharks called **sinacanthids** (meaning "spines from China") lived throughout the entire Silurian period (443–419 MYA). They are known only from their distinctive fossilized fin spines and scales. The largest of the sinacanthid fin spines suggest that these sharks reached a slightly larger size than their Ordovician ancestors, perhaps around a foot and a half long at most. They are typical acanthodian-like stem sharks whose spines occur in the same sites as mongolepid scales, so these groups might be closely related or the same group. We need to find a complete one to solve this mystery.

The oldest early shark known from an almost complete body fossil is *Shenacanthus* from Chongqing, South China, which lived around 437 MYA. Its discovery was quite serendipitous. Three young Chinese paleontologists were messing around play-fighting when one kung-fu kicked another into a roadside cliff face. Some rocks came tumbling down, and one split to reveal a hidden treasure of a spectacular fossil fish inside it. This turned out to be the oldest known complete fossil fish with jaws ever found! They couldn't believe their luck. With further digging they found a layer rich in complete skeletons of early fishes including *Shenacanthus*. It was a small fish, about an inch long, with bony plates covering the top of the skull and enveloping its chest area. It has pectoral fins without spines yet bears a stout dorsal fin spine similar to that of the acanthodians. Jaws with teeth have not been found on the specimen, so teeth might well have been absent. Strange tooth whorls that belong to a stem shark called *Qianodus* (see figure on p. 50) were also found in rock layers of this age in China. They are not the same as shark teeth with sharp-pointed cusps—these would not develop in sharks until the Devonian period.

*Shenacanthus* might bridge the gap between sharks and another early jawed fish group called **placoderms** ("plated skin"). Placoderms appear in the Early Silurian at exactly the same time. They are shark-like fishes with bony plates forming shields that completely cover the head and trunk area of the body. The condition seen in *Shenacanthus* might show that sharks and placoderms came from a bone-covered ancestor; alternatively, it could imply that the first sharks had an outer covering of bones, then rapidly lost them. More specimens of *Shena-canthus* with intact jaws and showing clear skull bone patterns are now required to show which of these two theories is more likely to be correct. (We will discuss the big question of shark evolutionary relationships in detail later on after more fossil evidence is revealed.)

**A restoration of the earliest known complete fossil of a "stem shark,"** *Shena-canthus,* **dated around 437 million years old** BY ZHANG HEMING, SUPPLIED BY THE INSTITUTE OF VERTEBRATE PALEONTOLOGY AND PALEOANTHROPOLOGY, CHINA

Now that we have covered the first 46 million years of shark history, let's recap what we've learned. In the Middle to Late Ordovician period (465–443 MYA) we saw the first appearance of fishes with rare shark-like placoid scales, such as *Tantalepis.* While they still lack the diagnostic features to confirm them as chondrichthyans, their scales are closer in structure to those of sharks than to those of any other group of fishes. Most paleontologists regard these creatures as the beginning of the line leading to modern sharks. Another line of stem sharks appeared by the end of the Ordovician period, and these would expand into a diverse group in the Silurian and later periods. Sharks with more complex scales evolved in the Silurian, like the mongolepids with neck canals that are found in the scales of living sharks, and the sinacanthids with thick dorsal fin spines made of shark dentine. The Silurian period (443–419 MYA) heralded the first complete body fossils of shark-like forms, including *Shenacanthus*. These shark-like fossils are so rare that none of them give clear clues as to the possible evolutionary origin of sharks, apart from vague similarities in their tissues to the scales of the jawless thelodonts. These earliest sharks were essentially the first seeds of a modern shark dynasty. While most of them would silently slip into oblivion, some of them would eventually become more like modern sharks. But first they needed something to give them an advantage over all their competitors.

Immediately after the Silurian period, the first sharks' teeth characterized by strong pointed cusps on a robust root start to appear, around 419 MYA. Sharks armed with large, sharp teeth swam about the shallow seas of what is now Spain. Teeth were a real game changer for sharks. The pace of our story now quickens, as the most dramatic arms race of the Devonian oceans is set to begin.

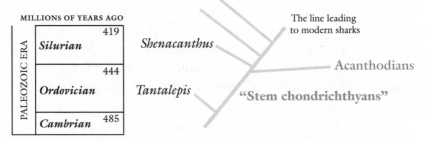

MILLIONS OF YEARS AGO

PALEOZOIC ERA

| | 419 |
| Silurian | *Shenacanthus* |
| | 444 |
| Ordovician | *Tantalepis* |
| Cambrian | 485 |

The line leading to modern sharks

Acanthodians

"Stem chondrichthyans"

**The early evolution of sharks from 465 to 419 MYA. As our journey continues, more of this tree will be revealed up to the present day** DIAGRAM BY AUTHOR

CHAPTER 3

# SHARKS BECOME PREDATORS

*Sharks' First Superpower: Deadly Teeth*

---

### Searching for the oldest known shark teeth

**SPAIN, JUNE 2022:** It was a scorching hot day as we passed through the jagged mountain peaks of the Iberian Chains of south Aragon to search for shark fossils. The landscape here was harsh yet serene. Forested mountains in the distance tower over 7,000 feet, framing the gently rounded hills that encircle a valley dotted with scraggly shrubs and stunted trees. A few small gnarly oaks clung desperately to the banks of a dry creek. The four-wheel-drive vehicles rattled along the old dirt tracks that bisected steep hillsides as we rambled past outcrops of jagged rock that date from the start of the Early Devonian.

Back then, atmospheric oxygen levels were about 8 percent higher than the air we breathe today, supercharging the metabolisms of the first jawed vertebrates, which in turn drove an explosion of fish diversity. These new creatures radiated into many bizarre kinds that frequented the shallow seaways and coastal estuaries and invaded freshwater rivers for the first time in Earth's history. Sharks slowly increased their diversity at this time, as shown by the emergence of a range of distinctive tooth types.

The fossil site was not far from the abandoned town of Nogueras, now a municipality of just eighteen souls. We drove past a pine forest, down a hidden track devoid of any signposts. Having arrived at our destination, we scrambled over the rocky outcrops searching for fossils.

The flat-lying outcrops of richly fossiliferous pale yellow limestones interfinger with thinly bedded dark shales, a sequence known as the Nogueras Formation. These rocks represent shallow marine sediments laid down in a coastal seaway between 415 and 408 MYA. They bristle with abundant fossils of creatures that lived in, on, or just above the seafloor. We found many rounded lampshells (brachiopods), the large, rounded domes of colonial corals, coiled snail shells, the lacework remains of sea mosses (**bryozoans**), and many other remains of invertebrate life. Eagerly we searched with lenses held up to our eyes for the telltale shape of a shark's tooth, for this is one of the sites where the oldest known sharks' teeth are found.

**Héctor Botella at one of the sites of the oldest known sharks' teeth in Aragon, Spain** AUTHOR PHOTO

Our field guide is Héctor Botella, whom I have known for some years now. He is a lanky, bearded Spaniard with a big grin and thirst for knowledge that rivals his appetite for joy and cold cervezas. Héctor is based at the University of Valencia, where he has made remarkable discoveries about the biology of these first toothed sharks. He has made thorough detailed microscopic and microscanning investigations into the tissue structure of the teeth, combined with complex statistical analyses of tooth sizes and shapes over time. Héctor tells us that in the early days of exploration, more than just isolated sharks' teeth and scales were found here. The remains of a shark showing associated teeth, scales, and fin spines from one individual were found here by a young German PhD student who collected samples in the area and extracted the first fossil sharks' teeth from the rocks back in the early 1980s. This specimen is now housed in a university museum in Tübingen, Germany.

While sharks' teeth are not uncommon here, they are hard to find when you have only a few hours at the site. To find them in abundance, we must first collect the right rocks—limestones with teeth or bone fragments visible in the field—then take them back to our lab. Limestone dissolves in certain mild acids that do not affect the teeth or bone inside them. We dissolve about a rucksack full of rock in a solution of 10 percent acetic or formic acid. Then we wash the remaining residues in water and then dry them in an oven. We must then painstakingly search through the insoluble rubble left in the bottom of the vat using a fine artist's paintbrush to pick out the tiny teeth. This job is done looking down a microscope, as the teeth are only about one-tenth of an inch long. This explains why that day we did not find any shark teeth in the field, but later on, Héctor took us to a neat local paleontology museum in the abandoned town of Santa Cruz de Nogueras. This town now has a population of only four families, yet fortunately for us, two of the residents are paleontologists who built the Museo de los Mares Paleozoicos (Paleozoic Seas Museum). After marveling at the wonderful displays of local Devonian fossils, we were shown to a clean and well-equipped laboratory to peer down microscopes at the spectacular fossil sharks' teeth found in the area. I was blown away by the excellent preservation of these oldest known sharks' teeth and other spectacular fish fossils collected from the region.

What I saw that day would revolutionize my understanding of what these early sharks were like. I could see in my imagination how they hunted their prey and used their hooked sharp teeth to peak advantage.

## Sharks' toothy origins

Teeth are extremely important to all sharks as they are vital for catching prey, or for cutting flesh from prey, or for reducing hard prey, like crabs, into bite-sized morsels to swallow. The appearance of true sharks' teeth in the fossil record marks a radical change in shark evolution, from the formerly presumed toothless protosharks known mostly from their scales and fin spines to creatures armed with a new superpower: the ability to make sharp teeth—many of them—and shed them quickly when new ones were needed. Only sharks can do this in such a highly effective way.

While typical bony fishes can grow new teeth up through their jaws to replace lost teeth, it is not at the same fast rate, and they cannot shed teeth routinely as sharks can. Many years ago I published a paper on a fossilized lower jaw of a large predatory Devonian fish from Antarctica called *Notorhizodon* with a severely bent fang still in the jaw. Sharks would never have this problem, for they invented the first kind of teeth that would never require a dentist (*hooray!*). This ability to have fresh, sharp teeth in their jaws really gave sharks the literal cutting edge over their competitors. Right from the start of this ability, they were dropping their teeth regularly into the sediments on the bottom of the seafloor, so their teeth became abundant as fossils.

So where did sharks get their teeth from? Here's another awesome fact about sharks—their teeth are enlarged mirror images of their tiny placoid scales, each with a dense dentine base embedded in the gums and a sharp biting part called the crown. Can you imagine being covered in thousands of tiny replicas of your teeth sticking out of your skin? The crown of a shark tooth might be simple, with one pointed cusp, like that of a mako shark (see figure on p. 16), or have many pointed cusps, like those of the sixgill shark (see figure on p. 229) that have teeth forming an elongated zigzagged tooth shape.

Sharks most likely evolved their teeth from modification of the scales around the mouth area, which eventually invaded the mouth and lined the jaws. Unlike scales, which grow only once, the teeth regenerate continuously, regulated by a different genetic process. Sometime around the start of the Devonian period, the genetic ability to keep growing mouth scales as teeth kicked in and was immediately successful. We can conclude this because before then shark teeth are almost nonexistent in the fossil record, while from the Devonian onward, they are abundant as fossils.

How shark teeth first developed might be linked to their taste buds. When a shark chomps on its prey, it tastes it through tiny taste buds that grow directly behind the last row of teeth in both their upper and lower jaws. A recent study led by Kyle Martin of the University of Sheffield showed a genetic link between how teeth and taste buds develop in the embryos of the small-spotted catshark. They found that the same embryonic stem cells form both the teeth and the taste buds. Then they looked at shark scales and identified a gene called **Sox2** that is not turned on when sharks make their scales (which are not replaced), yet it is active when making teeth and taste buds (both regularly replaced). They proposed a new theory that sharks got their genetic ability to replace teeth from the taste buds, which go back much farther in evolution than the first teeth, as taste buds are found in jawless lampreys. We know that jawed vertebrates had a genetic doubling of their genomes (total DNA in each cell) sometime before the Devonian period started, so this event might have caused the new genes to appear at this time. This genetic doubling helps explain the sudden appearance of the first jaws, teeth, and paired fins in sharks and other early jawed fishes.

Sharks show highly distinctive tooth shapes and structures, with each tooth formed from a variety of specialized tissues that give the tooth its strength and resilience. Some living sharks have homogeneous dentition, in which all the teeth are quite similar (e.g., the frilled shark). In nearly all sharks, the front teeth are taller than the much wider middle and rear teeth, so the exact jaw position of each tooth can be identified by its shape and proportions. Certain other sharks have **heterodont** dentition, in which the upper and lower jaw teeth are distinctly different. The sixgill shark, *Hexanchus,* for example, has wide, multicuspid lower teeth and much narrower upper jaw teeth with few cusps on them (see figure on p. 229).

The teeth of most sharks are made of a super-dense kind of dentine called **orthodentine**, defined by having a pulp cavity inside the tooth. Other sharks, like lamniforms such as white sharks, have **osteodentine** forming the tooth, lacking any pulp cavity inside, making it very strong. All shark teeth have a super-strong outer layer covering the biting part of the tooth, called **enameloid** (meaning "like enamel"). The enamel in our mammalian teeth has parallel bundles of dense hydroxyapatite crystals forming the structure, but these bundles of crystals are not as neatly arranged in shark enameloids. Enamel is the hardest tissue in our body, and modern shark teeth use multiple layers of enameloid over the dense dentine to make super-strong teeth. The better to eat you with, Grandma.

The ancient Spanish shark teeth belong to two distinct types. The first was named *Leonodus,* meaning "Leon tooth," after the region where they were first found. *Leonodus* had unusually complex teeth— they are double-pronged teeth resembling tiny grappling hooks, giving them exceptional capture power, because each tooth could pierce the prey twice and not let the struggling victim slip away. The second type of shark tooth came from the same layers and is called *Celtiberina,* after the Celtiberia region near Nogueras. *Celtiberina* teeth had wide roots with a single stout dagger-like cusp forming the main biting part of the tooth.

Both these early sharks were clearly formidable predators equipped with deadly dental arsenals. That day I stared down the microscope in the museum and was astounded at the robust build of the tiny *Leonodus* teeth, each pointed cusp bearing a series of sharp cutting ridges. These oldest sharks were no *Jaws* creatures—both were quite tiny. If scaled up to typical shark body lengths, the teeth suggest an average body length for *Leonodus* just over a foot and a half long, and a tad smaller for *Celtiberina.*

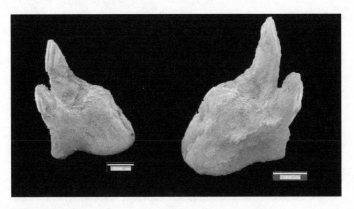

**Two *Leonodus* teeth**   SEM PHOTO BY HÉCTOR BOTELLA

We can imagine that such miniature sharks would have preyed upon fishes even smaller than themselves, as well as on other marine creatures like worms and shrimps. They were clearly not the top dogs in their oceans, so they would have had to keep a wary eye out for larger predators such as the three-foot-long lobe-finned fish *Porolepis* and some of the larger, armored placoderm fishes. Yet these sharks had not only real teeth, but another amazing capability—the ability to replace them regularly. Tooth replacement was their first true superpower.

## Sharks' first superpower: replacing teeth

It is incredibly surprising how much valuable information we can get from studying large collections of shark teeth. Héctor and his team did a series of studies using more than eleven hundred isolated teeth and broken cusps of *Leonodus*. He found that *Leonodus* teeth grew in rows, just like modern sharks' teeth, emerging from the dental lamina, an organ that works like a conveyor belt in its mouth to keep making teeth that pop up as required. This discovery of a dental lamina in *Leonodus* is important because it was the first hard evidence of an extinct shark that grew its teeth in exactly the same way that all modern sharks do. This is highly significant because up to now we have seen various teeth from much older rocks, like the odd tooth whorls called *Qianodus* from China, but they most likely belonged to archaic stem sharks (acanthodians). The way *Leonodus* grew and shed its teeth showed it was closer to the ancestral line leading to modern sharks than to any of the other stem sharks of the day.

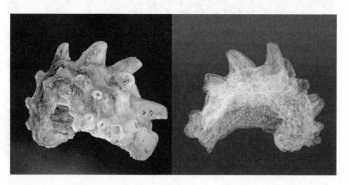

*Qianodus* tooth whorls from China (c.437 MYA); these are about ¹⁄₁₀ inch wide  SEM PHOTO BY PLAMEN ANDREEV, UNIVERSITY OF BIRMINGHAM, UK

Héctor and his team then tried to solve the burning question of whether these ancient Spanish sharks had the same rate of tooth replacement as living sharks. They set about calculating the degree of wear seen on each of the fossil teeth and then compared them to the wear patterns seen on the teeth of living sharks. He determined that *Leonodus* teeth were indeed replaced regularly, but about ten times more slowly than the average mean rate of replacement seen in most living sharks. We know the replacement rate in living sharks from various sources. For example, the sand tiger shark is a favorite species kept in aquaria around the world, and it has been observed to shed a tooth approximately every two days. *Leonodus* also grew and replaced teeth on a regular, if slightly slower, basis. Héctor's study also confirmed that the ability to replace teeth quickly is a highly specialized feature that would develop later in sharks to characterize modern forms. Teeth are the first key to knowing all sharks, past and present.

So, what was the world like in this new period we are now in, the Devonian? It was a majestic time in Earth's history, when new environments were open for sharks to thrive in.

## Welcome to the Early Devonian

We have a lot of information about the Devonian world, and this sets the stage for understanding the environmental changes that would drive major advances in the evolution of sharks and other fishes. During the Early Devonian (419–393 MYA), the lands were covered by small sprawling plants like lichens, liverworts, and mosses, with upright ferns that could barely stretch more than about three feet high. The ferns, along with simple stemmy plants called **psilophytes**, were confined to the edges of rivers and lakes, as they had not yet evolved the necessary structures to survive away from bodies of water. Most of life on Earth existed in the shallow seas embracing the margins of vast supercontinents and smaller islands that framed a very different map of the world from the one we know today. Jawed bony fishes and placoderms continued to diversify and occupy new habitats when they left the seas to invade large river systems.

Throughout the first 15 million years of this time, sharks underwent radical changes. The majority before this time were either protosharks known only from scales, or acanthodians (stem sharks). Some of these acanthodians had jaws with no teeth, or were lined with tooth whorls, while some had odd-shaped rounded teeth, or rows of sharp teeth fused to the jawbones—all these were nothing like the typical replace-

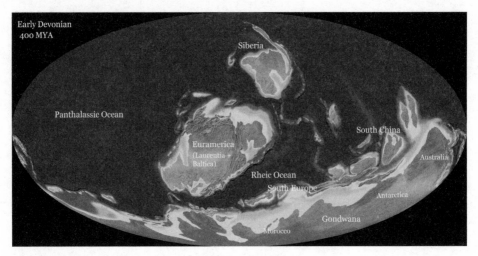

**Map of Earth during the Early Devonian period**   RON BLAKEY, DEEP TIME
MAPS (DEEPTIMEMAPS.COM)

ment teeth we see in modern sharks. A few do have rows of teeth (like
*Ptomacanthus*), and these are thought to be closer to modern shark
lines than the others. Suddenly we find that Early Devonian sharks
were growing teeth in the same way that modern sharks do and replacing them as they shed.

The story of shark evolution can be fleshed out more clearly as we
find more complete body fossils to see what drove their body shapes,
how their skulls reflect the evolution of their senses, and how they were
evolving the structure of their cartilages and tooth tissues. While we
have a couple nearly complete fossil shark specimens in the first half of
the Devonian, there is still some controversy over their status as early
examples of "real sharks" (those on the line to modern forms) rather
than "stem sharks" (acanthodians), which often lack characteristic features of living sharks. We don't get any definite complete shark fossils
on the modern line to today's sharks until the later part of the Devonian.

To succeed in life, sharks required more than just great teeth. They
needed streamlined bodies, wide pectoral fins for steering, and powerful tails to push them swiftly through the water. Up until the early
2000s, we had no idea what an entire shark from the Early Devonian
age might have looked like because we had only their teeth, scales, and
fin spines. Then, by pure serendipity, a nearly complete fossil shark was
finally found, in Canada—twice.

# The first nearly complete fossil shark

In 2014 I visited a beautiful and unique part of Canada, with my good friend and colleague Richard Cloutier of the University of Quebec at Rimouski. Richard Cloutier is a jovial fellow with a quirky and delightfully silly sense of humor and a perpetual smile at being able to do what he loves most: collect and study the fossils of this beautiful region in eastern Canada. When he was a postdoc working in London in 1992, he struck me as one of the hippest young paleo-dudes I'd ever met. Sporting a black leather jacket and a slicked-back hairdo, he was the epitome of Quebecois cool. He built his career working on the perfectly preserved complete fossil fishes of Miguasha, located to the east of the Restigouche River fossil sites on the northern shore.

We were on the way to the Miguasha Museum and stopped to admire the scenic Restigouche River near Atholville. It is a picturesque body of greenish-blue water that laps upon a rugged shoreline of brownish-gray mudstone and sandstone rock. Otters can be seen playing in the shallows as majestic herons glide overhead. In 1760 the river was the site of a historic naval battle when the English defeated the French before going on to conquer the rest of Quebec. While some find this recent history riveting, for me it's the rocky shoreline that is the region's most fascinating feature. These rocks are of Early Devonian age, deposited about 397 MYA, and like the pages of a book, each sedimentary layer reveals a story about the deep-time history of the region.

Discoveries of shark fossils at this site started more than 120 years ago. An amateur British fossil finder named Mr. Jex collected large, isolated fossil sharks' teeth, which he eventually donated to museums or sold to private collectors. Many of them ended up in major museum collections in England, Canada, and the United States. Some of these teeth had two slender cusps on each bony base and were a whopping half inch in height; others formed a tooth whorl with a series of sharp cusps curled back over themselves. In 1892, the curator of the Natural History Museum in London, Arthur Smith Woodward, made a detailed study of these fossil teeth and named two new forms, *Protodus* and *Diplodus*. Unfortunately, it was found that the latter name was already taken, so it had to be changed to *Doliodus* (meaning "deceitful tooth"). These early well-preserved shark teeth received no more detailed study until the early years of the twenty-first century. *Doliodus* suddenly became scientifically very interesting when more of it was finally found.

On July 4, 1997, a small group of keen fossil hunters led by Randy Miller of the New Brunswick Museum searched the rocky outcrops near Campbellton for fossils, hell-bent on finding something sharky. Sue Turner, a renowned fossil shark expert from Australia, who was then working there on the isolated teeth found from the region, yelled out to Randy, "Keep an eye out for sharks' teeth." That day they struck pay dirt. Jeff McGovern, of the New Brunswick Museum, and Heather Wilson, of the University of Manchester, were slowly excavating a layer of mudstone when they suddenly came upon a nearly complete fossil shark. It was a truly stunning specimen, sporting rows of pronged teeth at the front of the head. Unexpectedly, it had large pectoral fin spines—a feature never seen on any modern shark, living or dead, but well known in the archaic stem sharks.

The surprises from this site did not end there. Richard Cloutier's contribution to the shark story began during that same summer of 1997. A month after Jeff and Heather's find of the nearly complete shark that they suspected might be a *Doliodus*, Richard found a good chunk of a fossil shark braincase near the same site.

Richard soon told Randy about his recent discovery. Both pondered whether their specimens could be two parts of the same creature. A few weeks later they were able to meet up in New Brunswick, each bringing along his specimen. Imagine the joy of the moment as Richard slowly slipped the braincase block onto the broken surface of the main specimen. They were both stunned to see that the pieces fitted together perfectly. Only the end part of the tail was missing. The shark would have been about 2 feet long when complete.

Later it was CT-scanned to reveal incredible details of every tooth in its mouth as well as the structure of its braincase, jaws, gill arches, and cartilages forming the supports for the paired fins. It had robust fin spines preceding each of its pectoral fins, a feature previously thought to characterize the acanthodians. Inside its mouth lay a vital clue to its identity as a shark: rows of teeth identical to isolated ones from the same site earlier named as *Doliodus*.

These teeth were quite unusual in that upper and lower jaw dentitions were different. This meant that even at this early stage in shark evolution, it was not just the shape of the teeth that mattered, but also how the upper and lower jaw teeth differed (called **heterodont** dentition) to enable the shark to grab and manipulate its prey effectively. Countering this very shark-like feature is the fact that there is no prismatic calcified cartilage in the skeleton of *Doliodus*—the unique feature in all modern sharks. That remains as one mystery still to be solved before the true affinities of this fantastic early shark-like chondrichthyan can be confirmed.

The final big surprise was that the CT scans of the *Doliodus* braincase enabled a complete 3-D restoration of the shape of the inside of its nose, showing the olfactory bulbs that housed the organs for smelling. Creatures that have very large olfactory bulbs, like the turkey vulture, have an exquisite sense of smell that helps them find their food or navigate over long distances. In sharks these same organs are used to detect chemicals in the water, providing the shark with its keen sense of smell. *Doliodus* bore large olfactory bulbs, much bigger than those of other jawed fishes, which demonstrates that already at this early stage of their evolution sharks had developed what I like to call sharks' second superpower—a highly acute sense of smell.

The olfactory bulbs of *Doliodus*, as reconstructed from CT scans, are not only large but are also widely separated by a deep cavity in the front of the braincase—the precerebral fontanelle—a key feature seen on all modern shark braincases. This means it was capable of the specialized behavior used by sharks that smell while swimming by swaying their heads from side to side to zero in on the scent.

*Doliodus* has taught us a lot about early shark evolution. By the Early Devonian we have spiny "stem sharks" like *Doliodus* that are rapidly approaching a modern shark-like body with wide, triangular pectoral fins, dentitions very close to that of regular modern sharks, and a braincase similar to modern sharks in some respects. They grow teeth throughout life and shed them regularly. Some had already developed an ability to use their noses to the optimum ability by having wide olfactory bulbs to hunt by sweeping their heads from side to side to home in on prey.

So just how important is the sense of smell to sharks?

## Sharks' second superpower: smell

The keen ability of sharks to sense minute amounts of blood in the water from far away is the stuff of legend. The reality is that while their sense of smell is indeed very acute—some species are capable of detecting a single molecule of certain chemicals in a million parts of water— in the real world this does not equate to smelling a drop of blood a mile away. If the currents are just right, sharks can detect specific chemicals in blood emanating from wounded or struggling fishes from maybe a few hundred feet away, or at most a few miles if the current is strong. A shark can use its powerful sense of smell in conjunction with its lateral line system of sensory cells along the body to trace the current the smell is coming from, coordinate these bits of information in its brain, then home in on the prey from a long way away—something regular fishes can't do. Sharks' sense of smell is said to be thousands of times better than humans'.

While one much quoted scientific study published in 2010 found that sharks and bony fish have equal levels of smell sensitivity, the study was based on a limited number of chemical stimulants, mainly amino acids. Recent work on shark olfactory receptors—the proteins that bind odor molecules for smelling—concluded that sharks do have a keen sense of smell that has resulted from an evolutionary trade-off. While bony fishes have a much larger repertoire of receptor genes for smelling a wider range of different chemical odors than sharks, shark olfactory systems appear to be more finely tuned to maximize odor reception of fewer yet significant odors. This higher sensitivity gives sharks the edge over bony fishes in many cases, although each species of fish and shark is different in its ability to sense odors in the water. Eels and catfishes, for example, are among the top-ranked sniffers of the bony fish world, while white sharks are the clear champs of the shark world.

The classic case demonstrating the superlative smelling ability of the white shark was recorded in 1996 in South Australia off Dangerous Reef, where the *Jaws* live scenes were filmed. A cocktail of blood, macerated fish bits, and other chemicals was poured into the water, and, with a fluorescent dye added to it, created a visible plume stretching several miles from the boat. A large female white shark swam across the far end of the plume, about four miles from the source, and, detecting the highly diluted traces of the blood mixture, immediately turned a right angle to swim directly down the center of the plume toward the boat. This ability to smell minute particles of blood and other chemicals emanating from afar helps them find opportune food resources like dead whales floating in the open ocean.

The olfactory bulbs are what sharks use to smell as they process chemical signals in water that enters the nostrils. This makes sense when we see other animals that migrate over long distances also have unusually large olfactory bulbs for their keen sense of smell to keep on track during long migrations, as used by albatrosses. Some biologists have nicknamed white sharks "swimming noses." Their olfactory bulbs are the largest known in any shark, accounting for around 18 percent of the total brain weight. White sharks are seen to sometimes poke their noses out of the water to sniff smells that might tell them where they are with respect to the land.

Sharks can use their olfactory bulbs in a very special way to catch their prey by sensing potency differences in the odor source, although the potency of their sense of smell varies among different species. They do this by moving their head slowly from side to side as they swim through the water. Each olfactory bulb senses the odor on each side of the head independently, identifying on which side the smell is stronger. The shark can then slowly adjust its trajectory to home in on the source of the smell. Sharks with much wider nasal openings, like hammerheads, can calculate a smaller angle of attack using this method while cruising at higher swimming speeds. Sharks also use precise timing to turn toward the direction of the odor plume, rather than waiting the split second for the nostrils to confirm the odor's direction. This kind of behavior, coupled with an extraordinarily acute sense of smell, transforms shark olfaction into another real superpower compared with the olfactory abilities of other fishes. It has also been found recently that the internal structures of bonnethead sharks' olfactory bulbs are specially adapted for detecting faint scents in fast-flowing waters—a bit like us trying to smell a fresh cheese a hundred feet away in a field with the wind blowing toward us.

## Sharks' evolutionary experiments

Only a handful of other shark species lived during the Early Devonian alongside *Doliodus*, and these species are known only by teeth, scales, or spines. Some large three-pronged teeth from Algeria, named *Tassil-iodus*, suggest a moderate-sized shark maybe up to 3 feet long cruised the tropical waters of the northern Gondwana supercontinent about 400 MYA. The teeth suggest a predatory lifestyle adapted for gripping slippery prey like other fishes or squid-like creatures. Other large teeth of this age include the *Protodus* tusks from Canada, discussed earlier, suggesting a shark up to about 4 feet long, and with powerful, stout teeth adapted for grabbing struggling prey. Whatever it looked like, the bite of *Protodus* would have been deadly for most fishes. Perhaps *Protodus* is the first case of sharks ramping up their killing gear to tackle large prey, maybe creatures as big as themselves, such as the bone-covered placoderms and lobe-finned fishes around at the time. While these species are known only from teeth, other more substantial shark fossils representing skulls, jaws, shoulder bones and fins, have been found at the start of the next part of the period, the Middle Devonian (393–382 MYA).

Some of these intriguing early shark fossils come from South America and South Africa. These include some of the oldest shark skulls showing strange braincases and jaws with highly unusual teeth. They represent bizarre evolutionary experiments with shark body plans that give us an insight into how early stem sharks, similar to *Doliodus*, diversified. They also left us clues to decipher the stages leading to modern sharks.

I first found out about these remarkable fossils in 1995 when I came to work at the Muséum National d'Histoire Naturelle in Paris with Philippe Janvier. He is one of the great wise masters of ancient fossil fish research—a Gallic paleo-Yoda, keeper of Paleozoic vertebrate knowledge. Philippe has worked in the far-flung reaches of the world, often in dangerous and remote regions, searching for fossils. At one time in Bolivia, he was captured and ransomed by local farmers because he had unknowingly been collecting fossils from their "trilobite mine." (Read more about him on p. 407.)

The first strange shark-like fossil he worked on was *Pucapampella*, named after the region in Bolivia where it was found. It showed no evidence of shedding teeth like other sharks. At the time, Philippe concluded that it was "shark-like," but the jury was still out as to whether it could be confirmed as the oldest known shark ever found in South America. More of this fascinating beast needed to be found. Then suddenly, out of the blue, another set of specimens of this same enigmatic shark turned up halfway across the world, in South Africa.

The new South African specimens came from an Early Devonian site in the Southern Cape area (c.395 MYA), the same age as *Doliodus,* so slightly older than the Bolivian material. I spent some time in the mid-1990s working there exploring several areas searching for Devonian fish fossils in the snow-covered Cape Mountains. In 1999 we illustrated a skull of a strange shark from there and noted it was probably the oldest known shark cranium in the world at the time—just before the body fossil of *Doliodus* was found. Distinctive fin cartilages and a complete shoulder girdle came from the same site, both supporting its identity as a shark. Philippe became excited as he recognized it at once as being identical to his *Pucapampella* from Bolivia.

Philippe and colleague John Maisey (see bio on p. 421) then CT-scanned the new specimen to reveal a lot more details about the shape and internal features of the African *Pucapampella* skull. It showed many features that linked it to sharks, such as distinctive prismatic calcified cartilage, yet it had other features not seen in any known sharks. Its braincase had a slit dividing it into two parts, a feature found in some early stem sharks and in primitive bony fishes. *Pucapampella* also bore needle-like teeth directly attached to the jaw cartilages, not arranged in tooth rows for regular replacement as in modern sharks.

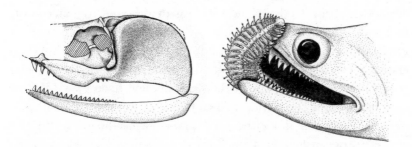

**Restored skull and jaws of *Pucapampella* (left) with reconstruction of it eating an arthropod (right)** DRAWINGS BY PHILIPPE JANVIER

Naturally, this caused quite a debate among paleontologists at the time as to whether it was a true shark, and as to exactly what features define "true sharks." Recent computer analysis of its features supports the idea that it is just another kind of "stem shark" that arose before true sharks appeared. It demonstrates that sharks were going through a big experimental phase of evolution at this time, trying out many options to tweak the structure and pattern of their skeletons, until some eventually settled on a fixed body plan that would work for them.

Another strange shark-like jawed fish Philippe studied from Bolivia is *Zamponiopteron,* known from a series of cartilage rods making up solid fin-shaped structures that resemble a set of pan pipes from the dawn of time. The calcified cartilage confirms that *Zamponiopteron* is indeed a primitive shark. These fossils suggest the fins were very important at this time for sharks to get their body shape just right, as they tried out a variety of fin rays of varying shapes. They were not successful, as later sharks all lack this feature. So while we can confirm that this is an early kind of weird shark of some sort, we know very little about it.

Philippe is also responsible for one last strange shark-like fossil from Bolivia. He found it in a Middle Devonian rock exposed along a road cutting in Cochabamba Province in 2005. When he split the rounded rock, he found what looked like a shrimp with many legs, so he gave it to another paleontologist to study. After it was cleaned in acid, the cavity in the rock revealed the truth of the specimen—it was an unusual kind of fish skull. Working with his protégé, Alan Pradel, they filled the cavities of both halves of the rock with a soft casting compound. This restored the lost beautiful form of the ancient skull, revealing its true shape to confirm that it was a complete braincase and associated jaws of another shark-like fish. The rock was then bound together as one unit and hand-carried to Texas, where it was CT-scanned at the University of Texas's high-resolution X-ray CT facility at Austin. This generated X-ray tomographic images to make models of the cavities representing the true shape of the shark's braincase and jaws. Following this, the rock (now with its own frequent flyer membership) journeyed to Grenoble in southeastern France to be the guest of honor at the European Synchrotron Radiation Facility. Here it was again scanned using a very powerful particle accelerator applying a different technique, one that measured differences in density of the specimen and its surroundings to generate images of where the cartilage had once existed.

Finally, after all this travel and technical analysis, Alan and his colleagues published their paper in 2009, naming the specimen *Ramiro-suarezia boliviana* in honor of a renowned Bolivian paleontologist. It turned out to be a strange new kind of shark-like fish that was so peculiar that it couldn't be assigned to either sharks, placoderms, or bony fishes, although its braincase and jaws most closely resembled those of a primitive shark. Recent analyses by various research groups all suggest that *Ramirosuarezia* is probably another kind of experimental stem shark on the line leading to modern sharks.

What might have been driving these weird experiments going on in shark evolution during this time? What was different about the middle part of the Devonian period?

Until recently, the middle part of the Devonian period (393–382 MYA) was thought to be a period of relatively low oxygen levels in the atmosphere, but new research is suggesting that until around 2019 we had it all wrong. A rapid oxygenation event causing levels of atmospheric oxygen to surge dramatically has recently been identified by geologists. This started around 395 MYA and continued for some 30 million years, having a truly profound effect on all life on Earth.

The seas and atmosphere were far richer in oxygen than today, and this drove plants to take over more of the land, growing larger and extending their ranges farther away from the waterways to whose environs they had been confined. The evidence for this theory is seen in a range of studies: the detailed measurement of increased tree height, fossil plant roots that grew deeper in ancient soils, the higher lignin content of plants (the woody part of the trunk), and in the dawn of seed production. This major boost of oxygen was as though all life was suddenly taking steroids, with higher activity levels easier to maintain as metabolic rates went up.

This global surge in oxygen had profound effects on life in the oceans and rivers at this time, creating aquatic environments able to support a larger biomass. This in effect kicked off a new wave of diversification for all fishes, including sharks. Higher global oxygen levels also enabled some lobe-finned fishes to evolve novel ways to supplement their oxygen, by swallowing air into their lungs through holes in the top of their head (called spiracles). They eventually diminished the use of their gills as lung breathing took over, invading land to explore new terrestrial habitats. These were the first **tetrapods**, or four-limbed animals, such as the earliest amphibians. The oldest evidence of tetrapods walking on land are the trackways found in Zachelmie, Poland, dated around 390 MYA, of large newt-like amphibians up to six feet in length. By the Middle Devonian, tetrapods had also left tracks over ancient Ireland and Scotland. By around 375 MYA, tetrapods had crossed the world and invaded Gondwana, leaving tracks in Australia.

How amphibians evolved from fishes is a great example of extreme evolution, in which a quantum leap in body form and physiology is necessary to transform a fish in water to an amphibian living on land. Such dramatic transformations that take place in a relatively short geological time usually occur when creatures invade a challenging new habitat.

This serves as a nice yardstick to measure how just much evolutionary change sharks were to undergo at this same time, as they faced the same challenge as they began invading new habitats. They would have to adapt to dramatic changes in salinity as well as navigating lagoons, estuaries, narrow river channels with strong currents, and large stagnant lakes. I was one of the scientists who made this discovery when I participated in field work collecting fossils from one of the most treacherous continents on Earth—Antarctica.

We started this book with a tale of collecting fossils in Antarctica; here we pick up the story to see what sharks were doing at this time in the vast river systems of east Gondwana, where Antarctica was then located.

## Cold killers: ancient sharks of Antarctica

I have collected many fossils from Middle Devonian rocks in Antarctica over four expeditions spanning the past thirty years. The multicolored fossil-bearing layers called the Aztec Siltstone represent a gigantic Middle Devonian river system with lakes dotting a landscape forested by low tree-like shrubs and ferns. Working in the footsteps of previous Australian, Scottish, and New Zealand scientists who had visited the region some two decades earlier, in 1992 I found a number of fossil sharks' teeth here, which later turned out to represent several new species.

Many people I speak to about these trips are surprised to learn that fossil shark teeth have been found in the wind-blasted, ice-covered rocky peaks of this harsh southern continent. The first Devonian shark's tooth found in Antarctica was a single small specimen named *McMur-*

*dodus featherensis,* collected in the late 1950s the hard way—by two intrepid New Zealanders, Bernie Gunn and Guyon Warren, on a three-month dogsledding journey through unexplored parts of the Skelton Névé region. The small collection of fossil fish they found was sent to the famous British Museum paleontologist Errol White. He studied the tooth and realized it was totally different from those of any known shark from the Devonian. It had a wide, flat root and bore many jagged cusps, a bit like a living bramble shark tooth. It was a mystery tooth, as nothing like it this old had ever been found before.

I examined this *McMurdodus* tooth in the British Museum in London in 1992. I estimated it had as many as fourteen small cusps, with the outer edges bearing the largest cusps, which are coarsely serrated with all the charm of a rusty steak knife. This kind of complex tooth represents the first appearance of any shark with serrations on its teeth, a trait specifically adapted for cutting flesh from large prey. It's a clear deviation from the *Doliodus* style of teeth that gripped small prey so they could swallow them whole. It signals a new, more aggressive direction: that sharks were beginning to gear up as serious predators by developing new tooth shapes for more specialized ways of hunting and killing.

I later found some new sharks' teeth similar to *McMurdodus* in the mountains near Mansfield in Australia. These teeth, up to one-third of an inch wide, differed slightly from the original *McMurdodus* tooth, so we gave them a new name, *Maiseyodus,* in honor of John Maisey from the American Museum of Natural History. Other *Maiseyodus* teeth had also been found previously in central Queensland from older rocks dated around 393 MYA. *McMurdodus* and *Maiseyodus* teeth represent a radical new design making these sharks capable of new ways of eating their prey—perhaps sawing through the flesh of large prey like some of the huge bony fishes that lived alongside them.

In 1971 a New Zealand expedition went back to Antarctica to search the same fossil layers where *McMurdodus* was found. Among the many superb fossil fishes found on that trip was a very nice articulated fossil shark from the Lashly Range found by Gavin Young of Australia (see bio on p. 413).

Gavin's shark from the Lashly Range was what is called a **Rosetta Stone** specimen—one that is complete enough to immediately clarify the identity, by comparison, of many isolated fossil teeth, scales, and fin spines from several other localities in southeastern Australia and other parts of the world. His specimen was a skull with the front half of the body and fins, covered by scales and fin spines. At the time it was found, it represented the oldest known shark head with the braincase preserved. He named it *Antarctilamna*, derived from the Greek word *lamna*, "shark," and "Antarctica." Its teeth and spines have since been found throughout the world, including in Bolivia, Iran, the United States, Canada, and South Africa. *Antarctilamna* was about 2 to 3 feet long with characteristic delicate teeth about a tenth of an inch wide. They bear two large, curved cusps with smaller cusps around their bases—somewhat similar to those of the living frilled shark, perfect for grabbing slippery prey like small fishes or worms.

Giant pronged teeth of *Portalodus* from Antarctica, representing a 10-foot-long Middle Devonian predatory shark   AUTHOR PHOTOGRAPHS

The fossil shark tooth specimens I collected from Antarctica over the 1991–92 field season represented three new sharks. I named a new species in honor of each of my fellow expeditioners: Margaret Bradshaw has *Portalodus bradshawae*, Fraka Harmsen has *Aztecodus harmsenae*, and Brian Staite has *Anareodus staitei*. *Portalodus* turned out to be the largest known shark's tooth so far found anywhere in the Middle Devonian. Later revision of this work by Michal Ginter of the University of Warsaw, Poland, culled this group to just two species, as he suggested *Anareodus* were likely the smaller back teeth of the larger *Aztecodus* teeth. These fossil teeth, together with Gavin Young's *Antarctilamna*, tell us sharks were heading in another new direction at this time.

# Big shifts in shark evolution

These discoveries from Antarctica signified that three big shifts in the evolution of sharks occurred by the start of the Middle Devonian. First, sharks had dramatically increased in size. From most sharks in the Early Devonian being small fishes about one to two feet long, with rare exceptions up to around 4 feet long, they swiftly moved to be apex predators like *Portalodus*, reaching maybe 10 feet long. This shark thrived alongside the similarly sized predatory bony fish *Notorhizodon*, for being the top dog in the ancient southern river system.

The second big shift is that sharks were invading freshwater habitats for the first time. This was a monumental step for sharks, because even today only a small number of sharks can live in freshwater. A few are very good at it, like the bull shark, which can venture up large tropical river systems to more than a thousand miles inland. The Aztec Siltstone represents sediments deposited by truly massive river systems—think Mississippi-scale rivers and adjacent oxbows and lakes. This has major implications for the evolution of the shark **osmoregulation** at this time: how they maintain the salt and water balance across the various membranes (through osmosis) inside their bodies. Sharks do not urinate in marine conditions; they expel water while retaining the vital chemicals in urea needed for osmosis to work. In order to do this, freshwater sharks have a special rectal gland that allows them to adjust the salinity level of their body fluids to match that of their environment. Perhaps this shift to freshwater by the Antarctic Devonian sharks indicates the precise time when rectal glands first evolved in sharks.

The third big shift is that we find a higher diversity of sharks—at least four species)—living together in the same ecosystem for the first time. This means that sharks were now able to coexist by occupying different niches without necessarily competing for the same food resources. For this to work they would have had to specialize so that each species hunted different kinds of prey living in the same waters. As the fish fauna from these layers is exceptionally diverse, representing some eighty species, this would not have been difficult.

The discoveries from Antarctica made me think about other fossil sharks from the Middle Devonian. What were sharks doing at this time in the seas and rivers of the Northern Hemisphere? Were they showing similar trends, or perhaps still experimenting with developing their body shapes and dentitions? How many other sharks existed at that time?

About a dozen or so other kinds of sharks swam in Middle Devonian oceans, all known from tiny teeth. Some like *Phoebodus* conform to earlier described forms, such as *Antarctilamna*, with small body sizes and grabbing piercing dentitions. At the end of the Middle Devonian we find a new kid on the block—*Protacrodus*. It represents the first time sharks developed domed teeth for crushing prey, like clams and snails. Up to this point all previous sharks were sharp-toothed forms, clearly predators. Sharks like *Protacrodus* would soon become much more common throughout the world. Sharks were now evolving at a more rapid rate. Several other new kinds of tooth shapes would soon appear, signaling the switch to a much broader range of prey and new ways to catch them.

Sharks had really come of age in the Middle Devonian, growing bigger, diversifying into new feeding strategies, and invading new freshwater habitats. But this was nothing compared to what sharks would achieve next. The Late Devonian saw sharks reach a new zenith of diversity and ferocity. They kept on going with their final push to take over the oceans and rivers, battling rising competition from other large predatory fishes.

CHAPTER 4

# THE FIRST RISE OF SHARKS

*Sharks Lash Out*

---

## Meanwhile, in the Late Devonian

The Late Devonian (383–359 MYA) represents 24 million dramatic years of Earth's history. Life on land took off in a spectacular way as the vegetation changed dramatically in scale. Middle Devonian forests had low to midlevel canopies, while in the Late Devonian we find the world's first towering forests made of gigantic plants. The horsetail clan of water plants, the **lycophytes**, morphed into trees over 100 feet high, jostling for light among the towering fern-trees like *Archaeopteris,* up to 80 feet high. Immense fungi, *Prototaxites,* formed bizarre rounded columns that soared almost 30 feet into the air. The first tetrapods, early amphibians sporting many fingers and toes, were venturing out of the water and invading the new frontier: land. The seas were teeming with plankton, more free-swimming than floating forms. The vast numbers of microorganisms fueled the complex food webs of the oceans. This rich energy source supported many species of placoderms, bony fishes, and sharks that continued to diversify and explore new ecological niches. Sharks flourished, primarily as small to medium-sized predatory fishes, most under 3 feet long and most armed with similar sharp, multipronged teeth. At the end of the Devonian, we see one shark group rapidly increases in size.

Throughout the Late Devonian there was a continual series of chaotic events shaking up life on Earth. Rather than one catastrophic hit, there was a maelstrom of around eight to ten smaller crises that choked the seas of oxygen, resulting in a prolonged mass extinction phase with two main killer pulses: the Kellwasser event at the end of the Frasnian stage, 382 MYA, which mainly devastated life in the seas, and the Hangenberg event at the end of the Devonian, 359 MYA, which culled all life in both marine and freshwater environments. Life became very challenging in both the oceans and the rivers at this time. This environmental disruption meant that creatures well adapted for change would survive, while others would succumb to extinction. How did sharks make it through these events?

To set the stage to answer this question, let's take a look at the kinds of sharks that were plentiful in the Late Devonian before the disruption. I found one in Western Australia some years ago; here's the story.

## Finding a new Devonian shark

The first half of the Late Devonian is known as the Frasnian stage (382–372 MYA). We know of a small scattering of stunning fossil sites from the Frasnian around the world, yielding spectacular fossils with excellent preservation of their skulls and jaws. Many more sites of this age have yielded large numbers of individual sharks' teeth, scales, and fin spines. Together, these finds give us the full picture of Frasnian shark diversity across the globe.

One of these exceptional sites is at Gogo, in the north of Western Australia. It's arguably the world's best fossil site for fishes of this age. The fossil fishes are encased in limestone nodules and can be acid-prepared out of the rock. This process reveals stunning 3-D preservation of their skeletons, as though they died yesterday, their bones still white in most cases. Recently we have found incredible preservation of muscles and soft internal organs.

Since the early 1960s, there have been many expeditions to Gogo by British and Australian teams resulting in excellent fish fossils. I have been collecting at these sites for almost forty years. However, until recently, not one shark fossil had ever been found here. It was a mystery: Why would such a diverse tropical reef environment, teeming with more than fifty species of fishes of all shapes and sizes, lack a major group like sharks?

Other marine sites around the world of this age had sharks. I had thought that sharks were simply not found at Gogo because they had delicate cartilage skeletons that did not preserve well—that time had slowly and irrevocably erased all trace of them. Then one day, out of the blue, one turned up, and this sublime moment of discovery is now forever stamped in my memory.

It was the morning of July 7, 2005. I was working alone in the long spinifex grass near a site called Stromatoporoid Camp. This beautiful, partially sheltered site is named after the ancient sponges called stromatoporoids that formed huge reefs here some 380 MYA. It is in a wide grass-filled valley overlooked by a ring of irregular limestone hills. As I walked slowly through the long grass, making noises to scare away the snakes, I wandered into small clearings with random patches of unbroken Gogo nodules—rounded concretions of limestone rock—concealing unknown and potentially priceless scientific treasures. I raised my sledgehammer and began to eagerly crack them open. A good strike will split the nodule in two parts straight through the center, instantly revealing if it has a fossil inside. Most don't. If you find one good fossil fish for a full day's work splitting rocks, then it's been worth the effort.

The nodules here were rather small and not the most interesting I'd ever seen. This led me to think that previous expeditioners probably walked right past them, searching for bigger nodules with potentially bigger fishes inside them. I cracked open an odd-shaped nodule, only about 4 inches long, and immediately saw it had a hazy outline of something in it. Under the glare of the hot sun, it was obvious that this was not a fish with bony plates like a placoderm, the commonest fish we find here. Curious, I crouched to pick it up, reaching into my pocket for my hand lens. On closer inspection, it became immediately clear that I'd found a shark.

My heart nearly exploded with excitement. I had always hoped—perhaps on some level I knew—that one day we would find a shark at Gogo. I screamed to my buddies searching for fossils in the long grass some 600 feet away: "Shark! Shark!" This warning cry usually sends terror through hearers, but not on this day. This was a resounding cry of joy. The small complex teeth with shiny enameloid blades were visible. Scattered teeth and scales enveloped a set of elongated rod-like structures that I suspected might have been made of mineralized cartilage, perhaps representing the two sides of the lower jaws. I felt elated by the thought that the rest of the head and braincase might still be inside the rock. There were cold beers around the campfire for everyone that night.

**The author minutes after the discovery of a shark from the Gogo site in Western Australia, July 2005**   PHOTO BY THE LATE LINDSAY HATCHER

Once back in my lab at Museum Victoria, I felt a little uncertain about what to do next. Examining the nodule bearing the shark, I could see the lower jaws and some other paper-thin cartilages exposed in the broken rock face. I was worried that the cartilage might break up in the acid bath. Science is a process of informed experimentation, and there was only one way to find the best approach. Or, as we say in the field: no guts, no glory.

I decided to test the fossil by immersing it in a much weaker acetic acid solution than normal for an hour or so to observe what happened. I set the timer on my phone and walked away to get some lunch. I couldn't eat, my nerves were jangling with anticipation. I returned to the lab and, to my relief, the cartilage stood proudly out of the etched rock, confirming that it was solid enough to withstand the acid treatment. Carefully I washed the sample in water to rinse off the acid residue. I then let it air-dry before applying a diluted plastic glue to the exposed cartilage and teeth. The glue seeps into the pores of the tissues and hardens, making the specimen robust and able to be handled.

Finally, I placed it back in a bath of dilute acid to continue the process overnight.

Preparing my own Gogo fossils still gives me a real buzz. It makes me want to rush in to work early every day to see what the acid baths have revealed overnight. Having left the shark alone in the bath overnight, the following morning I was overjoyed to see the dark brown jaw cartilages poking a little further out of the rock. As I examined and photographed it after each acid bath, I would note an emerging new bone, or a cluster of freshly exposed pointed teeth. After a few weeks of this process, two very large mysterious bony elements began poking out of the rock. I had no idea what they were until more was exposed. Then the penny dropped—they were the complete shoulder girdle bones that supported the thick triangular pectoral fins, with both sides in perfect 3-D shape. These bones are called **scapulocoracoids**, and all others known from sharks of the time were flattened 2-D specimens. I was looking at the first intact Devonian shark shoulder girdles ever to be prepared out of rock. A 3-D shoulder girdle showed its true shape, and that it was pierced by several small holes for the nerves and blood vessels of the pectoral fin.

The shoulder girdle bones were not the only treasures this shark

The Gogo shark as it was found: top; exposing jaw cartilages; below, after a week's immersion in weak acetic acid to dissolve the rock **away** AUTHOR PHOTOS

would reveal. The many teeth and scales that surrounded the head slowly dropped out of the rock in the acid residues. These were easy to pick out with a wet paintbrush and mount on microscope slides. It took me about six weeks of painstaking preparation for the entire rock to be dissolved away, liberating the shark's remains. Unfortunately, it wasn't a complete head, but comprised the two lower jaws, parts of the gills, the two shoulder girdles, and about eighty-two teeth and numerous scales. These represented much of the dentition from the lower jaws, together with scales from the head and body regions. I concluded that the fossil likely represented a decayed part of the carcass, where ligaments held these various parts together. The rest of the head and body must have been lost to a feeding frenzy of the scavengers so prevalent within reef habitats. Incomplete though it was, it remained an important and exciting find, for now we knew sharks were present in the Gogo reef deposits. There was enough of the shark preserved to describe it as a new species, which we named *Gogoselachus lynbeazleyae,* meaning "shark from Gogo named in honor of Lyn Beazley." At the time, Lyn was the Chief Scientist of Western Australia, and a dedicated supporter of our work at Gogo.

*Gogoselachus* is nothing special to look at; its skeletal features are like those of other fossil sharks with multipronged teeth, a bit similar to another common shark called *Cladoselache,* found in the Cleveland Shale sites of North America (I'll tell you more about this one in the next chapter).

The next big mystery to solve was whether *Gogoselachus* had a standard shark type of cartilage, as we rarely get thin cartilage shells like this preserved in such remarkable 3-D form. So I brought in my colleague from the Queensland Museum, Carole Burrow, a world expert on ancient sharks and stem shark (acanthodian) microfossils—their scales and teeth. She especially loves to dive deeply into the intricate complexities of the internal structure of their tissues. Carole ground down thin sections of some loose pieces of *Gogoselachus* jaw cartilage to a thickness of about 30 microns—thin enough to be transparent on a microscope slide.

I waited each day with great anticipation to hear what Carole would make of the samples I had sent her. Then one day, about a month later, the verdict came in. I was over the moon to hear from her that the internal structure of the mineralized cartilage making up this specimen's skeleton was radically different from that of any other shark known. *Gogoselachus* had a unique type of cartilage. While it had true shark features, like the globular and prismatic calcified cartilage that distinctly characterize the chondrichthyan group, there were also tiny holes in the matrix between the tesserae, called **lacunae**, each with radiating fine processes splaying in all directions. To a trained eye like Carole's, these resembled **osteoblasts**, or bone cells, located between the building blocks of prismatic cartilage. This was something very new, in a specimen very, very old.

No living shark has bone cells in its cartilage, as far as we know. It has long been known that cartilage is the precursor to bone in our human skeletons. This explains how a newborn baby has rubbery legs and cannot stand up; the limbs first form as cartilage rods, meaning there is no support until the bone forms around the cartilage core. Then the baby can walk. The nineteenth-century naturalists thought that sharks were primitive fishes because they hadn't evolved the ability to make bone. *Gogoselachus* was the smoking gun, telling us that at a certain point on their evolutionary path, sharks had once had bone in their skeletons. By the Late Devonian, around the time our *Gogoselachus* met its end, they were well on the way to losing it. Success for sharks meant casting bone aside in favor of developing more efficient, lightweight cartilage skeletons to supplement their sleek aerodynamic body shape. The most recent analyses of the evolutionary position of *Gogoselachus* show it as the ancestral species—what we call the "sister group"—to all modern sharks, rays, and chimaerids.

Other amazing new sharks, some with 3-D preservation and soft tissues preserved, have recently been found in later Devonian rocks of the Famennian stage (372–359 MYA) in another part of the southern continent of Gondwana. These have provided vast amounts of new information about the typical sharks of this age, and how they lived and hunted their prey. They were found on expeditions to northern Africa by an extreme-sports-loving German paleontologist.

## The mountaineer and his magnificent Moroccan sharks

Christian Klug grew up near Nuremberg, in Germany, fascinated by fossils. At the age of twelve his uncle took him out on a trip to search for "golden snails"—the fossilized shells of squid-like creatures called

**ammonites**. At fifteen, Christian found the oldest known crab fossil in the world, from the Early Jurassic (c.180 MYA). Today he is a professor of paleontology at the University of Zurich who loves rock climbing and exploring the Swiss Alps. Although he specializes in the study of invertebrates, he has recently ventured to the dark side—vertebrate paleontology—through his discoveries of some incredible fossil sharks on his expeditions to the deserts of Morocco.

**Christian Klug (left) with field team in the Devonian mountains of Morocco** PHOTO COURTESY OF CHRISTIAN KLUG

While on his first visit to Morocco in 2003 with his family, Christian stumbled upon a fossil shop selling limestone nodules with fossil fishes in them. He purchased one that had shark teeth around its edge, hoping it might contain a good skull of a Devonian shark. It did! As their relationship developed, the fossil dealer showed Christian where he could search for his own specimens. Christian scoured the spectacular desert terrain for fossil fish, and occasionally he purchased fossils from reputable fossil sellers, who eventually showed him the precise sites in the field so he could search for more. Over time he amassed a very significant collection of new kinds of Late Devonian sharks. Christian worked closely with the Moroccan authorities over several years, eventually reaching an agreement that allowed him to continue to collect or purchase scientifically important shark specimens as long as a fair representation of the new species are returned to universities or museums in Morocco.

The first big discovery Christian made in partnership with the government was of a nearly complete shark called *Phoebodus* that he had acquired back in 2001. *Phoebodus* was already well known by its teeth, represented by many species found around the world in rocks starting from the late Middle Devonian. Christian offered this beautiful specimen to his student Linda Frey to study for her doctoral dissertation. This 3-foot-long shark was named *Phoebodus saidselachus*, honoring her Moroccan field collaborator Said Oukherbouch, who found the original specimen. This animal was immediately distinguishable from the other Devonian *Phoebodus* species by the detailed features of its teeth.

dorsal fin spine

jaws, teeth

**Phoebodus from the Late Devonian of Morocco, showing exquisite preservation (above) and the restoration of the shark (below)** IMAGES BY CHRISTIAN KLUG

It turned out to be the earliest known shark to have a specialized eel-like body shape, very similar to that of the living frilled shark, *Chalamydoselachus*. The fragile jaws of the fossil shark suggested it had a bite force weaker than most other sharks of its age. The deposits where the fossil was found show that it inhabited continental shelf seas around 1,000 to 1,300 feet deep about 365 MYA. It bore some quite specialized gill arch bones that implied it fed using a swift snapping action of the jaws, combined with powerful suction, to capture its prey and swallow it whole. This further reinforces that the many Devonian species of *Phoebodus,* the most abundant of all sharks in the Devonian oceans, were living a lifestyle much like that of the living frilled shark. They were slow, opportunistic ambush hunters that probably fed on crustaceans, swimming worms, and small fishes in the middle of the water column.

This new, complete Devonian shark provided several critical pieces of anatomical information that allowed Christian's team to create a revised evolutionary tree for early sharks. The relationships of the new shark within a group of closely related forms is called its **phylogenetic position** (think of a phylogeny as a formalized kind of evolutionary tree). This can be determined by creating a data matrix of specific character states, each related to different parts of the anatomy, and scoring these features as present or absent for all the species in your matrix. The resulting data matrix of characters is run through a computer program to determine the likelihood of species being more or less closely related to other species. The results show us the likely point in time when sharks and rays diverged from chimaerids on the evolutionary tree. Their result pushed the timing for this big split in chondrichthyan evolution back by roughly another 10 million years, to 382.7 MYA, at the start of the Late Devonian. Like all findings based on new fossil discoveries, this date may be further adjusted by future discoveries of well-preserved shark fossils.

The next big find announced by Christian and his team came from the same region in Morocco and dated the same age. It was another totally mind-blowing fossil shark, almost complete, with soft tissues preserved, showing that its liver had two distinct lobes, its stomach and the spiral valve of the intestine all clearly seen. It was not possible to imagine just a few years ago that we could find such a well-preserved shark of this age and then image the inside of it with our new whiz-bang high-tech toolkit. It was CT-scanned so it could be digitally reconstructed to reveal the fine anatomy of its skeleton and all its soft tissues inside the rock. It had very large eyes, suggesting that it relied more on sight than its other senses to find prey in deep, dark waters. Even more intriguing was that it was preserved inside a rare red iron-rich concretion, so they named it *Ferromirum,* meaning the "Iron Wonder." This fantastic little shark grew to just over a foot long.

*Ferromirum* lived alongside a lot of strange, fat, shrimp-like crustaceans called **thylacocephalans**, which it may have hunted using a ram-

*Ferromirum*, the "Iron Wonder" from Morocco   ARTWORK BY
CHRISTIAN KLUG

feeding approach. Imagine a mouth full of fishhooks that when opened makes the hooks stick out in all directions—that's how *Ferromirum* hunted. The jaws were unique compared to those of any other sharks or fishes, living or fossil, because the way its jaw joints worked was counterintuitive to the way most jaws work. Instead of simply opening and closing the mouth, the jaw joints of *Ferromirum* worked by the lower jaws rotating outward as its mouth opened, displaying the rows of sharp hooked teeth, then pulling the teeth back inward as the jaws snapped shut. It likely swam with its mouth open into a swarm of crustaceans hoping to hook or snag one on its outward-facing curved teeth, then swiftly shutting its mouth. This motion was aided by strong gill arch bones and muscles that created a powerful suction feeding effect as the mouth closed. This is the first time in shark evolution we see this kind of specialized suction feeding coupled with a crazy jaw-closing mechanism.

It signifies that *Ferromirum* wasn't just another one-trick pony—it had developed its own very special way to catch prey, another first for sharks in our story so far. It is another example, one of many that will come later, showing sharks evolving distinctly specialized and often unique ways to catch their prey. This is the problem with studying sharks from their teeth alone—we can deduce so little from the tooth's shape. When we have complete fossilized heads with 3-D jaws and gill arch bones, we can build a working model for how the anatomy operates—how each lever and joint functions with the others—to make the jaws open and close in a specific way. This ability would be a driving force in the evolution of sharks and their kin from this point onward.

The updated shark family tree generated by the new information from *Ferromirum* also demonstrated that it was part of a diverse line of strange sharks that were more closely aligned to the **holocephalans** (ratfishes and chimaerids) than to the main line leading to modern sharks. This serves as a great example of how sharks of this age that resemble the typical sharks we know could go in one of two major evolutionary directions: toward modern sharks, or toward modern holocephalans. *Ferromirum,* while looking like your typical shark, was on the latter branch (see figure on p. 116), not far from the point of divergence, so it still retained an ancestral shark-like body.

Christian and his colleagues have recently described yet another well-preserved shark from the same age layers as *Ferromirum. Maghriboselache,* meaning Morocco shark, from an Arabic word, is known from several complete specimens, the largest being around 8 feet long. This makes it one of the largest sharks of this age anywhere. Like *Ferromirum, Maghriboselache* also evolved at the base of the holocephalan line, and is most closely related to *Cladoselache* from Ohio. It was indeed a weird-looking shark with a broad, flat head. Its eyes and nasal capsules are spaced widely apart, as though it was thinking of becoming a Devonian hammerhead. The brain shape with its wide nasal capsules are beautifully restored through CT scans. Imagine it swiping its wide head back and forth, taking in subtle differences in odor to better target its prey as it swam through the water.

The discoveries made by Christian's team have dramatically changed how we think about Devonian sharks. His three new sharks each show us distinct and specialized feeding mechanisms, highlighting that sharks were evolving more complex ways to catch food. The extreme body shape seen by the wide head of *Maghriboselache* also testifies that sharks were experimenting with new sensory ways to hunt and find prey. These significant discoveries also demonstrated the value of novel ways to study fossil sharks by 3-D modeling of their skeletal anatomy from scan data.

The complete *Phoebodus* specimens collected by Christian and his colleagues in Morocco were exceptions. Most species from this time are known only based on teeth. I collected some *Phoebodus* teeth on an

Reconstruction of the widenosed shark, *Maghriboselache,* from the
Late Devonian of Morocco   ARTWORK BY CHRISTIAN KLUG

expedition to northern Thailand many years ago that proved later to be
an important footnote in the shark story. It was also quite a wild trip.

## Fossil sharks of the Golden Triangle

Back in the late 1980s I worked at the University of Tasmania on a
project comparing the Devonian faunas of Southeast Asia that were
then part of the ancient Shanthai Terrane, or landmass, to those of
Australia and Antarctica that formed the East Gondwana region. My
goal was to find out whether there were similar species on the two land
areas to support a closer position of these continental blocks and figure
out when they split apart. We estimated that the best location for finding fossils was in the remote mountainous regions of northern Thailand, not far from the Myanmar border. This region is the infamous
Golden Triangle, well known for its drug lords and opium poppies.

By any standards, this was a memorable field trip. Despite our pre-
cautions we still had some close calls. One day our team had split up
and I was waiting by a road cut with Clive, my boss and fellow paleontologist, talking about some of our finds and joking about dinner. Suddenly we found ourselves confronted by hill tribesmen, their old
musket-like rifles pointed at us. We stayed calm and answered their
questions in broken Thai, but inwardly I was fearful for my life. After

several minutes (though it felt like hours), the potential bandits disappeared into the scrub upon the sudden arrival of our United Nations–bannered Land Cruiser. We later learned about a group of foreign tourists that had been robbed and murdered in these same hills. We had been lucky in more ways than one, making the daily risks of working in such remote and hostile territory worthwhile. We did find some choice fossils in our rock samples.

Among other fossils, I found fourteen *Phoebodus* teeth in Thailand. At home in my lab, I recognized that they were identical to teeth found in Australia, figured in a scientific paper by Sue Turner, who is warmly regarded as the mother of fossil shark research in Australia. In 1982 she published a landmark paper describing the first records of several fossil sharks from Australia, yet she did not create a new species for her *Phoebodus* teeth, arguing they were closest to an existing species from North America. Noting the distinct variations from the North American species, I now had enough evidence to name these teeth from Thailand and Australia as a new species. I must have run out of clever ideas that day, as I unimaginatively called it *Phoebodus australiensis*. The story that follows shows how scientists create new scientific names building on a first species named, and how that species can be upgraded to form new, higher levels of classification by others.

In 1992, after amassing a more extensive collection of Devonian fossil shark teeth from field trips across Europe and Morocco, Michal Ginter, from the University of Warsaw in Poland, recognized that there were variations in all the *Phoebodus* teeth of this age. One grouping of species were seen as typical *Phoebodus* with highly curved cusps coming off a wide root. My *Phoebodus australiensis* had cusps with distinctive thick ridges on them, and a differently shaped root. Michal thought these differences were significant, so he reclassified my species and put it into a new genus, renaming it after the person who first described the species (as is often done in paleontology). He called it *Jalodus,* using my initials, JAL (meaning "JAL's tooth").

When more species of *Jalodus* were found around the world in 2002, Michal and his colleagues decided to erect the new family Jalodontidae. Eventually, as other new genera and species were found allied to *Jalodus* now ranging over a 150-million-year period, the new order Jalodontiformes was created in 2021. I was thrilled to see the name climbing farther up the classification pyramid. Orders are very

Assorted teeth of *Jalodus* collected by the author from the
Golden Triangle in Thailand in 1989. Bar scales are
0.2 mm   AUTHOR PHOTOS

high-level groupings in biological classification. The order Primates, for example, contains around four hundred species of monkeys and apes, including us humans. It's humbling as a scientist to see how much later work was built upon the initial description of a thimbleful of tiny fossil teeth found in rocks collected from the steamy jungles of northern Thailand. A lucky trip indeed.

*Jalodus* was also a significant shark for determining the sea depths where sharks were living. Michal Ginter recognized three distinct shark faunas that represented different marine habitats in the Late Devonian oceans. Geologists can determine ocean depth from rock samples by studying the sediments and the fossils they contain. For example, shallow near-shore seas often have coarser sediments like sandstones or limestones with corals and other marine species that inhabit shallow depth ranges today—in the "photic zone" where light can penetrate. Deeper waters contain finer-grained sediments, like silts and muds, with the smallest particles like clays carried in the water longer and farther before the particles finally sink to the seafloor in the deepest ocean basins.

Using this logic, it's possible to work out that the first habitat was the shallow near-shore seas, which were dominated by the crushing tooth-plated sharks called *Protacrodus* and *Orodus*. These forms lived where there was an abundance of marine clams, snails, and crustaceans to feed on. It was heaven for sharks that loved a good clam chowder.

The second fauna of sharks was dominated by abundant *Phoebodus* and its close relative *Thrinacodus,* which lived in the open continental shelf areas that generally extend in depth from around 60 to 660 feet or so. These sharks loved their shrimp dinners, with a side of fresh fish.

The third shark fauna represented the deep ocean basins—the most expansive areas of our seas, which extend down to the seafloor about two and a half miles on average. This region was dominated by *Jalodus* teeth. These sharks probably preferred the soup of the day; they were specialists in small prey like plankton or wormy conodonts.

This new information about their habitats and depth ranges is vital for understanding how sharks would cope across the chaotic end-Devonian extinction events. Many species were now living well away from the shallow coastal seaways where sudden sea level changes could cause a catastrophic loss of habitat.

## Sharks scale down their diet

Some scientists spend their entire career working on isolated fossil sharks' teeth that are found by dissolving them out of rocks—the paleodentists of our world. Michal Ginter, whom we've just met, is one of the true heroes of this tiny tooth brigade. Michal and I recently reminisced about our first meeting in the early 1990s. He was a PhD student visiting Australia when he stayed a few days with me and my family in Perth. He told me that he still remembers the day he walked into my office at the museum in Perth and was elated because I had his first published paper on *Phoebodus* teeth sitting open on my desk.

We spent a good deal of time on that visit together staring down microscopes at various sharks' teeth that were unknown or undescribed at the time. Michal was really at the cusp (no pun intended) of a then rapidly expanding field of paleontological research, the study of shark microfossils. He specialized in the classification of shark species known only from isolated teeth. This kind of research is so important to our story because the majority of extinct sharks (perhaps around 98 percent) are known only from their teeth. Studying their teeth is our only way of accurately estimating shark diversity and their rates of evolution through time.

Today Michal is without doubt the foremost world expert on Devonian sharks' teeth. He was also one of the first scientists to recognize that by the Late Devonian, sharks were starting to try out new things, like a highly specialized way of feeding on very small prey. In 2008 he described some very unusual fossil teeth from Utah and Nevada, claiming they were heading in a different direction from all previously known sharks of the Devonian period.

These new fossil teeth were uncovered in shales and limestones deposited about 388 to 376 MYA in deep water, representing the open ocean. Michal named one species *Diademodus utahensis.* This species was about 14 million years younger than the complete *Diademodus* specimen from the Cleveland Shale in Ohio that was described by the Englishman John Harris at Cambridge University in 1951. Harris thought the strange, delicate teeth of this shark must have been fused together because they appeared far too wide for any shark of that age. Michal reexamined the original Ohio specimen and found that it was exactly like his new teeth from Utah, both having very wide bases with a comb of delicate needle-like cusps.

Michal suggested that the needle-like teeth meshed to create a sieve to filter out plankton and other small organisms from the seawater, as when you join your hands in a contemplative pose, fingers loosely interlaced. Predatory sharks that need to grip their struggling prey tightly usually have a swollen lump on the root for anchoring it to the next tooth in the row behind it. This arrangement prevents teeth from being pulled out of the mouth when biting down hard on wriggling prey. Michal observed that the *Diademodus* teeth lacked this special feature, supporting his idea that it was not biting hard on prey, but using its teeth as a sieve for filter feeding. Today all filter-feeding sharks are massive; whale sharks, basking sharks, megamouth sharks, and manta rays are classic examples.

All these living forms feed by using their gills to sift out the plankton from the water passing through their mouths and out the gill openings. None of them use teeth to feed, so most have minute teeth that might have a function only when mating. I can imagine the warm Late Devonian seas above current-day Ohio where large schools of 3-foot-long *Diademodus* swam, ramming through clouds of plankton with open grinning mouths, sieving the tasty tiny morsels out of the water by trapping them between their teeth. They might have been the first creatures on Earth to take advantage of the abundant plankton now proliferating in the open seas.

Michal identified a second species of filter-feeding shark from the same age with similar teeth with wide, slightly more robust cusps,

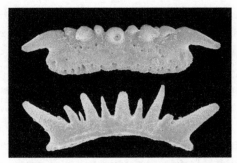

The widely spaced, delicately cusped teeth of *Diademodus* sug-gest that it was a filter feeder   SEM IMAGE COURTESY OF MICHAL GINTER

which he named *Lesnilomia*. These teeth have been found in the United States as well as Moravia in Europe. The teeth of *Lesnilomia* are slightly bigger than those of *Diademodus*, so the entire shark might have been up to 4 feet long. The sharp ridges on its teeth suggest it also retained some ability to catch soft prey, like squids, or plump **nu-dibranchs** (also called "butterflies of the sea"—a kind of sea slug), while still being capable of sieving out small plankton-sized morsels from the water.

Since our first meeting three decades ago, Michal's work has made major contributions to our knowledge of shark diversity and their en-vironmental ranges through time. Collectively, his discoveries have tripled the number of species of sharks known from the Devonian pe-riod since he began his quest. These discoveries confirm that at the very start of the Late Devonian, some sharks were becoming inventive in securing their food. As the body shape of *Diademodus* had not radi-cally changed in comparison to other Devonian sharks, we know they achieved this change by simply experimenting with their tooth shapes. Once more, we can see how they were changing with the environmen-tal resources and altering the food chains around them.

While filter feeding is a radical breakaway from the catch-and-kill style of predation, other sharks at this time went even further. Building on the success of *Protacrodus*, other forms appeared throughout the Late Devonian with rounded, arched teeth specifically adapted for crushing hard prey, including *Orodus*, known from the Cleveland Shale sites in Ohio, and *Deihim*, another shark known by crushing teeth from the end of the Late Devonian of Iran. *Deihim* has round, elongated teeth with little rows of secondary rounded cusps to make two crushing surfaces on each tooth. It's an insane style of tooth for a shark, operating like a mammalian molar in the way it has complex ridges and valleys for grinding its hard prey.

There was another group of sharks that took this adaptation of a crushing tooth to the next extreme. They would alter the shark dental plan so much that they could break away from all sharks of the day and form a radical new lineage.

## Sharks reinvent themselves, for the first time

Sharks have radically reinvented their body plans twice over the course of their long evolution. The first time they did this was in the Late Devonian. The entire body of evidence for this great transition to an entirely new shark lineage is—as crazy as it seems—just one incomplete tooth found in Poland and studied by Michal Ginter. Truly. This game-changing tooth comes from *Psephodus*, the first known holocephalan—the group containing chimaerids, ratfishes, and spookfishes, whose fossils are commonly found in younger Carboniferous rocks. The *Psephodus* tooth is a totally new design for sharks; it has a smooth, well-rounded crown and is the first tooth they evolved without any sharp pointy bits—imagine how good that must have felt compared to having a mouth full of razor-sharp teeth. Similar crushing teeth later evolved in sharks when they invented themselves for the second time, much later on, as rays. (Rays have pavements of flat teeth used to grind up their crunchy prey, such as crabs, clams, or snails.)

This toothy experiment worked. The proof is the great diversity of sharks with rounded crushing teeth that proliferated straight after the Devonian ended. They were suddenly everywhere, nosing around the shallow seas of the world, represented by many strange-looking species. From hereon I refer to this radical group of sharks with rounded crushing teeth by their proper name: holocephalans. Holocephalans are known by about fifty living species that are mostly deep-water inhabitants. These living forms differ from all other sharks and rays by having a set of large crushing tooth plates in the lower jaw and another set of tooth plates located in the palate, with the upper jaw cartilages fused firmly to the braincase. Unlike sharks' teeth, which regularly get replaced, the tooth buds coalesced to grow within a large single unit, sort of like a mini tooth factory contained within one rock-hard structure. In most sharks, the jaws are loose, allowing them to project them forward when feeding, a neat feeding trick holocephalans cannot perform. Holocephalans also have something else that sharks and rays lack: a flap of skin covering their gills, which gives them protection from parasites.

While modern-day holocephalans have these few large elongated tooth plates for crushing prey, the first holocephalans—like our Devonian *Psephodus*—had many rows of individual crushing teeth. Later these rows would slowly coalesce to grow as one fused unit, forming the modern holocephalan tooth plates.

So far our story has focused on the slow, steady rise of sharks, but we haven't given much attention to the incredibly fierce competition they were up against. The biggest survival challenge facing sharks wasn't from other sharks, but from the rapid rise of the armored placoderm fishes. Some of them were enormous and had powerful blade-like jaws. How did sharks manage to cope with them? And what was the evolutionary relationship between sharks and placoderms?

In the next chapter we will look at placoderms to see why they were such a threat to sharks, and what they can teach us about the evolution of certain shark features. We will also examine what sharks were doing in the last few million years of the Devonian period, and how both sharks and placoderms faced another series of devastating mass extinctions at the end of the period.

CHAPTER 5

# SHARKS' ARMORED RIVAL
## *Sharks Versus Placoderms*

---

### Ancient American sea, 359 million years ago

Warm equatorial waters lapped the coast of the vast Panthalassa Sea that once covered large expanses of the states of Ohio and New York some 359 MYA. Imagine two hundred feet below the surface, where the last faint rays of sunlight barely illuminate the muddy seafloor. A sleek, streamlined shark about 4 feet long is circling, hunting for its next meal, perhaps a small fish, a shrimp, or a worm-like conodont animal. Looking much like a modern-day dogfish, the only clue that this sleek hunter, *Cladoselache,* is an early shark is its mouthful of pronged teeth. These teeth interlock tightly when the mouth is closed. It is well equipped to seize small, slippery prey, like the fast-moving trout-like fishes inhabiting this twilight world.

The shark uses its keen sense of smell to locate its prey. As it sweeps the seabed, a cloud of mud wisps up around it. It quickens its pace and sweeps left and right, with each sway of its head homing in to where the smell is strongest. Within seconds it seizes a fat 3-inch shrimp that pops open inside its mouth. It swallows the mangled morsel in one gulp.

Not far away from this action, just above the seafloor, another *Cladoselache* is slowly moving along searching for its meal. Its nose senses a welcoming whiff, the familiar odor of death. Homing in on the source, it spies a large decaying shark carcass and moves in closer to feed, tearing chunks away from the festering remains of its departed kin, thrashing its head from side to side. In the Devonian, a meal is a meal; deceased friends and relations are not spared.

In the shallower waters just beneath the sunlight-drenched surface, a group of five *Cladoselache* are cruising around on the hunt for something fresh and delectable. This tight pack closes in on a school of small, bony, thickly scaled ray-finned fishes called *Kentuckia,* busily feeding on a pack of conodonts, free-swimming worm-shaped animals. Slowly, then in increasingly faster circles, the sharks edge closer and closer to the small *Kentuckia* shoal, until one loses control and darts in quickly with its mouth open, narrowly missing a fish with its snapping jaws. But all is not lost, for it has wounded another with its razor-sharp fin spines. The blood of the wounded fish immediately sends a sharp olfactory signal to the shark's brain, making it jackknife around to snap up the bleeding fish in its mouth. Once the fish is caught between its rows of jagged teeth, there is no escape. The *Kentuckia* is gulped down in a flash as the shark turns and lines up another angle of attack.

Immediately other *Cladoselache* sharks dart in and out among the panicking *Kentuckia,* slashing through the shoal with their open mouths and sharp dorsal fin spines, maiming as many of the fleeing fish as possible. The sharks dart in for one last pass, picking off slow strays as they scatter in different directions. The *Cladoselaches* are focused so intently on catching dinner that they fail to notice the large, dark shadow slowly approaching from the stygian waters below.

Without warning, a dark behemoth lifts its massive head and, with a powerful crack of its tail, pushes its chunky 15-foot body up through the chaos of scattering sharks. In a sudden rush it breaches the surface, sunlight flashing off a stunned and bleeding *Cladoselache* trapped between its massive, jagged jaws. With an explosive splash, the giant predator crashes down into the sea again to begin chomping hard on the captured shark. The life of the *Cladoselache* is extinguished with a couple powerful bites.

Meet *Dunkleosteus,* the dark lord of the Devonian deep and the shark's mortal enemy. This massive armored fish had the most powerful jaws of any creature on Earth at this time, twelve hundred pounds of bite force, a record that would be displaced only by the appearance of *Otodus megalodon,* the biggest shark of all, 340 million years later. Nothing on the Devonian planet was more ferocious than this terror.

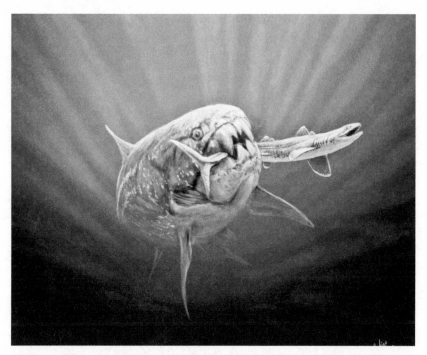

**A 15-foot-long *Dunkleosteus* placoderm grabs a 4-foot-long *Cladoselache* shark in the Late Devonian seas around Ohio, USA** ARTWORK BY JULIUS CSOTONYI

Its thick, rounded bony shield enveloped its head like Darth Vader's helmet. Behind the bone-clad shoulder was a chunky shark-like tail, as seen in all members of this archaic fish group called placoderms (meaning "plated skin"). Take away the bony shields of placoderms like *Dunkleosteus,* and you essentially have a brutish shark-like fish. Think of them as the panzer tank divisions in the Devonian war in which the sharks were battling for survival.

The rapid rise of the placoderms throughout the Devonian period was without doubt the greatest challenge sharks had faced so far in their history. Placoderms were the most diverse group of fishes on Earth at this time, and their ranks included some of the largest and most grotesque predators, like our *Dunkleosteus,* and the aptly named, similarly sized *Gorgonichthys.*

At the very end of the Devonian, we also have a large number of shark species living alongside the largest and most dangerous placoderms. It is a very important time to study sharks, as it gives us a baseline for shark diversity just before a major round of extinction events at the end of the Devonian. It also shows us how some sharks were evolving their own defense strategy against these large predatory placoderms. The scene was now set for a final battle to attain the top of the food chain. These last few million years of the Devonian period are perhaps the most significant time so far in our history of sharks. It is when they rise to a new zenith of both diversity and body size—their way of dealing with the placoderm competition. First we need to meet the players, then explore the relationship between sharks and placoderms to reveal the evolutionary legacy that placoderms bestowed upon both sharks and us.

## The fabulous fossil sharks of Ohio

The Cleveland Shale provides a rare window into the last 3 million years at the very end of the Devonian period (362–360 MYA). It is one of the world's most important sites for the study of Late Devonian sharks because of their complete, nearly perfect preservation, sometimes even showing muscles and soft tissues in the fossils, and their great abundance. Let's look at this site in detail, highlighting how these sharks were preserved and which were the top sharks at that time, and weigh up the scientific evidence for our reconstructed scene.

These sharks are so beautifully preserved because the seas here were not far from river deltas that brought in large amounts of decaying plant material, causing a depletion of oxygen. This results in what geologists who study modern deltas call a "dead zone" where nothing lives on the seafloor. Instead of being torn apart by scavengers, the delicate remains of dead sharks received a softly ceremonial burial. Over time, the chemistry of the decaying shark caused the sediments building up over them to bond and harden. This process continued until eventually a solid rocky concretion formed, a natural coffin around the body. This strange, slow natural process not only signifies the place of their burial, but also completely encloses and preserves the remains intact, much as they were in life.

After millennia of natural catastrophes, drifting continents, and cataclysmic environmental change, the uplift of mountains pushed these rocks, formed in what was the bottom of a seaway, to the surface on dry land. Because of the chemistry of decay, the rocky nodules were much harder than the soft shales around them. They would erode out of cliffs, high and exposed and far from their marine origins. The skeletons thus remained protected and intact for hundreds of millions of years. They sometimes retain even the delicate faint impressions of the soft tissues, like the outline of muscles, or even their last meals. They are a unique portal into a vanished ecosystem opened across a vast time.

Paleontologists study these exquisite remains like forensic detectives to establish what killed them, and what their last meal was, by analyzing their stomach contents. They can even determine how these creatures behaved in life, analyzing the subtle evidence on their carcasses of scars and bite marks, indications of aggressive or defensive interactions with other predators.

The sharks of the Cleveland Shale in Ohio were the first complete sharks found from the Devonian period, and they have been studied meticulously for the past 140 years. The American paleontologist John Strong Newberry first studied complete fossils of sharks found in Cuyahoga County, Ohio, in 1889. His intricate work showed the fine detail preserved in these sharks—the outlines of beautiful fins, clear details of the jaws, teeth, and gill arches, the delicate little scales forming rings framing the eyes. Some years later another stellar American paleontologist, Bashford Dean, at the American Museum of Natural History, referred Newberry's sharks to a new genus *Cladoselache,* meaning "branch shark," because of the branching cartilage rays that support its fins (see photo of *Cladoselache* on p. 89).

Newberry is one of the pioneer shark paleontologists I admire the most. Although he trained in medicine and served in this capacity during the Civil War, his real love was paleontology. He was one of the most important American scientists of his time, rising to become the president of the American Association for the Advancement of Science and of the New York Academy of Sciences, both very prestigious positions. Newberry made immense contributions to our understanding of the anatomy of placoderms and fossil sharks. Over his long and illustrious career, he named more than fifty new species of fossil sharks, placoderms, and bony fishes from the Devonian and Carboniferous of the central and eastern United States, including the formidable *Dunkleosteus,* about whom we will hear more shortly. He was a keen observer who relished describing and illustrating the finest details of their skeletal structures.

Newberry noted that *Cladoselache* looks like a typical shark, not very unlike the modern dogfish (*Squalus,* see p. 273), with dorsal fin spines in front of each fin, except the spines are shorter and more robust in our fossil form. *Cladoselache* was around 4 to 6 feet long and certainly built for speed. Its broad triangular pectoral fins and a high-angle tail fin resemble the tail profiles of fast-swimming modern sharks like white sharks and makos. Its multipronged sharp teeth were perfect for grabbing and securing the fast-moving little bony fishes and eel-like conodont animals it preyed upon. However, while stomach remains reveal that small fishes were a key source of nourishment for these sharks, they also tell a more sinister tale.

**John Strong Newberry**   PUBLIC DOMAIN

Mike Williams, who also used to work at the Cleveland Museum, identified fossil *Cladoselache* specimens with stomach or intestine contents preserved. Mike was a kind and jovial man whom I'd met a couple times at various paleo conferences. He was a legend at the Cleveland Museum, renowned for his impromptu bluegrass performances on his banjo. His detailed analysis of fifty-three fossil sharks with prey inside them showed that 64 percent of them fed on the small trout-like ray-finned fishes (*Kentuckia*), 28 percent fed on the shrimps (*Concavicaris*), and about 9 percent fed on the wormy conodonts. One specimen had six sharks' teeth inside it, suggesting it was from a carrion feed. Unfortunately, this *Cladoselache*-like specimen was missing its head, so we cannot confirm that it was the oldest known case of sharks eating their own species (cannibalism). However, it is still the first evidence we have of sharks eating other sharks.

While most scientists regard *Cladoselache* as the typical bog-standard Devonian shark, recent research suggests that *Cladoselache* is an early chondrichthyan on the beginning of the line leading to the modern holocephalans. Like several of the sharks we met from Morocco, including *Cladoselache*'s closest relative, *Maghriboselache*, it looks like a typical shark because it occurs very close to the time of divergence from typical sharks. The critical differences that demonstrate that it is closer to holocephalans than to modern sharks are in the subtle anatomical features of the skull and braincase shape—not the body—which is why it retains an essential shark body shape.

*Ctenacanthus* was another Cleveland Shale shark known from relatively complete fossils. Its lineage, the **ctenacanthiforms**, represents the first definite line of true sharks leading to modern forms (selachians). It bore two powerful dorsal fin spines and a robust skeleton. The spines have many ridges on them, each with swollen blobs of shiny dentine, strengthening them. Its jaws were lined with many rows of strong multicuspid teeth. While the museum specimen indicates it might have been around 12 feet long in life, an isolated lower jaw collected by Peter Bungart of the Cleveland Museum of Natural History in 1924 hints that if it belongs to this form, this shark may have grown to around 16 to 20 feet long, making it the largest shark of the Devonian period. Period.

Unlike the more delicate teeth we see in most other Devonian sharks, these were heavily built, with large, pointed central cusps lined with several fine linear ridges to help puncture its prey. They were clearly the teeth of an aggressive hunter that probably fed on other smaller sharks and placoderms. I can imagine *Ctenacanthus* feeding on juvenile *Dunkleosteus* and other placoderms, although mature sharks at the top of their size range could have given even large *Dunkleosteus* a fair run for their money. The giant *Ctenacanthus* was the apotheosis of shark evolution during the twilight years of the Devonian. With brute force it powered its way through the devastating extinction events that punctuated the end of that period.

There are many other kinds of Cleveland Shale sharks, too many to discuss here. But there is one very strange shark found nearby in Pennsylvania that represents yet another new direction for how sharks fed, so I think it's important to include in our story at this point. Discovered by my colleague Ted Daeschler of the Philadelphia Academy of Sciences, it lived in the twilight years of the Devonian.

Ted Daeschler is a fish paleontologist who has spent his life finding really cool stuff in the Devonian sites of North America. He spent many field seasons working with Neil Shubin of the University of Chicago in remote parts of Arctic Canada, which resulted in their discovery of *Tiktaalik,* one of the most spectacular finds in paleontology in recent years—a bony lobe-finned fish that bridges the gap between fishes and land animals. Ted has been described by the *New York Times* writer Carl Zimmer as having an "otter-like face" and a gentle personality, and I must defer to Carl's accuracy here. I spent two long field seasons working in Antarctica with Ted and Neil, so we know one another pretty well these days.

Ted is nothing if not determined. He took me to the Pennsylvania site some years ago. It's on a large highway road cutting near Hyner, where a thick pile of red Devonian sandstone and shales are nicely exposed. Despite the ground-shaking semi-trailers and other assorted vehicles whizzing closely past, Ted just kept plugging away relentlessly for years at this site named Red Hill. His efforts resulted in some fantastic finds of early four-limbed animals (tetrapods), as well as many other new species of placoderms, bony fishes, and sharks. The shark teeth are particularly important to our story at this time, as the Red Hill site is well dated as being the same age as the Cleveland Shale, but it represents a fauna that lived in a different environment—freshwater river systems.

Our story concerns a peculiar shark named *Ageleodus* that Ted found

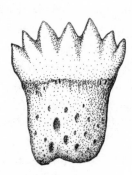

The strange tooth of *Ageleodus* from the Late Devonian, a new type of tooth for sharks
ARTWORK BY AUTHOR

at this site. This small shark has tiny comb-like teeth, each armed with many sharp little triangular cusps on a flat, deep root system, in some ways resembling the jagged lower jaws of the little cookiecutter shark. These sharks can latch on to large prey like whales and cut a clean plug of flesh out of the side of the host by rotating their razor-sharp zig-zagged lower teeth around, working it out like a can opener. Some undersea telecommunications cables have even fallen victim to these sharks, which I doubt would have liked the taste of the polyethylene covering. Perhaps *Ageleodus* was using much the same action, feeding parasitically on large bony fishes that lived in the same big river system. *Ageleodus* teeth are also known from younger Carboniferous rocks, showing that it survived the end-Devonian extinction event without a hitch. This fits with the pattern of other sharks we have seen, showing a remarkable capacity for adaptation and survival.

I will now introduce you to *Dunkleosteus* to examine how placo-derms differ from sharks, and to explore whether there's any merit in the idea that they held sharks back from diversifying throughout the Devonian period.

## Meet Deadly Dunk

If you remember the name of just one placoderm, make it *Dunkleosteus*. Big Dunk was one of the really bad boys of the Late Devonian seas, frighteningly large and massive of form. Like all placoderms, it bore thick bony plates that interlocked around its head like a helmet. Other bony plates protected its body and stomach like a living suit of armor.

Its armor plating—the external bony shield around its head, called a "head shield," and around its upper body, called a "trunk shield"—distinguishes placoderms from all other fishes. This massive covering of bone served to protect them and anchor powerful muscles. While bony fishes like trout have bones similar to placoderms' in the corresponding places, their bones are generally lighter and lack the rigid interlocking joints seen in placoderms. Most placoderms have rela-tively thick plates, so their bones fossilize easily; as a result, they are known from hundreds of fossil sites around the globe.

Some recent estimates place Dunk at up to 29 feet long, the size of a large orca; others suggest it was a smaller yet very chunky 14 to 15 feet long—slightly larger yet much heavier than the biggest tuna ever

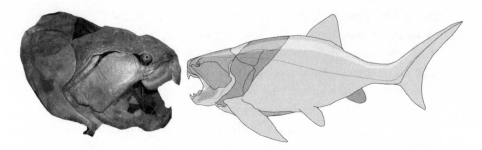

***Dunkleosteus*** **skeleton on display in the Cleveland Museum of Natural History, Ohio, with restoration by author**   PHOTO BY GLENN STORRS, ARTWORK BY AUTHOR

caught (13 feet, 2,000 pounds), earning it the nickname Chunky Dunky. Either way, it was clearly one of the biggest predators around at this time.

The lower jaws of *Dunkleosteus* were long blades of thick bone with a few sharp teeth that merged to form large triangular blades. Two sets of equally fierce toothed bones formed its upper jaws. Together they functioned like two opposing guillotines that could slice the heads off sharks with one bite.

Even though it has been extinct for 360 million years, we know a lot about Dunk from its bony plates and jaws found in the United States and North Africa. While we don't know what the tail and fins of *Dunkleosteus* looked like, we can guess from the remains of closely related placoderms that are preserved as whole organisms, like the smaller *Coccosteus* from Scotland. This fish, just over a foot long, somewhat resembled a small sixgill shark with a bony cover over the front third of its body. It bore a pair of large triangular shark-like pectoral fins that emerged from holes within the armor behind the head. They also sported a pair of similarly shaped pelvic fins located halfway along their bodies at the point where the body armor ended. Along its back, *Coccosteus* bore a single dorsal fin, differing from those of sharks as it was more elongated and rounded. It bore a long tail, slightly upturned, like those seen in many living sharks, of which the upper tail lobe is long and the lower lobe short. I suspect *Dunkleosteus* most likely had a similar arrangement of fins along its body, with a shorter, more muscular tail making it look more an iron-pumping armored tuna.

Placoderms like *Dunkleosteus* had protection not only over their heads. The eyes of all placoderms were also armored with thick, curved,

*Coccosteus,* a common placoderm found in Scotland. Its body shape is not unlike that of a typical shark ARTWORK BY AUTHOR

bony scleral plates that enveloped the outer part of the eyeball, leaving only a tiny central hole about an inch wide to allow the light in. Paired nostrils opened at the front of the face, or just in front of the eyes in some forms.

Basically, all placoderms resembled sharks with armored plates around the front third of their bodies, protecting their most vital organs while allowing for the smooth operation of those nightmarish jaws hinged by smooth cartilaginous joints. They differed from sharks in some key anatomical features, like having a small bone called the **parasphenoid** lining the palate, which all bony fishes have but sharks lack. They also lack the principal features defining sharks, like special calcified cartilage and shedding-type teeth, so they are technically not a branch of archaic sharks, but a separate group of early fishes.

John Strong Newberry was the first scientist to describe the giant placoderm *Dunkleosteus* from the Cleveland Shale sites in Ohio in 1868. I love the way he writes about this fish in the gripping introduction to his huge monograph *The Paleozoic Fishes of North America,* published in 1889, in which he refers to the mighty power of a giant fish he named *Dinichthys* (its name would later be changed to *Dunkleosteus*). Newberry writes: *"placoderms included some of the most highly specialized, largest and most formidable of all fishes; for example,* Dinichthys . . . *was about 15 feet in length and could have severed the body of a man as easily as he bites off a radish!"*

Speaking of their choppers, placoderms were long thought not to have real teeth, just jagged bony blades on their jaws. In recent years we have confirmed from CT scanning that they do possess true teeth with their own kind of dentine and pulp canals running through them. A recent study of some early placoderm jaws from Europe has even suggested that sharks might have got their teeth from placoderm ancestors.

Valéria Vaškaninová from Charles University in Prague is renowned around the world for her amazingly detailed research into placoderms. Placoderms had been known from the Early Devonian rocks around Prague for more than a hundred years (the "Pragian stage" is a subdivision of the Devonian period named after these rocks). Her study of primitive placoderms found there revealed something quite extraordinary about the origins of teeth that is especially relevant to our understanding of how sharks developed their teeth.

Valéria concluded that the way placoderms developed teeth was closer to how bony fishes do it, because both grow teeth from their special jaw bones. These ossified jaw bones wrap around the primary cartilage frame of the jaw, called **Meckel's cartilage**. Her study implied that sharks develop their teeth in a more advanced way than placoderms, as sharks do not require special bones for teeth to grow from. This means sharks are not likely to be ancestors of placoderms, but the reverse is possible: that sharks could have evolved from placoderms. More about this intriguing evolutionary mystery later in this chapter.

Placoderms competed with sharks for around 100 million years in all known aquatic environments, from the open oceans and shallow seaways to the rivers, estuaries, and lakes. As far as we can tell, typical placoderms like *Dunkleosteus* and *Coccosteus,* which belong to the group called **arthrodires,** fed by hunting live prey or scavenging. Other placoderms called **ptyctodontids** evolved powerful crushing tooth plates for chowing down on clams and snails. Remarkably, they did this at least 30 million years before the first shark did the same thing. Others, like the weak-jawed *Bothriolepis,* belonging to the **antiarch** group, might have been sediment sifters looking for tiny organisms in the mud. Overall, placoderms show a much larger range of feeding strategies than sharks living at the same time. Placoderms are very common in Devonian-age rocks on all continents, even though they started much earlier, in the Silurian, around 438 MYA.

The unprecedented success of the placoderms is a real enigma for us scientists. Placoderms were far more diverse than sharks, with at least six hundred species in the Devonian (about six times more species than sharks), and they were far more abundant—many more placoderm specimens are found in Devonian fossil sites than sharks. Placoderms also occupied a wider range of ecological niches than sharks did throughout most of the Devonian, as shown by their more varied feeding strategies. Yet sharks were lighter, more streamlined, armed with deadlier, replaceable teeth, and generally thought to be much faster swimmers than placoderms. If sharks were the sporty Corvettes of the ancient seas, placoderms were your Ram pickup trucks. Sharks could likely easily outswim any placoderm, but could they beat them in a head-to-head battle? How did placoderms manage to keep sharks at bay for so long?

I think the shark-placoderm rivalry mirrors what happened later with dinosaurs and mammals. While the first mammals evolved at much the same time as the first dinosaurs, they lived in their shadows, unable to radiate into the many modern groups until after the dinosaurs went extinct. This is because the extinction of dinosaurs opened up many new ecological niches for mammals. The same scenario likely occurred with sharks. It has long been thought that sharks were less adaptable, so they could not truly "break free" and diversify into many varied, wonderful forms until the placoderms died out at the end of the Devonian. The more likely story is that placoderms began their diversification well before sharks, so they got in quickly to occupy the main ecological niches first. Sharks then had to earn their place at the top—an evolutionary challenge that would take them almost 60 million years—and, more important, had to fight to maintain their superiority in the oceans.

I still haven't answered the question of how sharks and placoderms are related to each other and to other living bony fishes. This thorny question has vexed scientists for the past two centuries, but finally we are getting nearer to a solution. Let's meet a nineteenth-century American scientist who tried to answer this question by studying both sharks and placoderms, along with medieval suits of armor.

## Sharks, placoderms, and medieval armor

Throughout the nineteenth and twentieth centuries, most scientists thought that placoderms and sharks were closely related, some even placing them together in a superclass they called Elasmobranchiomorphi (try saying that after two martinis). Yet as early as the 1910s, one American scientist argued they belonged in distinct groups, which is how we think of placoderms today. Bashford Dean, the dashing and legendary curator of reptiles and fishes at the American Museum of

Natural History in New York, was also an expert in the seemingly un-related area of medieval armory. Is it any wonder that Dean was par-ticularly drawn to study armored fishes?

In 1873, at the age of six, Bashford Dean became fascinated with helmets and medieval armor. He was an exceptionally prodigious stu-dent who enrolled in the College of the City of New York at age four-teen, graduating only five years later. In 1890, aged just twenty-three, he received his PhD in zoology and paleontology from Columbia Uni-versity. As if his talent bucket wasn't already overflowing, it's worth noting that Dean was also a brilliant artist who illustrated most of his published works. He was the original multitasker, a true Renaissance man. Between 1890 and the early 1900s his research took him to Eu-rope, Russia, Japan, and up and down the Pacific coast of the United States. Dean held three serious academic positions in New York City simultaneously. He was a professor, teaching and doing research at Columbia University, then later became the founding curator of the Department of Reptiles and Fishes at the American Museum of Natu-ral History, and also served as the honorary curator of arms and armor at the Metropolitan Museum of Art.

A polymath like Dean was always doomed to leave some of his

Bashford Dean, circa the 1920s. Dean worked on fossil sharks and placoderms and held prestigious curatorial positions at the Ameri-can Museum of Natural History and the Metropolitan Museum of Art simultaneously  PUBLIC DOMAIN

dreams and visions unfulfilled. In the late 1920s he spent a great deal of time and energy planning a grand exhibition for the public, showcasing for the first time the story of shark and fish evolution alongside the wonderful diversity of modern fishes. This vision would be realized through his pièce de résistance, his curation of a spectacular new Hall of Fishes for the American Museum of Natural History. Sadly, he died from a botched operation on December 6, 1928, the day before his new gallery was opened to the public.

Dean's paleontological work focused on producing detailed descriptions of the superb fossil sharks and placoderms from the Cleveland Shale. He also wrote detailed anatomical monographs about extant sharks and drew upon this knowledge to inform his studies of Devonian forms. His expert knowledge concluded that placoderms were not close relatives of sharks, but a "dead-end" lineage with no modern relatives. So, was Dean right or wrong?

We have recently discovered that placoderms did in fact leave their imprint on the world of modern fishes and so have a close evolutionary relationship to sharks. This has all been revealed over the past decade through the extremely significant discoveries of the oldest complete placoderms from the Silurian period of China. They have reversed everything we thought we knew about placoderms, and frankly, they are beyond Dean's wildest dreams.

## Revolutionary Chinese placoderms

YUNNAN, CHINA, MARCH 2018: In the mountainous region outside Qujing, we drove along narrow bitumen roads that passed through a series of small mountain villages. Farther beyond the city limits, we began to encounter flocks of chickens, herds of goats, and gatherings of small children. We passed an old man carrying wood, bent over with his load and stumbling along the pavement. A young woman with a child or two on a scooter buzzed by us, leaving a trail of smelly exhaust fumes.

We parked the vehicles, got out our gear, then walked single file between some farmhouses to the nearby rice terraces. We crossed a babbling creek and emerged onto a muddy goat track that snaked its way up the gently rounded hills flanking the village. Eventually we reached a rather nondescript quarry site. Several large piles of massive gray limestones from the Kuanti Formation, dated at around 430 MYA, were the focus of attention for a small but industrious team of workers. Among them was my good friend and close colleague Professor Zhu Min. Min leads a team that spends half the year hunting for fossils and the other half of the year preparing their finds.

**Zhu Min (standing) at the site near Kuanti in Yunnan, China, where the placoderm *Entelognathus* was found** AUTHOR PHOTO

I've visited this beautiful site with Min and the team twice in recent years and am amazed (and perhaps a little envious) of the scale of their operation. Some layers of dull gray limestone outcrop naturally, exposing the rocks for ease of access. In other places, swaths of the area have been opened up by bulldozers to give the fossil hunters ample fresh material to search. I am always amazed at how rare it is to find a complete fossil fish. It's only through massive amounts of sweat equity that any fossil is found at all. However, when complete fishes are found at Yunnan, they always turn out to be something very, very special.

On a visit there a few years ago, we were splitting rocks for some time and finding very little. After a hard day's labor under the bright mountain sun, we laid down our tools to search for beer and sample the local culinary special at a café in Qujing. Over a steaming bowl of pungent black mountain goat soup, Min reminisced with me about how their many years of perseverance at the site had finally paid off. They seemed to have hit a rich seam, a particular layer within the rocks that held a great range of fish fossils. He was excited to share with me his plans for how they would study the delicate fossils without damaging their structures.

Min's team have at their disposal a special custom-built CT scanner that scans flat 2-D fossils in slabs of rock, then their software adjusts the compression of the fossil to restore it to 3-D form. This enables scientists to breathe virtual life back into the squashed fossil specimens without risking their physical integrity. Min and his team have used these methods to study what were then the oldest known complete fossils of both bony fish and placoderms, yielding discoveries that resulted in a major shake-up at the base of the vertebrate family tree.

Min's big discovery of the oldest known relatively complete placoderm was announced in 2013, a fish called *Entelognathus,* meaning "jaw without teeth." *Entelognathus* offered a never-before-seen look at a very early placoderm skeleton. This new find disproved once and for all that placoderms were an evolutionary "dead end." I regard this specimen as perhaps the most significant fossil found in the past century, given the huge impact it has had on changing our views about our own early origins, as well as the origins of sharks.

Think of *Entelognathus* as an Enigma machine that decoded the origins of many parts of our own skeleton for the very first time. For example, the dentary bone is the one in your lower jaw, and the maxillary and premaxillary bones are the tooth-bearing bones of your upper jaw. These bones were thought to be absent in placoderms before this discovery. *Entelognathus* also allowed us to identify certain paired skull bones of placoderms as being the same bones found in the human skull—the parietal bones, for example. It proved for the first time that these early placoderms, 425 MYA, were true "deep-time ancestors" of humans.

The discovery of *Entelognathus* caused a massive and frantic rewrite

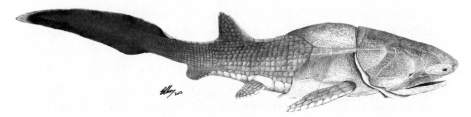

*Entelognathus* **from the Silurian of China revolutionized the study of placoderms, linking them back to us humans**   ARTWORK BY BRIAN CHOO

of every biology textbook. Placoderms now had to be introduced into the biological lexicon as a significant group of animals, one that we all should know. More recent finds from the same Qujing sites have only reinforced these theories. Other placoderms have since been found, such as *Qilinyu*, that show further stages in the development of the jaw bones, representing intermediate stages between placoderms and bony fishes (the line leading to us). Even older complete placoderms like *Xiushanosteus* were recently found in the Early Silurian of China, dated at around 438 MYA. This series of remarkable discoveries from Min and his team have kicked off what I have dubbed the Placoderm Renaissance that now constitutes a new chapter in the book of deep human ancestry. These new placoderm discoveries also have major ramifications for our understanding of the origins of sharks.

This is because the evolutionary position of *Entelognathus* and the other maxillate placoderms is not where we would intuitively expect to find them—at the base of the placoderm tree. Instead, they are right at the top—they are the most advanced placoderms, yet they are enigmatically among the oldest. This sometimes happens when the fossil record of fishes is poor at certain time intervals, as it is for most of the Silurian period. Placoderms therefore sit on the evolutionary tree just below the node where sharks and bony fishes diverge. They are now regarded as the ancestral group that gave rise to both bony fishes and sharks. Because bony fishes gave rise to four-limbed land animals, we humans also fit into this picture.

Bony fishes evolved from placoderms by retaining many shared bones like the parietal, premaxilla, maxilla, dentary, and others around the shoulder area. Sharks evolved from placoderm ancestors by sequentially losing all the main dermal bones, developing a different way of making teeth, and opting for a lighter, more flexible cartilaginous skeleton. The precise timing of this divergence—when sharks and bony fishes arose from the placoderms, or which of these two groups came first—has not yet been resolved. The recent discovery of the armored shark-like *Shenacanthus* from the Early Silurian also reinforces the view that sharks could have emerged from placoderm ancestors. The line of modern sharks likely came from the archaic stem sharks that simply lost their original suits of bony armor to make them more agile, and faster off the blocks.

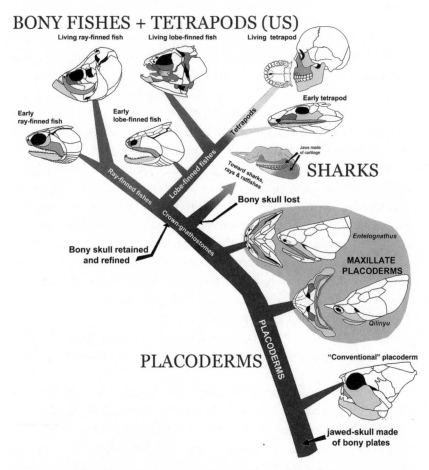

**BONY FISHES + TETRAPODS (US)**

Living ray-finned fish

Living lobe-finned fish

Living tetrapod

Early tetrapod

Early
ray-finned fish

Early
lobe-finned fish

Tetrapods

Jaws made
of cartilage

**SHARKS**

Ray-finned fishes

Lobe-finned fishes

Toward sharks,
rays & ratfishes

**Bony skull lost**

Crown-gnathostomes

*Entelognathus*

**Bony skull retained
and refined**

**MAXILLATE
PLACODERMS**

PLACODERMS

*Qilinyu*

**PLACODERMS**

"Conventional" placoderm

**jawed-skull made
of bony plates**

Evolution of skull bones from placoderms to sharks and bony fishes. Sharks basically lost most of the bones present in placoderms to opt for a lighter, more agile skeleton   ARTWORK BY BRIAN CHOO

These earliest complete placoderms from China give us a rare glimpse into the shared ancestry of placoderms, sharks, and bony fishes. However, to truly understand how placoderms became successful and outcompeted sharks for most of the Devonian period, we need to look at the superbly preserved placoderms from a site in Australia where fossils of perfect uncrushed skeletons and even soft internal organs of placoderms are found. Some of these finds revealed where sharks got their saucy style of reproduction.

# How placoderms made sharks sexy

Certain placoderms from the Gogo site in Australia, where I found *Gogoselachus,* can explain some shark behaviors. In 2007, I was working on new placoderms we had collected from Gogo with a crack squad of Aussie researchers: Kate Trinajstic, Gavin Young, and Tim Senden. One specimen gave us a shock when we found it contained an embryo of an unborn fish, the first ever found in a placoderm! We named it *Materpiscis attenboroughi,* meaning "Attenborough's mother fish." (Sir David Attenborough told me when we met in Adelaide some years later that he was absolutely delighted to have a Gogo fish named after him, as he'd visited the Gogo site while filming *Life on Earth* back in 1979.) *Materpiscis* was one of the ptyctodontid placoderms, a group known for their powerful crushing tooth plates and long, tapering bodies. At the time, only a very few of them had been studied from well-preserved 3-D examples, and all had come from Gogo. None of the earlier studies had given us any hints about their sex lives.

In 2008, when we published our study of *Materpiscis,* the practice of CT-scanning fossils was uncommon. Tim, however, was not a paleontologist, he was a professor of physical sciences, and he had just developed a powerful new micro-CT scanner for his lab. This collaboration enabled us to scan the specimen and reveal the minute details of its fossilized umbilical cord. Kate subjected it to a powerful scanning electron microscope and identified features that were found only on the placenta of living creatures, such as tiny attachment scars for ligaments. These powerful new scanning tools proved beyond any doubt that the little bones inside the fish were not its last meal, but its embryo, a mini replica of the adult still connected by a long cord that ended with a yolk sac. It may be the most poignant discovery I've ever made: the fossil of a mother that died carrying her unborn young, entombed together and preserved in perpetuity.

In one stroke we had discovered the origin of live birth in vertebrates, pushing the known fossil record of this method of reproduction (called **viviparity**) back by almost 180 million years. This fact helped me achieve something I never thought possible: Our discovery made it into the 2010 edition of the *Guinness Book of World Records* as the "world's oldest live birth" (I'm holding a model of this fish on p. 55 of that edition).

Two mating *Microbrachius dicki,* with the male (right) showing the spiky claspers used to inseminate the female    ARTWORK BY BRIAN CHOO

*Materpiscis* confirmed that placoderms copulated. They did "it," 380 MYA. This was an unexpectedly advanced form of reproductive behavior, as copulation has consequences, as most of us know well. It means that the young subsequently develop inside the female. This kind of reproductive strategy is called "internal fertilization," the opposite strategy to that of the vast majority of living fishes that simply spawn in the water, like salmon. It meant that placoderm mothers, like *Materpiscis,* had to invest a lot of their energy into caring for their young. They also took a big risk by increasing their own vulnerability to predators while carrying large unborn young inside them. This was the first time in vertebrate evolution that this advanced kind of reproduction appeared. After finding the first placoderm embryo, we rushed back to the museum collections in Australia and the UK to search for more. We soon located another Gogo ptyctodontid that I'd found earlier and had named *Austroptyctodus.* This specimen revealed three embryos inside it!

We were keen to see if the most abundant group of placoderms, the **arthrodires** (the group including *Dunkleosteus*), also showed evidence of internal fertilization. One specimen of interest was a Gogo arthro-dire called *Incisoscutum*. This fabulous little fish bore a cluster of tiny bones inside it, interpreted by its paper's authors in 1981 as the remains of its final meal. We could confirm our suspicion only by hunting it down in the Natural History Museum in London.

In late 2008 we called on a close colleague, Zerina Johanson, the curator of fossil fishes there. I had known Zerina, a Canadian, from her time spent working at the Australian Museum in Sydney before she took up her job in the UK. Zerina examined the specimen and quickly confirmed that it indeed contained an embryo. She also found a sec-ond *Incisoscutum* specimen in the NHM collections with an embryo, which really nailed it for us—the most diverse placoderm group, the arthrodires, also reproduced using internal fertilization. Strangely, none of the many hundreds of well-preserved arthrodires in museum collections we had seen showed any signs of males bearing the neces-sary telltale shark-like claspers. This posed our next big question: How the heck were these arthrodires doing it?

If our theory was correct, we should find the smoking gun—the male claspers on arthrodires—by going back to the museums and look-ing carefully at the best-preserved specimens to find them. (Life as a paleontologist can lead you down many unexpected roads.) A few months later, Kate was working with another colleague, Per Ahlberg from the University of Uppsala. Together they stumbled across the infamous missing male clasper in a placoderm from Gogo at the West-ern Australian Museum. Embarrassingly, it was a specimen I'd found earlier that I had given to one of my students to study for her disserta-tion. At the time, we thought we had a pelvic girdle that supported the pelvic fin, as seen in many other Gogo placoderms. Per and Kate spot-ted that the bone in question bore tiny ridges on the head of the clasper, like the kinky hooked and barbed claspers of some living sharks. Finally the elusive arthrodire clasper had been found, and it had been right under our noses for years.

The next discovery in this sexy prehistory tale was made by Kate and Zerina. Together they conducted an exhaustive survey of all the placo-derm specimens they could access in UK museum collections. They found from close inspection that all placoderm claspers were not con-nected to the internal pelvic girdles, as we see in sharks. They are in-stead a separate set of bony and cartilage structures that developed in the same way as the paired pectoral and pelvic fins and girdles. We proposed in our 2015 paper that placoderms had six pairs of paired limbs: the two pectoral fins, the two pelvic fins, and the paired claspers in males and paired pelvic plates in females.

The final consummate act in this prolonged saga of placoderm sex came unexpectedly in late 2014. I was working on a study of fossil placoderms from Estonia and Russia with the great Estonian paleontologist Elga Mark-Kurik at the Tallin University of Technology in 2013. Elga, who sadly passed away recently, had access not only to amazing collections of Baltic placoderms, but also to many excellent specimens collected by geologists working at sites across most of Russia, sent to her when Estonia was part of the Soviet Union. After completing much of the work I'd traveled there to do, I spent some time looking through the rest of her collections to learn about the incredible placoderms from the Baltic states.

Then it happened. I had a total Eureka moment, filling me with joy and utter excitement. Inside a drawer I found a small wooden box which contained a tiny, strange, hooked bone. Using my hand lens I could see it was a clasper attached to one trunk plate of a tiny antiarch placoderm called *Microbrachius* (only about 3 inches long when alive). This was the first time a clasper had been found in the entire antiarch group of placoderms—another major placoderm group with several hundreds of species. This chance find led to more searching and eventually finding some complete examples of *Microbrachius* from the barren, wind-swept Orkney Islands. I participated in two expeditions to these remote and beautiful islands off the north coast of Scotland with some friends from the UK, collecting more specimens of these intriguing fossils for our research.

Most antiarch placoderms bore peculiar bony arms, looking a bit like crab legs, as their pectoral fins. For many decades, scientists pondered what these strange arms could be used for in a fish. The new specimens of *Microbrachius dicki* (named after the Scottish fossil collector Robert Dick, actually) revealed that the males had a pair of large hook-shaped bony claspers coming off the rear of their trunk shield. The claspers bore a deep groove for passing sperm along on the internal face. They also had many sharp spikes on the external side of the fish, possibly the male placoderm equivalent of a peacock's tail to show off and attract females. Alternatively, the spikes might have served to anchor the fish into the lakebed, keeping it stable while it was copulating with its enormous claspers—each mature clasper, if straightened out, would be nearly half the fish's length! *Microbrachius* also displayed the first special mating structures found in any female vertebrate. These were small cheese-grater-like plates near the cloacal opening that helped to lock the male claspers into position for mating.

Our discovery solved a long-standing riddle about why some antiarchs had long, jointed, bony appendages as their pectoral fins. They were necessary for the males to get purchase while inserting the rigid claspers into the females by using a side-by-side mating position—the only sexual position possible given the size of the claspers. They were hooking up using their jointed arms intertwined "square dance style," or what I call the "do-si-do" mating position. This is the very first known sexual position for all vertebrates, and the first of many in the still-being-written Placoderm Kama Sutra (and really, nothing could surprise me now). When our paper came out in *Nature* in late 2014, the discovery was featured in much of all the world's media, even popping up as a short chat on *Saturday Night Live* in the United States and *QI* in Britain. It totally redefined the term "hooking up."

Aside from the inevitable puns and giggles, these finds bore real significance for our understanding of evolution and humans' relationship with sharks. The discovery of sexual reproduction in antiarchs is important because they sit at the base of the placoderm evolutionary tree. This discovery confirmed that it was highly likely that *all* placoderms used sexual intercourse for reproduction. This meant that internal fertilization, common in early jawed fishes like placoderms, using external genital organs, was passed on to sharks, who maintained it, but it was lost in bony fishes.

When bony fishes left the water and invaded land as tetrapods, they reevolved internal fertilization. The genes to do this were already in place from what placoderms had first developed, so it wasn't hard to do. All land tetrapods reproduce this way, obviously including us. We have placoderms to thank for both the joy of sex and the labors of childbirth.

Sharks, rays, and chimaerids all mate by the males placing packages of sperm inside the female via insertion of their claspers—long lobes coming off the base of the pelvic fins. Placoderms gave sharks this intimate way of reproducing by passing on the ability to develop paired claspers in the males. The claspers are separated from the pelvic fins in placoderms, so all sharks had to do was unite the claspers with the pelvic fin, and this is exactly what we find today in all sharks, rays, and chimaerids. All chondrichthyans reproduce this way, although species vary in exactly how the females deliver their young.

## Placoderm "rats of the Devonian"

Some placoderms competed with sharks not by being large, scary predators like *Dunkleosteus*, but by simply being everywhere. The global success of the antiarch placoderms might also explain why many sharks stayed on the sidelines at this time. These little fishes were ubiquitous in the Devonian world, being the commonest vertebrate fossils found in any Devonian sedimentary rocks. The most common of all antiarchs, *Bothriolepis*, is known from many species found on every continent (even Antarctica), so it was once common across the entire Devonian world. It lived primarily in freshwater habitats like rivers and lakes, with a few species also living in shallow marine near-shore habitats. This suggests that it might have dispersed by its larvae drifting on the ocean currents, then invading estuaries and river systems.

*Bothriolepis* is the fossil fish I first cut my paleontological teeth on as a grad student back in 1980. It really was the rat of the Devonian, found absolutely freaking everywhere. My undergrad thesis identified four new *Bothriolepis* species from sites in central Victoria, Australia. They are the first vertebrate genus to diversify into well over a hundred valid species, just like some catfishes in the Amazon basin today, where forms that somewhat resemble *Bothriolepis*, like *Plecostomus*, are known by more than 150 species. Such high species abundance meant that *Bothriolepis* clearly dominated the riverbeds and lakebeds. It had puny, twisted, toothless jaws, and its stomach contents confirm that some species ate tiny shrimp-like critters called **conchostracans**; others might have been mud grubbers, sifting organic particles out of lake floors and river bottoms. This total world domination by *Bothriolepis* likely prevented sharks from venturing into the same ecological niches. I feel that mud grubbing was a vocation too low even for sharks, so the majority of sharks kept on surviving through the Devonian period by sticking to what they did best—being active predators.

As we've seen so far, from the earliest known relatively complete fos-

The placoderm *Bothriolepis* was everywhere in the Devonian period, and its success might have precluded sharks from occupying certain ecological niches
ARTWORK BY PETER SCHOUTEN

sils from around 438 MYA, sharks lived in fierce competition with placoderms, their main rivals for territory and survival. Placoderms actively prevented sharks from branching out into trying new food resources because placoderms were able to modify their jaws and teeth for a wide range of prey types well before sharks could. Placoderms were the most diverse, most abundant, and largest of all fishes during the Devonian period, successfully invading new waterways, developing a diverse range of species, and inventing copulatory sexual reproduction, a strategy that, going by the numbers, worked to ensure that new generations grew to adulthood. However, all that was about to change.

In the latter part of the Late Devonian, things suddenly got a lot better for sharks. They were finally able to give their armored cousins a real run for their money. What was the game changer for sharks? Finally, at the minute-to-midnight moment as the Devonian drew to its tragic end, sharks challenged the placoderms to a knockout match in which the winner takes all and the loser goes extinct.

## Sharks and the Devonian extinction events

The Late Devonian extinction event unfolded over two deadly phases: the Kellwasser event at the end of the Frasnian stage (373 MYA), which mainly devastated life in the seas, and the Hangenberg event at the end of the Devonian (359 MYA), which seriously culled life in both marine and freshwater environments.

Scientists regularly debate what caused these events and likely agree that there was not just one definitive cause, but more of a cluster bomb of tragic circumstances. We know from the geological evidence of carbon and oxygen isotopes in the rocks that at around 373 MYA, many of the world's coastal seaways suffered a series of oxygen depletion events. These were accompanied by dramatic sea level rises and falls caused by periods of glaciation sucking up the water into polar ice, and then warmer periods releasing that water. Such massive environmental shifts could well have been instigated by the success of land plants. As they evolved, plants developed specialized vascular tissues to store water and produce large woody trunks, enabling them to grow to enormous sizes.

End-Devonian forests were dominated by horsetail trees (lycophytes) up to 100 feet tall. In the Middle Devonian, plants rarely ever grew above 30 feet or so tall. The advent of the first prototype seeds enabled these plants to invade large areas of the continents, their prior dependence on living near large bodies of water becoming redundant as they spread inland. The crash of falling trees in these primeval forests meant that huge amounts of organic material ended up in the rivers, which carried the woody debris out to sea. Concentrated plant accumulations today can cause algal blooms in shallow seaways that suck the oxygen from the seawater, killing all life in the water below. The oceans of the Devonian were equally susceptible.

Others argue that it was a gigantic volcanic eruption centered in Siberia, resulting in a formation called the Viluy Traps, which spewed forth around three hundred thousand cubic miles of lava around the same time as the Kellwasser event was taking place. This eruption may have raised carbon dioxide and sulfur dioxide levels, destabilizing the atmospheric system and causing rapid cooling, with subsequent sea level falls as water contracted and froze.

Another explanation favors a massive supernova explosion that destroyed Earth's ozone layer. This invisible shield protects life from the effects of harmful UV rays that if unchecked cause widespread genetic mutations and pathologies. We find evidence of these effects in a high percentage of deranged, mutated pollen and spores from this time period.

Finally, one other possible cause is one I published in 2016, working with an international team of geochemists. We looked at essential trace element concentrations in deep-ocean sediments, based on some three thousand samples taken around the world. Our research revealed that the Late Devonian oceans went through a major depletion of essential trace elements, including selenium, zinc, magnesium, cobalt, and cadmium, which caused havoc with marine food chains. The depletion took the form of a rapid drop in many of the essential trace element levels to more than a hundredfold lower concentrations than in modern oceans. Certain trace elements like selenium are necessary for life. This deadly decline started just before the Kellwasser event began, and continued through the end of the Devonian, rising again just after the Hangenberg event.

To my mind, there is convincing evidence for all these causes working in sync, with some offset in timing between their maximum effects. The combined effects would have placed tremendous environmental pressure on all life over a prolonged time. Some things had to break.

Immediately following the Kellwasser extinction event, when sea levels dramatically dropped and the shallow seaway became starved of oxygen at the end of the Frasnian, the extent of reefs around the world diminished by a factor of five thousandfold. The major reef-building organisms, the sponge-like **stromatoporoids**, never recovered. This time was a significant challenge for all reef-dwelling organisms.

Many of the specialized families of placoderms living on Frasnian reefs like those at Gogo were lost. Sharks, however, were not that reef dependent at the time, as evidenced by their scarcity at sites like Gogo, and the high abundance of shark teeth found in deeper, open ocean marine deposits. Sharks just sailed through this first mass extinction pulse without any noticeable effects, quite probably swimming out to seek shelter in the high seas, away from the tumult of crumbling reef ecosystems. The devastation of habitats and loss of many families of placoderms at this stage meant that many niches opened up in the Famennian stage of the Late Devonian (382–359 MYA). Sharks rushed right in.

## Finale: Devonian Sharks triumph over placoderms

The many shark fossils of the Cleveland Shale in Ohio support the idea that sharks eventually gained the upper hand over placoderms before the final end-Devonian extinction event struck. This site holds the highest diversity of sharks from any one geological formation for this time (within the final 3 million years of the Devonian period), with more than thirty shark species now recorded. It also shows that some of the last sharks of the Devonian were the largest yet, with *Ctenacanthus* reaching up to 20 feet in length. A similar-sized giant shark, probably another *Ctenacanthus*, living at the same time was found recently in Morocco by Christian Klug and his team, but this find is yet to be published. One recent study scaled down the size for the placoderm *Dunkleosteus* to a maximum of 14 feet—in line with Newberry's original estimation of its size at 15 feet. If we take this estimate into account, it's likely that at least one shark, *Ctenacanthus*, might have been larger, and much more massive in bulk, than the largest predatory placoderms around at this time.

The most recent tally of Cleveland Shale placoderms shows that they were easily outnumbered by sharks, and this dominance of sharks over placoderms is also reflected in the microscopic fish remains found in northern Italy at this time. This finding heralds the first time in Earth's history that sharks became the most diverse group of fishes in our oceans. If we include the numerous other shark species known only from teeth living in this late Famennian time, while also considering all known marine placoderm species globally, we get a fresh view of life in the terminal Devonian seas. It is overwhelmingly a picture of high shark diversity, just before the last great extinction pulse struck chaos around the globe.

I think the reason sharks became more diverse than placoderms at this time was that they finally showed a wide range of highly specialized feeding strategies, while placoderms were declining after the Kellwasser event chaos at 373 MYA. Furthermore, sharks were not wedded to the shallow seaways like most placoderms. We have numerous fast-swimming active predatory sharks like *Cladoselache, Stethacanthus,* and *Ctenacanthus* living alongside a guild of filter feeders like *Diademodus,* with shell crunchers and all-around omnivores that can eat anything, like *Orodus* and *Protacrodus.* Add to this mix the arrival of a new type of shark body plan as seen in the early holocephalans like *Psephodus,* and we have an entirely new opera ready to start after the end-Devonian chorus was silenced. Suddenly, in a geological flash of lightning, placoderms are on the wane and sharks are on the rise.

It's true to say that for the first time in Earth's history, from their mysterious origins about 465 MYA, sharks are now in a very strong position. The environmental chaos at the end of the Devonian, coupled with the growing competition from sharks, proved too much for placoderms. They disappeared forever from the face of the Earth about 359 MYA.

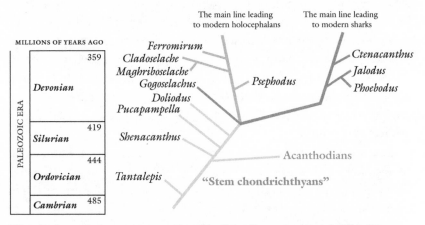

The shark evolutionary tree up to end of the Devonian (359 MYA). Note the first big split where sharks reinvented themselves for the first time on a line leading to modern holocephalans   DIAGRAM BY AUTHOR

Lauren Sallan is a very talented young American paleontologist based at the Okinawa Institute of Science and Technology in Japan. She studied the effects of the Devonian mass extinctions on vertebrates for her PhD at the University of Chicago. She meticulously compiled a comprehensive new database of all fishes of the Devonian period, detailing precisely when and where each lived. Her groundbreaking papers based on her mining of this database revealed some astute new findings that have changed how we think of sharks at this time. First, she noticed that fishes were less affected by the Kellwasser event 373 MYA than previously thought. While some groups went extinct, the actual composition of fishes in the oceans after that extinction was not dramatically changed. Second, she found that the end-Devonian event was much more devastating for fishes than previously thought. About 50 percent of all vertebrates succumbed, and large fishes like placoderms were particularly vulnerable. Third, she found that the size of fishes decreased radically after the Hangenberg extinctions (359 MYA), so much that the term "Lilliput effect" is used to describe this ecological state in which species are on average much smaller in the recovery phase after the mass extinction event than they were just before it.

The Carboniferous period was now ripe for a new invasion. Sharks and their newly minted offspring, the holocephalans, would march straight in and try to take over. How would they do this, in a freshly postapocalyptic world where another major group, the ray-finned fishes, was also beginning their rapid diversification? Today, the ray-finned fishes, like salmon and marlin, are the most diverse group of vertebrates on Earth, with more than thirty thousand species.

So how did sharks cope with this new rising competition? We will explore the answers to this question, as well as how sharks got very weird very quickly, in the next chapter.

# PART 2
# SHARKS RULE

*Akmonistion*

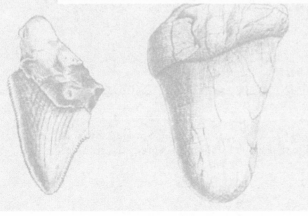

CHAPTER 6

# THE FIRST GOLDEN AGE
# OF SHARKS

*Sharks Take Over the World*

---

## An ancient Montana bay, 318 million years ago

Above the turquoise waters of the open bay, the summer storm clouds are gathering from the west. It is monsoon season, and this shallow sea embayment about 12 degrees north of the equator soon begins to feel the heavy rain building into a torrential downpour. Large mats of green seaweed float in the bay, with clusters of spiny lampshells and mussels dangling from their fronds. Translucent tiny fish larvae dart in and out of these green floating masses, chasing little shrimps and worms. Below the surface, the daily rituals of life carry on as always. The pouring rain above doesn't worry the numerous sharks and other fishes that live in these warm shallows, even though for some it could spell disaster.

The map of the world at this time featured two major supercontinents, Gondwana in the south and Euramerica in the north, comprising North America and parts of the Baltic states. Our warm seaway is located close to the western margin of this smaller supercontinent, just north of the equator. The atmosphere was nothing like it is today, as oxygen levels soared; life was high on the rich air, and the seas were supercharged with nutrients.

Most of the fishes in this large bay of around forty square miles live in the brightly lit surface and midwater zones, although a few prefer nuzzling for tasty worms or weeds on the muddy seafloor. Two odd sharks, a male and female *Falcatus,* are swimming slowly side by side, their large eyes alert for any signs of danger ahead. Both are small, just under half a foot long. The male *Falcatus* is easily recognized by his long slender dorsal fin spine that extends over his head, looking like the sword of Damocles, and by his paired claspers, or male genital organs, that trail off his pelvic fins.

There are also many larger sharks, some up to 12 feet long, cruising these waters. Below the pair of *Falcatus,* a strange long-snouted eel-like shark called *Thrinacoselache* darts out of a clump of algae, jagging a passing squid with its tiny grappling hook–shaped teeth that line its mouth. Hidden by a cloud of black ink released by the doomed squid, the shark twists and thrashes its body around, throwing the squid down its gullet in a couple satisfied gulps. Flitting along the muddy seafloor is a very strange-looking shark with wide stingray-like pectoral fins. This is *Squatinactis,* poking about for small worms hidden in the fog of fine sediment, using its keen sense of smell to detect the hidden prey.

Farther along, hovering just above the seafloor, a pair of bizarre shark-like fishes, *Rainerichthys,* slowly flap their large wing-like pectoral fins emerging from near their necks like dystopian underwater butterflies. To our eyes, their wings would resemble a pair of moose antlers from above. The pair slowly pull themselves along through the water, their stumpy bodies ending with weak little club-shaped tails. Their large eyes fill the skull. Small, jagged, wheel-shaped rows of teeth project from the tips of the jaws. The *Rainerichthys* maneuver through the water toward a large drifting jellyfish, which is unable to escape. The sharks carve it up into thick chunks with a scissoring action of their jaws, the razor-sharp lower tooth whorls jerking up and down in a flurry of hunger and speed as their mouths open and suck in the tasty morsels.

Above our pair of cruising *Falcatus* sharks, a school of deep-bodied ray-finned fishes, *Discoserra,* pass by, the sunshine glinting off their silvery scales as they chase a school of shimmering shrimps. Closing in quickly behind them comes a large male shark, about six feet long, *Stethacanthus,* sporting his anvil-shaped dorsal fin bristling with teeth-like spines. With a powerful swish of his forked tail, he pushes swiftly through the water, tearing a deadly swath through the shrimps, blood now tainting the water. He emerges from the maelstrom of the panicked and wounded with a bunch of shrimps caught firmly in his jaws. He did not notice the two smaller *Falcatus* swimming below.

A small patch reef of weeds harboring communities of clams and snails lies ahead. Around this verdant patch swims a weird menagerie of peculiar-looking shark-like fishes. One has a large head with big round eyes, high serrated dorsal fin spines protruding from its back, and a series of peculiar little "antlers" protruding from ridges above its eyes. This is a male *Echinochimaera*, an early holocephalan about 6 inches long. The females of this species are half his size and have a single set of curved horns above the eyes. The male peers into the weeds, delicately pulling coiled snails off one by one and smashing them up with his robust tooth plates.

A peculiar shark-like fish slowly moves through the weeds in search of a quick meal. It is *Belanstea,* a deep-bodied form with a large head and two fan-like dorsal fins extending all along the body—resembling a modern lumpsucker fish (*Cyclopterus*). It sports massive paired pelvic fins under the shoulder girdle, at the same level as the pectoral fins. It is unlike any shark, ray, or chimaerid alive today (see images, p. 132). *Belanstea* perhaps represents the most extreme chondrichthyan body plan ever evolved.

Back in the bay, the male *Falcatus* swims erratically in front of his partner, lightly grabbing her pectoral fin with the tip of his jaws. She responds with a small shudder, wriggling her body to shake him off. He holds on for few seconds, then lets go and moves ahead, waiting to see if his romantic overture has worked. She moves forward with a quick flick of her large tail and catches up to him, moving above him to swim together in perfect synchrony. She then grabs his dorsal fin spine in her mouth, griping it with her teeth, using special small tooth whorls located in her cheeks at the junction of the jaws for the job.

Above the water, the torrential downpour of monsoonal rain has shifted massive loads of sediment rich in plant debris from nearby land into the bay, forming fast-moving currents of organic-rich turgid water. With his spine now tightly held in her jaws, he senses the time is right to mate. He twists his body around as his claspers become erect, moving them toward her pelvic fins, which are now open to expose her cloacal opening.

Suddenly, the sediment-laden warm surface water moves swiftly over the lower layer of cooler water forming the bay, then sinks, co-alescing the two layers.

He inserts one of the claspers inside her and wriggles, edging the clasper deeper into her body. They begin their writhing dance of love, the male pumping packages of sperm inside her while she keeps biting down on his dorsal spine. As he withdraws, and before she can relax her grip, they are suddenly covered by a massive fall of toxic water carrying deadly sediment that hits them like the hammer of Thor, pushing them to the bottom of the bay. The organic-rich waters suck all the oxygen out of these deeper waters, where our couple are killed instantly, still entwined like a Carboniferous Romeo and Juliet.

As the storm passes, the two dead sharks lie buried under a thick pile of sediment. The female died with her mouth wrapped around the male's dorsal fin spine. After compaction flattens the sharks, tectonic forces and the sheer weight of later sediment accumulations will compact and transform the limy mud of their tomb into limestone layers. Eventually other tectonic forces, like the movement of continental plates, will push the doomed pair up toward the surface of the Earth. Eons later, at a site named Bear Gulch, a human knowing exactly what to look for will use a small crowbar to pry a thin slab of limestone apart and be awed to find the two shark lovers still together, bound in the same position as the day they died 318 million years earlier.

## Welcome to the age of coal swamps

The Carboniferous period (359–299 MYA) was when Earth first became covered in lush forests of tree-sized lycophytes (horsetails) and tree ferns (*Archaeopteris*), with the first seed plants appearing, the **progymnosperms**. It was a time when vast swamplands spread over the world's shallow land areas because climates had rapidly cooled since the end of the Devonian. These sodden marshlands filled with leaf litter and dead trees eventually formed much of our massive coal deposits—the name Carboniferous meaning "coal-bearing." Gigantic 7-foot-long millipedes, like *Arthropleura,* sifted through the leaf litter, avoiding the 3-foot-long scorpions that scuttled noisily over the forest floor. For the first time, Earth was home to many kinds of terrestrial amphibians—huge newt-like creatures, some up to 15 feet long, that lurked like slow-moving crocodiles in the rivers and lakes hunting fishes. The atmosphere contained 29 to 30 percent oxygen, the highest level our planet ever attained. Forest fires were common.

The Carboniferous was the time when sharks underwent their most rapid burst of diversification in their entire history. The seas, rivers, and lakes of the world would never be the same again. The age of the placoderm was over, the time of the shark had come.

After the great double-punch extinction events at the end of the Devonian, the seas were quiet, allowing life to slowly recover. Although seawater temperatures plummeted by as much as 27°F at the end of the Devonian, they rapidly bounced back again by the end of the first half of the Carboniferous to nearly the previous levels. This epoch is known as the Mississippian (359–323 MYA), named after the extensive lime-stone deposits found in the Mississippi Valley. The second half of the Carboniferous period is named the Pennsylvanian epoch (323–299 MYA), for the rich coal deposits of Pennsylvania, as such deposits are found around the world.

In the shallower, cooler continental seaways of the Euramerican su-percontinent and the coastal margins of the Gondwana landmass, we see a spectacular change in the fish faunas at this time. We find a very high abundance of sharks and a range of bizarre shark-like forms of uncertain affinity, so we will simply call them **stem holocephalans**. Other fishes, like the ray-fins (trout-like forms), are also found in higher abundance and diversity, much higher than in the Late Devo-nian oceans. However, they are still in the minority compared to the huge diversity of the sharks, holocephalans, and their kin.

Finally, after a hundred million years of competition and evolution since they first appeared, sharks have entered their first golden age in the Carboniferous period. They would rule the seas unchallenged for the next 130 million years, thanks largely to their next superpower: the ability to craft and shape new tooth types with new tissues.

To explore the great variety of sharks living at this time and hear how they use their dental superpowers, we need to visit Bear Gulch. First let's find out how this remarkable site was discovered, then learn why the fossils were so well preserved and see what they tell us about the new advances in shark evolution at this time.

## Dick Lund, the shark whisperer of Bear Gulch

For nigh on fifty years, a small, stocky man who chews baccy and dresses like a cowboy of old, Dr. Richard "Dick" Lund, has been the "shark whisperer" of Montana. "Dig we must" is Dick's motto, and it has served him well.

In 1967, Bill Melton of the University of Montana received some superb fossil specimens from a student whose uncle was a rancher who owned the land out at Bear Gulch. These finds contradicted earlier geological reports stating that the limestone in the area contained no fossils. So Bill organized an expedition that summer, accompanied by a keen young grad student, Jack Horner, who mapped the geology of the area. He kept finding fossil fishes and sharks everywhere the rocks outcropped. Jack went on to become one of the world's most famous dinosaur paleontologists. In the following summer of 1968, they were joined by another newly graduated offsider, Dick Lund. That field season was the beginning of Dick's long relationship with the Bear Gulch site. Dick eventually landed a job at Adelphi University and continued working the site for more than five decades, digging every year, aided by teams of willing students and volunteers.

In a 1976 newspaper interview, Dick's infectious enthusiasm for paleontology shines through. When asked what the most exciting fossil was he ever found at Bear Gulch, Dick replied, "The next one."

*"When you split a rock," he explained, "open it up, and see a beautiful fish before you that has not seen the brilliant light of a Montana day for three hundred and twenty million years, the sight is breathtaking and the feelings are of awe, transcending all else."*

Every year they found spectacular, complete sharks and other fish fossils, as well as a diversity of other fossils—squids, ammonoids, nautiloids, shrimps, brachiopods, horseshoe crabs, worms, and seaweeds. They could reconstruct the entire ecology of the ancient bay where all these organisms lived. It was indeed a rare piece of land, where the fossils were preserved not only whole, but also in great numbers—what we scientists call a *lagerstätte* (in German, "place of storage"), that is, a fossil site of great significance owing to its great numbers and extraordinary preservation. Without doubt, Bear Gulch is the single richest locality known in the world for well-preserved fossil sharks of this age.

**Dick Lund working at the Bear Gulch, Montana, site in the early 1990s** PHOTO BY DICK'S WIFE, EILEEN GROGAN

Dick and his team have now collected more than 5,900 fossil fishes. The sharks and their cartilaginous kin—stem holocephalans and related forms—dominate the fauna with more than eighty species recorded, alongside some fifty species of ray-finned fishes, about a half dozen coelacanth fishes, and a lamprey. This marks the first time in Earth's history that we have a marine fauna in which the diversity of sharks and other chondrichthyans is far higher than the total diversity of all other fishes combined. The bulk of them are now safely curated in the Carnegie Museum of Natural History in Pittsburgh, with several other U.S. and European museums now also sharing in the site's bounty.

Dr. Eileen Grogan from Saint Joseph's University in Philadelphia has been working with Dick on the Bear Gulch chondrichthyans for the past three decades, and they are also partners in life. She is a leading expert on holocephalans. Together, Eileen and Dick have studied the fishes as well as investigating the forensics of the site—how these sharks died and why they were well preserved. As always, there is a geological story underpinning how such intriguing fossil deposits were formed.

The sharks that lived in the ancient Bear Gulch sea included some we regard as "true sharks" (selachians) as well as a wide variety of stem holocephalans and others that are side branches to the evolutionary line to modern holocephalans. Most were small, squat, or elongated forms. Very few shared similar body proportions to a modern-day shark. Together these sharks and their close kin present a pageant of bizarre and unusual body shapes, some of which were unique to this time period.

How do we know all this? Let us plunge into the science underlying the earlier reconstruction of the scene in the ancient Bear Gulch Bay, take a detailed look at some of the extraordinary sharks, and weigh up the evidence for how we know about their world and their lifestyles.

## The Bear Gulch shark menagerie

The Bear Gulch chondrichthyans look like a motley crew of outcasts when compared side by side to modern-day sharks and holocephalans.

The hero and heroine of our re-created scene are the *Falcatus* couple. Known from more than a hundred fine examples, some were found as groups of individuals buried together. This suggests they were social creatures that liked to hang out with one another. For me, the most dramatic fossil ever found at Bear Gulch was the slab of rock perfectly preserving the two *Falcatus* sharks lying one above the other. The male, identified by his claspers coming off the pelvic fins, is below the female, whose mouth is wrapped around the prominent dorsal fin-spine inclined above his head. This is one stage of an intimate courtship ritual; certain living sharks often bite each other, especially on the pectoral fin, to get a good purchase when mating. It's impossible to tell if their courtship was actually consummated. I've chosen to present them here as being captured forever in the postcoital position.

Mating *Falcatus* specimen, male below showing dorsal fin spine protruding over his head with the female above him clasping her jaws onto it   IMAGE COURTESY OF DICK LUND

How did these sharks get buried in their life positions so rapidly? The geology of the region has been meticulously studied to provide the answer. The bay where these sharks once lived supported various habitats, from the deep center to the shallower margins and channels leading into the bay. The best fossils, the most complete ones with soft tissues preserved, are those that were rapidly buried in the center of the bay. Here the sharks were killed by a deadly double punch: simultaneous burial and asphyxiation from lack of oxygen. This was caused by a **pycnocline**, a phenomenon in which bodies of water of different densities sit on top of each other in layers created by differences in salinity levels, temperature variations, and other factors. Sharks lived in both the top and bottom layers depending on what kinds of prey they hunted. Summer monsoons bring waters full of sediment, which were washed from the shores, far into the bay. These sediment-laden waters then fall rapidly, mixing up the two water layers of the pycnocline. The organically rich sediments in the falling column of water are full of decaying plant matter, which absorbs oxygen as the sediment-rich water rapidly descends to the bottom, both suffocating and entombing all living things in its path.

Another Bear Gulch shark mentioned in our scene, *Thrinacoselache*, was like nothing living today, with its extreme eel-like body shape, lacking any dorsal fins and bearing a very long tail fin underneath its body. It had a long, pointed snout and rows of grappling hook teeth, perfectly suited for jigging squids, which were abundant in this bay, or hunting fishes. The oldest definite example of the squid family, *Gordoniconus*, was described from here. *Thrinacoselache*'s body shape suggests that they might have hunted like moray eels, darting out of crevices to ambush passing prey. One Bear Gulch example shows how effective *Thrinacoselache* was as a predator. A specimen of *Falcatus* was found inside the digestive system of a *Thrinacoselache*, neatly sliced into three parts, the result of two clean bites, even severing its sickle-shaped dorsal fin spine. This marks another significant evolutionary first for sharks: They can be seen to have developed long, slender bodies devoid of dorsal fins, adapted for completely new ways of ambushing their prey.

*Squatinactis* is another captivating early shark featured in our opening scene. It possessed a rather peculiar pair of wide pectoral fins, like a stingray, implying that it flitted about above the seafloor using gentle

***Thrinacoselache*** from the Bear Gulch site—a shark trying to be an eel

undulations of these large fins. Its delicate teeth were armed with many finely pointed prongs ideal for clutching small prey, like worms or small swimming sea butterflies (nudibranchs). *Squatinactis* represents yet another completely new body form—one of many—developed by sharks of the Carboniferous, and one that would reevolve several times, culminating in a new group with wide pectoral fins about 100 million years later, when the first rays appeared.

The strange *Rainerichthys* represents another totally new and weird group, the **iniopterygians**, known by many good examples. The iniopterygians were stem holocephalans—an offshoot from the evolutionary branch leading to living holocephalans. Barbara Stahl of Saint Anselm College in Manchester, New Hampshire, was the first scientist to solve a big mystery about these strange creatures. She recognized that the pectoral fins were high up on their bodies, and used as wings, ray style. She also noticed that the upper jaws were not fused to the braincase in some forms, distinguishing them as being more primitive than all the modern species of holocephalans, in which fusion is the norm.

*Rainerichthys* clearly occupied a highly specialized niche in the Bear Gulch food chain. It had a sharp tooth whorl on the tip of its lower jaw that met a similar pair of whorls in the upper jaw, while the rest of the jaws lacked teeth. The tooth whorls would shear when they met like a pair of sharp scissors, aided by a joint through the skull that enabled a rapid closing movement using its flexible braincase. They were also unique among the early holocephalans (chimaerids and kin) in their high-performance suction feeding. Stomach contents found in some of

*Rainerichthys*'s close relatives testify that it ate a broad diet, feeding on shrimps, conodont animals, and plants. Unexpectedly, this suction-feeding method they used was found to be closer to how some turtles and salamanders feed than to the suction feeding method used by living sharks.

**Restoration of the weird *Rainerichthys*, an iniopterygian that flapped through the water with its wing-like pectoral fins. Sharks trying to be birds ahead of their time** ARTWORK BY NOBU TEMURA

*Echinochimaera,* meaning "spiny chimaera," is the oldest definite complete fossil of a holocephalan sharing the same body plan as modern chimaerids. As the first dorsal fins of some living forms have venomous spines, it's quite possible that *Echinochimaera* may also have used a toxic defense to protect itself from the larger predatory sharks of the bay like *Stethacanthus.*

*Belantsea,* perhaps the weirdest of the Bear Gulch holocephalans, is named after the First Nations Crow people's name for themselves. *Belantsea* is known for a truly beautiful fossil, bearing dark stains indicating the position of its liver. It wasn't built for speed but more for specialized feeding on prey that required considerable manipulation using its flat, jagged teeth. These teeth are shaped like a hand-chipped early human stone tool, with coarse grooves along the biting edges. These teeth highlight the way sharks activated their next dental superpower: the ability to evolve teeth of many strange shapes using new dental tissues (more detail about that soon). Its teeth also confirm that it was a **petalodont**—a group of stem holocephalans known mostly from their distinctive flat teeth with only seven teeth in each jaw. As no living fishes have teeth like it, *Belantsea* was either feeding on some specific food type that's not around anymore or using its teeth to open clams in a way that's hard to visualize.

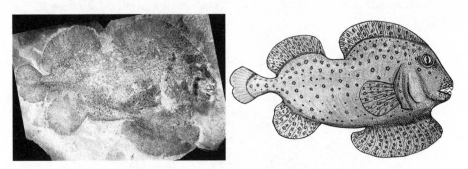

*Belanstea,* **heads to right, with weird pelvic fins below the head—perhaps the most extreme body shape that ever evolved for a chondrichthyan fish** PHOTO COURTESY OF DICK LUND, ARTWORK BY THE AUTHOR

Of the eighty or so species of sharks and their kin that once lived in this bay, I've selected just a few of my favorites to highlight the great variety of sharks and their kin at this time. Through half a century of hard work by Dick Lund and his colleagues, the fossils of Bear Gulch have given us an IMAX view of the life of the past, showing us in amazing detail the first great diversification event of sharks and holocephalans on Earth.

## Thoughts from a 300-million-year-old brain

While Bear Gulch has shown us some incredibly well preserved fossil sharks, there are rare examples of even better preservation in which something extremely delicate is preserved in a fossil chondrichthyan. No book on the prehistory of sharks would be complete without telling the story of an almost unbelievable fossil found in Kansas. While the Bear Gulch fossils are preserved as flat outlines, certain rock concretions (or nodules) from the heartland states contain complete three-dimensional shark and holocephalan skulls and jaws. One nodule from Kansas revealed an unexpected treasure of paleontology: a real fossilized brain. It was found by a French student, Alan Pradel, while he was working on his doctoral dissertation with John Maisey at the American Museum of Natural History in New York.

Using the European synchrotron to scan the inside of the rock containing the fossil, Alan revealed that the rock contained the skull, braincase, and actual brain of an iniopterygian, a close relative of our friend *Rainerichthys*. This was not a cast of the brain made when sediment filled in the space where the brain once existed. It is the mineralization of the soft brain tissue, preserved shrunken inside the much larger cavity of the braincase. It is complete, showing the delicate spaghetti-like cranial nerves emerging from it. Strangely, none of the other soft tissues of the head were preserved in this fossil, just the brain. The internal chamber of the braincase provided a unique chemical environment for the brain tissue to be fossilized before it could decompose, because the braincase chamber created an environment devoid of oxygen. Decomposing fatty tissues of the head caused a drop in the acidity levels inside the chamber, allowing calcium phosphate to mineralize the brain's tissues. The saturation of the brain chamber with phosphatic minerals is also mirrored by the preservation of these fossil skulls in nodules held together by the same phosphate minerals.

So, what does a 300-million-year-old fossilized brain tell us about these strange iniopterygian sharks and their lifestyle? Alan observed it had a small cerebellum (the main thinking part), in contrast to the brains of fast-swimming predatory sharks, like makos, whose cerebellums were much larger. This suggests that this iniopterygian was more of a slow mover. The shape of its inner ear canals (used for balance) resembles those of living fishes that like to dwell on or just above the seafloor. The geology of the fossil site tells us it lived in a shallow subtropical seaway. The lifestyle of the iniopterygians was therefore similar to the way modern holocephalans, like the deep-sea ratfishes and chi-

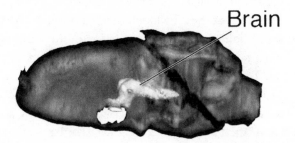

**A 300-million-year-old holocephalan brain imaged inside its braincase. The specimen is still encased in rock** SYNCHROTRON TOMOGRAM BY ALAN PRADEL

maerids, live their lives, except that most of the recent forms now prefer much deeper ocean floor habitats.

Here's an interesting afterthought: A second early fossil fish brain was announced recently, this time in a ray-finned fish called *Coccocephalus*, from England (dated at 319 MYA). This find suggests that perhaps more fossil brains might be discovered if only we would take the time to search for them using powerful micro-CT scanners and synchrotrons.

Other discoveries of sharks from the slightly older layers of the Carboniferous period suggest to me that the great diversification of sharks and holocephalans taking place at this time was caused by a reactivation of their dental superpowers—the ability to evolve many new tooth shapes and to develop new tooth tissues.

## Sharks activate their dental superpower

Sharks have unique teeth that are regularly replaced and made of superstrong material. However, the way they develop and use their teeth is the key to understanding what drove their evolution from these early forms through to the modern shark groups. The two big factors at play here are the shape of the teeth and the tissues making up the teeth. Most sharks had sharp-pointed teeth for predatory lifestyles, like having a mouthful of knives with which to stab or pierce your prey. Many others developed flat crushing or grinding tooth plates for eating hard prey like clams or crabs—the steamroller approach to feeding. There are also many in-between cases in which some have sharp teeth at the front of the mouth and rounded crushing teeth at the back of the jaws—the Swiss Army knife approach to feeding.

There were very few shark species with crushing tooth plates in the Late Devonian seas, yet these included the first holocephalans and their kin, which sailed right through the extinction events, proving that this new type of dental tool was perfect for adapting to rapidly changing circumstances. Today's living holocephalans—chimaerids, ratfishes, and spookfishes—all bear distinctive tooth plates and are specialists that thrive on low-nutrient foods. In the early years of the Carboniferous period they flourished in shallow coastal seaways, bays, and lagoons.

Sharks and their offshoot, the holocephalans, activated their new dental superpower in a big way in the Carboniferous. Holocephalans developed new super-strong tissues to harden their tooth plates, making them incredibly robust, perhaps the hardest teeth of any creature. Simple rounded tooth plates of *Psephodus* show the tooth-bearing rows of pillars made of super-strong hypermineralized dentine. This is a kind of tooth unique in the entire animal kingdom, for it gives such teeth incredible strength for powerful bites, likely opening up new food resources that no previous species could crack open, like larger clams or lampshells (brachiopods) that were abundant at this time.

Unfortunately, most of these species of early tooth-plated sharks and holocephalans are known only from their teeth. This created a huge problem for paleontologists in the nineteenth century. In these early days of fossil shark research it was common practice for scientists to give new species names to every single shark tooth, or fin spine, that differed even slightly from another named type.

But this is not how nature works. Not every variety of tooth indicates a different species. Think of your own teeth—there are several different shapes and sizes, all named as human teeth (one species), rather than a new species for every tooth shape. The teeth in the jaws of any shark can also vary greatly from top to bottom, and from the front of the mouth to the back. Similarly, a shark can also have multiple fin spines with different shapes. This provides an insight into the true nature of paleontology, with its occasional attendant frustrations. It's not just about discovering new fossil species all the time; sometimes there is a lot of detailed, time-consuming clean-up work to be done, fixing the taxonomic messes made with the best of intentions by overzealous earlier scientists.

Aside from the benefits of the internet, which has allowed us to search museum databases and compare specimens quickly across continents, what was missing most in those heady nascent days of fossil shark work was a Rosetta Stone shark—a complete specimen with intact jaws so that the many teeth and isolated remains could be studied and combined into just one species.

As an example, a common form of shark tooth found throughout the Early Carboniferous are called **cladodont** teeth, named after *Cladodus*. This tooth type has three or more prominent pointed cusps, with the central cusp being the largest, on a wide root. We know it is part of the ctenacanth shark family, as one form, *Ctenacanthus,* is known from nearly complete specimens from Ohio showing very similar teeth. As

*Cladodus* teeth were collected in great numbers in the early days of paleontology, the genus expanded to having more than seventy named species, all based on isolated teeth discovered by a range of passionate fossil hunters. Research done in the 1990s was able to collapse these seventy down to just five valid species. Currently we still have several hundred questionable species of fossil sharks in the literature. Fossil shark specialists know that the real number of valid species will be far lower once these problematic forms are restudied and culled to fewer species, but it will take decades to do this work.

Some of the best Rosetta Stone specimens that helped solve these mysteries came from Scotland. The oldest complete fossil sharks of the Carboniferous period were found at Wardie Bay near Edinburgh (dated at 335 MYA) and Bearsden near Glasgow (dated at 329 MYA). Like Bear Gulch, these sites formed in shallow marine lagoons and bays. While the Wardie Bay fossil sites were first collected over a hundred fifty years ago, it is thanks to the extraordinary work of one canny Scot that many spectacular fossil sharks were uncovered in the past few decades. Finding these fossils wasn't luck; it was a real triumph of human will and the dogged determination of a passionate enthusiast. Let me introduce you to the remarkable Stan Wood and the story of how he found his spectacular fossil sharks.

## Mr. Wood's incredible Scottish fossil sharks

In September 1982 I was a grad student attending my first overseas conference at Cambridge University. I had just presented my talk when a tall, lanky Scotsman with a broad smile approached me. Stan Wood congratulated me on my discoveries and extended an irresistible invitation—if I was heading up to Scotland, I was welcome to stay with him and his family and see his fossils. So I did.

Stan became interested in fossils in 1969, at age twenty-nine, when he visited an archaeologist at the Royal Scottish Museum who convinced him that he should switch his interest from ancient Roman artifacts and try searching for fossil fishes. At the time, there were several important fossil sites that needed to be rediscovered because the local knowledge of the exact rock layers that the fossils came from had been lost. Stan started with a site within the Edinburgh city region easily accessed by public transport, Wardie Bay. In 1992 he published his account of working at Wardie Bay, laconically entitled *A Challenging Edinburgh Fossil Site*. Its poignant opening cuts to the chase, stripping bare the soul of a fanatical fossil hunter:

> Once upon a time a local Edinburgh lad of 30 years itching for further stimulus found that searching for fossils in and around Scotland's capital provided an ideal "run off" for physical and mental excesses. The challenge of a new locality, the anticipation of finds, their moment of discovery, the triumphant return home; all for the cost of a local bus fare. Happily that value for money is just as true today as it was then. If, as I did, you too become immersed in this intriguing hobby, then even the pain of a back bent for more hours than advisable, the penetrating cold from North Sea winds and the throbbing ache of unintentionally hammered knuckles, will only add depth to the satisfaction of a new specimen won through adversity.

For two years, Stan scoured these local beaches where complete fossil sharks and fishes had been found hundreds of years before, and found zilch. Then, as he read more about the history and geology of the sites, he turned his focus to certain thin layers of rock. With persistence and hard work, he soon found some outstanding specimens, some of which would later become new species; one was named *Diplodoselache woodi* in his honor. It represents the oldest known definite member of a new group of sharks, called **xenacanths**, that would soon dominate the freshwater rivers, lakes, and lagoons of the world and thrive well into the age of dinosaurs. They are so important to the shark story that I treat them in detail in the next chapter.

Stan Wood back in 2012 carrying a fossil he'd just dug up from a site in Scotland
PHOTO BY GLYN SATTERLEY (GLYNSATTERLEY.COM)

Stan's new *Diplodoselache* species represented the first complete specimens ever found, as previously it was known only from teeth. Even at up to 7 feet long, it wasn't the top dog in this ancient Scottish lagoon. That place was held by the 20-foot-long *Rhizodus,* a monstrous bony rhizodont fish with 6-inch-long dagger-shaped teeth. This beast was the largest bony fish of the entire Paleozoic era, and it likely fed on *Diplodoselache* when it could ambush them.

Stan's two other significant fossil sharks from the Wardie Bay sites are *Onychoselache* and *Tristychius.* Each heralds a new direction for sharks, ultimately leading to the rise of all modern sharks, because *Onychoselache* and *Tristychius* are the oldest known definite members of the group containing modern sharks and rays, the **elasmobranchs** (albeit as early or "stem" members). They belong to an extinct shark group called **hybodontiforms**, the group that would dominate the seas and rivers in the first half of the Mesozoic era, when dinosaurs ruled the land.

*Onychoselache* was a small shark at around a foot long and is known from only three examples. Its dentition shows tiny wide crushing teeth that lacked distinctive cusps, the tools of a generalist feeder that could eat almost anything it could fit into its mouth. A specimen Stan found from another site revealed new information on the skull and fin structures of *Onychoselache,* including its extraordinary similarity to the specialized mobile fins of today's epaulette sharks (*Hemiscyllium ocellatum*), which use their paired fins to walk across exposed coral reefs. Perhaps, when in a tight spot, *Onychoselache* could shift itself out of the water, walking up the lagoon bank and away from the terrifying mouths of large rhizodontid fishes.

*Tristychius* is the most common shark from the Wardie site, with more than a hundred specimens found. Like most of the best fish fossils from Wardie, it was encased within reddish to mauve ironstone concretions. Manually preparing these fossils out of the rock is a real pain, taking a lot of care and time and blunting your preparatory needles. However, to make a big discovery of this kind it requires a team effort. Stan finds the fossils and can beautifully prepare them with fine needles and drills. His colleague and friend Mike Coates, a British shark paleontologist based at the University of Chicago, can then take the discovery to the next level by using high-tech tools. Mike used CT scanning to save Stan a lot of time and effort preparing some of the *Tristychius* fossils inside ironstone tombs.

I first met Mike in 1982, at the same paleontology meeting where I met Stan. Mike and I were both poverty-stricken grad students studying fossil ray-finned fishes for our PhDs. Mike invited me to stay at his place when I visited Newcastle upon Tyne a few weeks after the conference. He welcomed me to the ramshackle student house he called home. I was glad to sleep on the floor, but in the dead of night I found myself attacked mercilessly by an apparently starving cat. Nonetheless, we bonded over tepid pints of Newcastle Brown Ale.

Thirty-six years later, Mike's CT scans of *Tristychius* were used to generate 3-D printed models of its head and shoulder anatomy—the braincase, jaws, gill arches, and shoulder girdle cartilages. These were manipulated to determine the range of movement possible as the jaws opened. Next, using a combination of video and computer programs to digitally model these movements, Mike showed that *Tristychius* had an unexpected and exceptional power that had not been seen in any fossil shark before: As *Tristychius* approached its prey, it could use its gill arches to open its mouth cavity so wide, and in such a special way, that it created a powerful suction that could pull its prey into its mouth. Imagine the shark slowly approaching a shrimp. Under normal conditions, our tiny crustacean friend might dart quickly away once it spots the predator coming. The shrimp might hide in a small crevice and think it's safe. Not so: *Tristychius*'s suction power could wrench it out of its hidey-hole and into its mouth in a split second.

Such a specialized way of feeding is what drove the success of the modern ray-finned fishes (**actinopterygians**). Today, nearly all of them feed in this "suck-and-crunch" way—just watch your goldfish eat its food. Yet *Tristychius* had mastered this trick in the Early Carboniferous, 50 million years before the first ray-fins learned how to do it. This innovation destroys any idea that Carboniferous sharks were all "primitive" and was a precursor to the feeding strategy used by some living sharks, like nurse sharks that would evolve independently again, much later in time.

The takeaway from this study is that we can learn so much more about fossil sharks using high-tech methods to re-create the accurate anatomy of the shark. Printing the fossils as 3-D models allows us to digitally explore the shark's movements without jeopardizing the original specimen. We can also directly compare the results with living forms that feed in the same way; by measuring their feeding speeds and the water volume changes in the mouth, we can extrapolate back to how the fossil shark hunted its prey and captured it.

Ultimately, the advanced suction feeding style of *Tristychius* did not create a huge evolutionary advantage for sharks. As useful as this new capability was, we don't find a subsequent large number of species using the new way of suction feeding. We don't know why. It may be that the many other experimental feeding methods used by sharks at the time, like new crushing dentitions and a wide variety of odd tooth shapes, were simply too successful. This next spectacular discovery from Scotland shows that some sharks had the ability to make large tooth-like structures in strange places, even outside their mouth.

## *Akmonistion,* the Bearsden Shark

Without doubt Stan Wood's most famous fossil find is the Bearsden Shark, now lauded in verse as one of the great hero fossils of Scotland. Poet Laureate Edwin Morgan's 2010 poem titled "The Bearsden Shark" called it a "Heroic long-dead creature, waiting in death."

Indeed, the nearly perfect, complete fossil shark waited some 330 million years for Stan to find it. He uncovered it shortly after he moved to the town of Bearsden, six miles from Glasgow city center, in 1982. He was walking his dog when he spotted a fossil poking out of the rocky layers of the bank of the local creek, the Manse Burn. Further excavations at the site revealed a shark with a massive dorsal fin spine containing hundreds of tooth-like structures, similar to the *Stethacanthus* from Bear Gulch that we have already met. Stan kept digging at the site and found more good examples of this shark. He asked his friend Mike Coates to undertake the study of his prize specimens.

It turned out to be a real Rosetta Stone fossil. When Mike Coates and his student Sandy Sequeira began studying these sharks, they came up against a wall of confusion—many names of different fossil shark bits all belonged to this one shark, uncovered by past researchers working from fragments. For example, the name *Stethacanthus* was originally based on its robust fin spines, *Cladodus pattersoni* from one of its scales, *Stemmatias simplex* from a throat denticle, and *Lamndodus hamulus* from another denticle found in the dorsal spine brush. To avoid any future confusion, they simply decided to give it a new name, *Akmonistion,* meaning "anvil sail," alluding to its massive dorsal fin spine structure.

*Akmonistion* was about 2 feet in length, sporting a narrow body with the lower lobe of the tail longer than the upper lobe, a shape suggesting that it spent time resting on the bottom of its lake. The head bore a conspicuous cap of thick, pointed denticles that were basically like teeth sticking out of its skull, matching those on top of the dorsal fin brush. Its teeth had large, robust cusps with stout pairs of pointy side cusps, signaling that it was a predator. Unlike most sharks, it lacked placoid scales covering its body. Its ungainly dorsal fin spine probably functioned to stabilize the shark when swimming, resisting sideways wobbles. We guess it was capable of short bursts of speed to ambush prey because we find the remains of agile shrimps in its gut.

*Akmonistion* with massive dorsal spine brush covered in tooth-like denticles. See reconstruction on p. 119    PHOTO BY J. K. INGHAM, UNIVERSITY OF GLASGOW

The unusual cranial denticles and dorsal fin spine brushes that look like teeth are developed on *Akmonistion* and *Stethacanthus* and all their family (called **stethacanthids**). These characteristics exemplify the early shark superpower of **dental plasticity**—the ability to modify teeth or scales into tooth-like structures inside and outside the mouth. Teeth are replaced throughout life in sharks, whereas scales are made just once. Some sharks could occasionally modify their teeth and scales in totally unexpected places, like on top of their head, above their eyes, or sticking out of their dorsal fins. It's a neat party trick and something not often seen in the natural world. The Carboniferous period, however, is typified by lots of bizarre sharks with a variety of teeth and denticles, even on the top of their head.

Mike regards his work on *Akmonistion* as one of his most important achievements because it inspired everyone in this field to think deeply about where these stethacanthid sharks fit into the big family tree of sharks. This specimen generated lots of new research papers that added and refined ideas about early shark evolution. The original *Akmonistion* study of 2001 hinted at a radical idea—that *Akmonistion,* along with the common Late Devonian *Cladoselache* and other typical "sharky sharks" known as **symmoriiforms** (including *Phoebodus* and *Ferromirum* from Morocco), were not actually true sharks on the line to living sharks. Instead, we should think of them at the base of another branch of sharks leading to the modern holocephalans. As controversial as this idea was at the time, it is now the most accepted idea. The important features that support this idea are seen in the large eyes, large inner ear chambers, and other parts of the brain structure. They still look like typical sharks, as they have just diverged from the main evolutionary branch containing sharks, so they haven't started changing to the holocephalan body shape just yet.

Sadly, Stan Wood passed away in 2012. His legacy is that he discovered the sacrosanct holotype specimens—the museum reference specimens designated when naming a new species—of thirty-two new fossil species, including five new sharks and many other bony fishes, plus two scorpions, a horseshoe crab, a plant, and twelve stunning new kinds of early tetrapods. Five fossil species were named in his honor, including a stout woody plant that appropriately bears the name *Stanwoodia.*

Strangely, these Scottish sites and the Bear Gulch site lack sharks that were really huge, like the enormous Late Devonian *Ctenacanthus.* This raises some interesting questions: Were there any very large predatory sharks around at this time, and if so, what do we know about them? To answer these questions, we will need to get down on our hands and knees to crawl through a tunnel in a cave in Kentucky.

## Giant predatory sharks rule the seas

One of the most exciting large shark fossils of this age was found many years ago deep inside a cave in Kentucky, but its true significance has only recently been made clear, thanks to the efforts of a young shark paleontologist named John-Paul Hodnett (JP to his friends). He grew up in Arizona. As a young boy, JP began collecting fossils while visiting his dad at Calvert Cliffs in Maryland, or sometimes buying them from shops. Fossil sharks' teeth were among his favorites. At Northern Arizona University he found a neat new kind of shark tooth in a rock exposed on campus grounds (more on that find later), prompting him to study paleontology, and this training led to him becoming a national park ranger.

The U.S. National Park Service initiated an inventory of the fossils in the Mammoth Cave National Park in Kentucky in 2019 and brought JP in to help. Mammoth Cave formed in limestone that was deposited in a shallow sea between 345 and 325 MYA, a little older than the Bear Gulch site. The rangers working on the site had known about these shark fossils for a long time, so they led JP to the secret location.

After crawling a quarter mile through narrow tunnels on their hands and knees, the park scientists and JP emerged into a rocky room deep inside Mammoth Cave. They shone their headlamps upward onto a set of large fossilized shark cartilages poking out of the cave wall. The lower jaw was exposed as a cross-section, an impressive 2.5 feet long. This was a very large shark. Because of its size and fragility, and the hard limestone entombing it, it will never be removed from the cave wall—the risk of damage to the specimen is too great.

JP studied the fossil as best he could in situ. Its teeth confirmed that it belonged to a form previously known from isolated teeth found in Britain, *Saivodus*. The Kentucky specimen came from a shark about 17 feet long. JP used photogrammetry techniques, taking a series of still photos that can be merged into a 3-D image, to record the beast. While it was big, this one turned out to be just a relatively small example of this monster shark.

*Saivodus* is the biggest shark so far in our story, an Early Carboniferous predatory behemoth. Its teeth, up to 2.5 inches high, are similar to *Ctenacanthus* from Ohio, so we know it belongs in the ctenacanthid shark family. Using the largest known teeth, we can calculate its maximum size at a whopping 24 to 28 feet long—about 4 to 8 feet longer, and probably double the weight, of the largest white sharks. Its teeth have sharp ridges on each of their large central dagger-like cusps, which are perfect for impaling squids, ammonoids, small fishes, or other sharks.

Some braincase fossils of similar-sized sharks from 300 MYA are known from Texas (nicknamed the Finis Shale supersharks), although there isn't enough of the remains to identify them precisely. John

Maisey studied them and confirmed that they belong to enormous ctenacanthid sharks around 23 feet in length and weighing up to 2.75 tons. It's quite possible they might belong to *Saivodus*, but there are no teeth to confirm this guess.

*Saivodus* **was the largest predatory shark of its time, reaching up to 28 feet long. This *Saivodus* tooth is about an inch wide**
AUTHOR PHOTO, COURTESY OF THE NATURAL HISTORY MUSEUM, LONDON

Ctenacanthids like *Saivodus* were clearly the top dogs of the Carboniferous marine food chain. Although we know little about the big species, we do have a superbly preserved fossil of one of the smaller ctenacanth species. It provides valuable new information about how these ancient predators looked and might have hunted. And JP discovered it.

May 21, 2013, is a day that is "burned forever in his memory"—when JP found his best fossil ever. He was on a field trip to the Kinney Brick quarry near Albuquerque, New Mexico, with a bunch of other paleontologists. In the heat and dust of the quarry site, he found something large that resembled a cross-section of a big thigh bone. He thought he'd found a large fossil amphibian. Other more seasoned paleontologists took over and excavated the specimen from the thin shale layer on the quarry floor. Next day, one of these guys walked in with a huge grin on his face. He informed JP that he had found something very special. It was a very large fossil ctenacanthid shark. Perfect. Complete. JP was asked if he'd like to work on it for his master's thesis, which he was doing with the legendary Dick Lund. He jumped at the opportunity.

They named it *Dracopristis,* from the Greek words for "dragon" and "sawfish," because its face was covered in sharp spines that made it look like a dragon. *Dracopristis* was about 6 feet long, with characteristic dorsal fin spines measuring nearly 2 feet. The teeth are broad and very thick, clearly able to exert great crushing power when piercing its prey. The pectoral fins of *Dracopristis* are large and triangular, resembling those of slow-cruising living sharks. The ctenacanth shark body shape was not suited to fast swimming to catch prey. JP suggested that it might have even been the first of the ctenacanth sharks adapted to live and hunt as an ambush predator on the seafloor, a coastal estuary specialist like today's aggressive bull sharks.

Dracopristis

**Dracopristis, the "dragon shark," a 6-foot-long predatory ctenacanth from New Mexico** ARTWORK BY JULIUS CSOTONYI

*Dracopristis* was not alone. Another large ctenacanth shark, known only by its teeth, also occurs at the same locality in New Mexico— *Glikmanius*, reaching about 12 feet in length. Its teeth are so rare at the site that it was likely an occasional visitor to the lagoon rather than a resident. The large sharp dorsal fin spines of *Dracopristis* would have deterred the much larger *Glikmanius* from eating it. All this is telling us there were several large ctenacanths at the top of the food chains in different parts of the world at this time. However, they were all feeding on smaller prey that they could grab and swallow whole. Another kind of killer shark was probably hunting large prey in the same seas, and it might even have stalked ctenacanthids.

*Carcharopsis* had teeth serrated like a steak knife, and surprisingly similar to those of the white shark. It's a tooth befitting a true super-predator, a voracious beast capable of preying not just on other large fishes but on creatures larger than itself. That's how I define a super-predator; just think how orcas can bring down a much larger right whale, or how lions can team up to kill an elephant or wolves to overcome an elk. *Carcharopsis* teeth have been found in England, China, and Russia, as well as a really nice specimen, revealing more than just teeth, from the United States.

A detailed study of a partial skull of *Carcharopsis* led by Allison Bronson of the American Museum of Natural History showed that this strange shark grew to a very large size. The largest teeth measure over 1.3 inches high, which suggests that *Carcharopsis* grew to around 17 feet. The teeth bear large, coarse serrations, with some bearing even smaller serrations on their edges. This was undoubtedly a ferocious flesh-cutting predator that fed on large prey, perhaps even *Saivodus*.

Based on the study of its braincase, Allison suggested that it was also a highly advanced form on the line leading to modern sharks. This was first recognized by the American paleontologists John Strong Newberry and Amos Henry Worthen, who suggested back in 1866 that *Carcharopsis* teeth might represent an extinct species intermediate between the ancient hybodontiforms and the living white shark.

## Carboniferous shark wrap-up

We have now seen how sharks and their kin, the holocephalans, diversified as never before in the first few million years of the Carboniferous period, when hundreds of new species, representing both predatory and shell-crushing species, suddenly appeared. Many different new shark body shapes appeared, from elongated eel-like forms to deep-bodied squat shapes and ray-like forms. Bear Gulch shows this diversity very clearly. Most intriguing is the shift in both how sharks swim—some by using pectoral fins as wings—and how they make teeth—using super-hard new dental tissues.

Other sharks evolved super-suction feeding methods, both in sharks like *Tristychius* and in stem holocephalans like the iniopterygian clan. Some sharks, like *Akmonistion* and the stethacanthids, developed massive dorsal fin spines with tooth-like structures on their heads. We also see gigantic ctenacanth sharks now at the top of the food chain, in the seas of the Laurasian landmass, some maybe reaching 28 feet long, while other giants, like *Carcharopsis*, had coarsely serrated teeth for cutting flesh, so it might have preyed on sharks larger than itself. Finally, we see the first hybodontiform sharks appearing, like *Tristychius*, that would ultimately lead to the rise of modern shark groups. Yet all these new evolutionary directions are driven by one common superpower: the ability of sharks and their kin to evolve new tooth shapes and tooth tissues and adapt them for innovative feeding methods.

While sharks were going wild and growing large in the warm Carboniferous seas, other sharks began hunting around estuaries, coal swamps, and lagoons, eventually invading freshwater rivers and lakes. There was also a particularly strange shark that developed a long snout, an indication that sharks were starting to activate a new superpower: finding food in murky rivers. To find out more about these electrifying river sharks, we will have to dive into a Late Carboniferous river in Illinois.

CHAPTER 7

# SWAMP SHARKS

*Sharks Take Over Rivers and Swamps*

---

### An Illinois delta, 310 million years ago

A large shark, around 10 feet long, sporting an elongated flat paddle-like snout, is slowly swimming its way down the river channel, sweeping its long snout just above the sandy floor. Around it the river is awash with mats of floating plant debris. There is a gentle swish of shrimps and small fishes fleeting past the shark in the slow current. The shark is the largest creature in this river, yet it is finding it hard to find anything through the murky water. She is a female, and her swollen body shape tells us she is gravid, ready to lay her eggs when she finds her special breeding area.

Suddenly she slows down, feeling a weak signal deep inside her brain, coming from her long snout packed full of special cells that detect electric fields. There is something alive in the sandy bed below it—all living things have an electric field around them, from the electrical activity of cells making muscles work. She circles and hovers her long paddle-shaped snout around the spot where she can sense something. Seconds later, she nuzzles her long snout through the sandy bed to home in on a stronger signal. Her brain is detecting electrical pulses of a small living creature, a kind of primitive crab that thinks it's safe down in the burrow beneath the sediment. The shark feels the pulsing electricity of its prey and charges deeper into the sand, eventually hitting a small solid moving object in the small depression where it had stirred up a cloud of silty muck. She quickly opens her jaws and sucks the small crab-like critter into her mouth, crunching it up in one bite before swallowing.

*Bandringa* continues on her journey down the river, emerging into a wider stretch of the river delta, close to the sea. She is on her annual migration. Eventually she feels the water changing its salinity; the salt-water has infiltrated this delta. The soup-like water here is full of insects, centipedes, and the odd struggling newt-like amphibian; all have been washed down from recent flood surges upstream.

She passes through the river mouth out into the shallow coastal seaway. A range of typical sea creatures are swimming around her in the salty water, a variety of shrimps and crab-like creatures, some sea scorpions, shoals of deep-bodied ray-finned fishes, and lots of jellyfishes. A very odd creature passes by. It looks like a squid with eyes on long stalks and strange pincer-like jaws protruding from a long narrow tentacle coming off its head. It eyes her suspiciously (it is a mysterious beast known enigmatically as the Tully Monster). Below her, on or near the seafloor, are scattered clams with their shells open as they sieve the water for tiny organisms. Coiled snails move at a tedious pace through the sandy bottom as their close relatives, squid-like creatures in elaborately coiled shells called **ammonoids**, drift over them searching for small fishes. There will be lots to eat when she has finished her mission.

Suddenly a tiny eel-like lamprey, *Hardistiella*, swims by and latches on to her tail flank, its circular mouth full of keratinous teeth that cut into her flesh. This parasite will hang on and feed, sucking her blood while she continues her journey. Another kind of small shark bearing dorsal fins with stout spines, *Dabasacanthus*, swims slowly along the sea bottom looking for worms and small clams, clearly oblivious to the much larger *Bandringa*.

*Bandringa* eventually finds the special place where she always goes, year after year. It's a quiet area with lots of weeds that will give her hatchling young a good chance of survival. Already there are several sets of eggs deposited by her sisters. Some have already hatched. A tiny, long-snouted *Bandringa* pup spots her and quickly darts back to the safety of the weeds. She swims down and positions herself with her cloaca in the weedy area. As she lays her eggs, they come out as coiled packages with spiral ridges around them. The trailing tendrils on each egg entangle with the weeds and secure them safely. Her job is done. She will now turn her attention to feeding before migrating back upstream to the wide, sweet freshwater stretches of her river.

## The amazing Mazon Creek fossils

This scene took place in what is now Illinois somewhere between 311 and 305 MYA. Today these creatures are all known from superb fossils preserved in concretions known as the Mazon Creek Fauna. Before we learn about these sharks from Mazon Creek, we need to understand the big picture—what was happening with the world's oceans and climates, specifically a major shake-up that took place halfway through the Carboniferous period, which had devastating effects on some, but not all, sharks at that time.

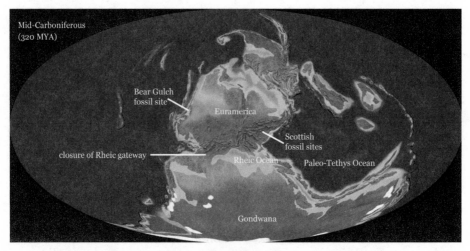

**The world during the middle of the Carboniferous period, 320 MYA**
RON BLAKEY, DEEP TIME MAPS (DEEPTIMEMAPS.COM)

Around 325 MYA, the supercontinent Gondwana in the south collided with Euramerica in the north, closing a long-open seaway that had connected two major oceans, the Paleo-Tethys Ocean in the east and the Rheic Ocean in the west. The closure of this Rheic–Tethys gateway brought about major changes in ocean currents. This initiated the glacial conditions in the Southern Hemisphere and severe climatic fluctuations throughout the world that would steadily increase in severity during the middle and end of the Carboniferous period. This event is called the Serpukhovian crisis, named after the time stage in the Carboniferous.

Iris Feichtinger of the Natural History Museum of Vienna did some amazing research that showed that the Serpukhovian crisis saw a 28 percent decline in shark genera and many other marine organisms. The decline was more severe in Eurasia than around the ancient North American region. One review of world mass extinctions and biotic crises ranked the Serpukhovian crisis as more devastating for life than the end-Ordovician extinction event. *Cladodus* went extinct at this time, and the giant *Saivodus* might have gone extinct in Eurasia, yet it survived in Euramerica. *Glikmanius,* another large predatory ctenacanth shark with teeth up to an inch or more across, also survived the event and lived through to the Early Permian.

The invasion by sharks into freshwater rivers, estuaries, and nearby coastal swamps was one way for sharks to cope with these severe and threatening environmental conditions. This leads us back to Mazon Creek, where the fossils occur in two differing environmental settings: one representing either a freshwater river or shallow marine setting receiving a considerable freshwater flow, the other in open marine conditions. One shark, our *Bandringa,* is the only fish known from the area that lived in both environments.

Our opening scene stars the long-snouted shark *Bandringa* and a range of other creatures all found preserved as complete organisms in the ironstone concretions of the Mazon Creek Fauna, from the famous fossil sites in eastern Illinois in the central United States. These exquisite fossils are from the second half of the Carboniferous period, dated between 311 and 305 MYA. They were first recognized for their wonderful preservation as far back as 1866. Collecting the fossils really kicked off in a big way only when intensive strip mining for coal began in the 1940s. This resulted in massive amounts of overburden, full of concretions containing fossils being moved and left in mountainous spoil piles. Local collectors descended upon them once they realized you could find neat fossils in them. Since then, hundreds of thousands of fossil-filled concretions have been collected and searched, many of them ending up in museums and private fossil collections around the world. Once the concretions were seriously studied, a diverse fauna and flora of ancient life emerged, including our very snouty shark, *Bandringa*.

These sites preserve fishes, amphibians, reptiles, and many kinds of invertebrates and plants as complete creatures or parts of plants, sometimes with soft tissues evident. The Mazon Creek Fauna is especially important as it gives us a rare window into life in the rivers, swamps, and shallow seas at the same time, about 310 MYA, in that two of its main sites represent different environments.

During the second half of the Carboniferous period there were many fluctuations in sea levels. Swamps filled up with fallen logs that became inundated with seawater (these would eventually form coal). Sea levels would drop as phases of cooler climate sucked up all the water into the growing southern polar ice cap. The rivers would then once again wash over a landscape of shallow marine deposits, replacing them with new coal swamps and sandy river deposits. Each of these cycles lasted around a hundred thousand years, as climate variation and sea level change deposited a discrete package of marine and freshwater sediments and the freshwater coal lakes were rapidly buried by marine incursions as the sea levels rose to complete each cycle.

This package of lake and river deposits with marine layers above them is called a **cyclothem**, and they provide evidence for what was alive in the water and surrounds, recording traces like a geological fingerprint. The cyclothems were repeated throughout the Carboniferous period. At least fifty cycles are recorded in the great Illinois Basin, where most of the biggest coal deposits formed. Another forty or so corresponding cycles have been documented from the similarly aged British coal measures. This explains why so many of the best Pennsylvanian fossil deposits containing sharks occur in layers just above the coal seams.

One group of Mazon Creek sites contains the Essex Fauna, named after the village of Essex. It presents a variety of common marine invertebrates—snails, shrimps, horseshoe crabs, sea scorpions, worms, and sea anemones, all perfectly preserved inside the rounded concretions. The sea anemones called *Essexella* are the most common fossil found here, so much so that they are known to all the local collectors as "the blobs." They were thought to represent jellyfishes for many decades, until the mystery was finally solved in 2023. Several kinds of

*Bandringa,* **foreground, in its shallow seaway, with a group of ray-finned fishes in the background** PAINTING BY JOHN MEGAHAN, UNIVERSITY OF MICHIGAN

ray-finned fishes are also found here, as well as our hero shark, *Bandringa.*

About ten miles north of another nearby town, Braidwood, one can find concretions with a different assemblage of aquatic animals and plants that once lived in or near a river. We find velvet worms, centipedes, millipedes, insects (more than two hundred species!), fishes, amphibians, and reptiles, along with a spectacular bouquet of the typical land plants of the day. Our shark *Bandringa* lived here too, being the only fish in the fauna that occurs in both environs.

These superb fossils are preserved by rapid ironstone formation around the decaying organisms, yet original tissues like bone and cartilage are often missing. A 3-D mold forms where the bones or cartilage once existed. This method of replacement of original tissues by other mineral deposits results in the replication of soft tissues as dark stains. It's really amazing to see these visible around the eyes, and some specimens show color pigments remarkably preserved.

*Bandringa rayi* was named back in 1969 by the great Rainer Zangerl of the Field Museum in Chicago for its finder, Ray Bandringa. Rainer has been dubbed "the father of modern U.S. paleoichthyology" (the study of fossil fishes; read his bio on p. 417). He was a pioneer shark paleontologist who did outstanding early research on the fossil sharks from the black shales of Illinois and neighboring states. He recognized that all the *Bandringa* specimens from Mazon Creek were tiny, under 6 inches long, so they must have been juveniles just hatched out of their egg cases. However, pieces of large paddle-shaped snouts were known from other sites in the Illinois Basin. Based on these remains, and knowing the growth changes in living sawsharks, from newborn pups to adults, researchers are confident that fully grown *Bandringa* sharks might have reached 8 to 10 feet long. There have been other studies of this curious little shark fossil, including a recent revision of its anatomy and environmental preferences. This study revealed a lot of new details about *Bandringa* that support the idea that it had a new superpower that all living sharks have, yet up to now had not been confirmed in any fossil shark: electroreception.

## Sharks' next new superpower: electroreception

Sharks have a particularly well-developed system for sensing the weak electric fields of other creatures, and use this power in several ways. They can find food by electrosensing creatures buried in the seafloor or living in muddy waters. Embryos of sharks still in their eggs can use their electrosensory ability to sense predators approaching and remain still inside their eggs. Other sharks, like scalloped bonnetheads (*Sphyrna corona*), can follow magnetic trails along the seafloor using this system to find their way between nocturnal feeding sites and daytime resting sites. Some rays can use electricity to stun their prey—sending powerful electric shocks—and we now think at least one extinct shark had this power too (more on that one later). The main structures visible on sharks that give them this ability are the rows of pores around the head, particularly on the snout. Each pore is a flask-shaped cell filled with conductive jelly to pass electric signals to the brain via a sensory layer of nerves. These cells are called the **ampullae of Lorenzini**.

The ability of sharks and rays to detect electricity was first confirmed in the early 1960s by experiments that demonstrated behavioral changes to sharks when electricity was applied to the water. We now know that every known species of shark, ray, and holocephalan has these special electrosensory cells, as well as a few primitive bony fishes, like sturgeons and paddlefishes.

We find the ampullae of Lorenzini in all living sharks and rays in greatest numbers around their heads. These organs are especially concentrated around the snout, as this is the prime area for sensing prey. Electroreception most likely evolved in sharks much earlier than *Bandringa,* but we don't yet have the evidence to prove it. This superpower must have been present in a common ancestor of both bony fishes and sharks, unless one of these groups evolved this ability first and then passed it on to the other group. This chain of events implies that electroreception could have evolved as far back as the Silurian period, possibly in the placoderms or earliest stem sharks.

*Bandringa* is the first shark to show an elongated, flat snout, called a **rostrum**. The main reason for developing a paddle-shaped snout is to spread your electrosensory cells across a wide surface area to better pick up the electric fields of living prey. Every creature, from plankton to human, generates its own weak electric field that can be detected with the right array of sensory cells. The Mississippi paddlefish (*Polyodon*) and various living sawsharks and sawfishes still have similar long, flat rostrums that are packed full of electroreceptors. Paddlefishes can use their rostrum to detect the very weak electric fields emanating from swarms of tiny plankton. It's like a metal detector that can find gold dust scattered widely over the ground. Sawfishes and sawsharks use their long snouts to stun prey, or to immobilize prey by holding it down on the seafloor. *Bandringa* probably used its flat snout more like a mine detector, sweeping it across the large river or seafloor to detect the electric fields of small crab-like critters living on the riverbed, hidden by muddy waters, or buried just below the surface. A recent study of *Bandringa* fossils revealed the contents of their stomachs, so we know this shark fed on small crustaceans. The researchers' study of its jaw structures revealed it could draw them into its mouth using a suction feeding action, just as we saw earlier in *Tristychius*.

## Sharks take over the rivers

Why are freshwater and estuarine habitats important for sharks? Today, rivers, lakes, and estuaries make up only around one percent of the Earth's surface, yet 25 percent of all the backboned creatures (vertebrates) on our planet inhabit these environs, including around 50 percent of all the known fishes. Sharks began invading freshwater habitats back in the Early to Middle Devonian, as we saw in chapter 3 with the eclectic bunch of sharks found in the river and lake deposits of the Aztec Siltstone of Antarctica. As Devonian shark fossils remain a relatively rare find in definite freshwater deposits, we can tell that the serious conquest of freshwater by sharks went up a gear in the Carboniferous period.

20 cm

**The complete skeleton of a *Xenacanthus* from the Lower Permian of Germany shows the eel-like body and sharp neck spine that typify the xenacanths** PHOTO BY OLIVER HAMPE, MUSEUM FÜR NATURKUNDE, BERLIN

The Early Carboniferous rivers and lakes were still dominated by gargantuan predatory bony fishes such as *Rhizodus,* a 24-foot killer with very powerful jaws and banana-shaped teeth. The sudden decline of rhizodontids in these rivers and lakes in the Late Carboniferous, from around 310 MYA, was the big opportunity sharks had been waiting for. They gracefully glided in and took over.

One large group of freshwater sharks rose up to become the most successful apex predator in freshwater environs around the entire globe at this time—the xenacanths. These were fearsome, serpentine beasts like the sea monsters of old, characterized by their deadly jagged neck-spines, which might have been venomous, and distinctly pronged teeth. Xenacanth means "foreign spine," alluding to the unusual position of this deadly spine from the back of their heads. They were mostly small to medium-sized predatory sharks, between 2 and 4 feet long, with the largest, like *Lebachacanthus,* reaching at least 10 feet, while one other species may have been a filter feeder up to 16 feet long.

We have already met one of the oldest known xenacanths, *Diplodoselache,* from Scotland, that lived at the start of the Carboniferous period (350 MYA). One of the other oldest members of the xenacanths is *Bransonella,* represented by small three-pronged teeth with coarse wavy ridges on them. *Bransonella* lived in the sea, showing that later xenacanths had their origins in marine stock. It was a hardy shark that ranged from the base of the Carboniferous through the middle of the Permian period, but it is known only by its teeth.

There's only so much xenacanth teeth can tell us. The first superb complete specimens of xenacanths were found in Bohemia (now Czechia) as the coal mining industry expanded and miners kept accidentally finding fossils. One very talented scientist studied xenacanth sharks and did pioneering work setting up his country's first national museum. Let's meet the extraordinary Anonín Frič.

## Antonín Frič and the Bohemian xenacanths

Antonín Frič (1832–1913) is held today in great respect by the people of Czechia, not just because he was one of the world's leading paleontologists, but because he was truly a great scholar and a gentle, caring human being. Antonín was the scion of a deeply intellectual family; his father, Josef, was the lawyer who first introduced the Czech language to the judiciary, which previously had to hold court in German. Antonín himself led an exemplary life of public service and duty. He was far ahead of his time in his fervent love of nature, espousing how being outdoors benefits human health. He was also an outspoken critic of the use of tobacco.

Antonín was appointed to a professorship at the Charles University in Prague sometime around 1863, and soon after became the director

**Antonín Frič** CZECH ACADEMY OF SCIENCES, PUBLIC DOMAIN

of the nascent National Museum in Prague. In addition to setting up the country's first great natural history museum, he was fanatical about correct care and conservation of museum specimens, making sure every specimen was properly curated and cataloged. He had trained as a taxidermist in his youth and imparted his skills to his museum staff.

His brother Václav worked closely with him to help build up the collections of the national museum. Václav used the surplus natural history specimens from the region to trade, or sell to other natural history museums around the world. This was the stuff they had loads of, like the well-preserved xenacanth sharks from Bohemia in the western half of the country. They had a large store of excellent fossil specimens, including the best xenacanth sharks around. It was a time when many of the world's largest cities were setting up museums and galleries as sites of public education, so there was high demand for interesting specimens and artifacts from around the globe. Geological specimens and many other natural history objects travel well and thus could be traded easily.

Václav quickly realized that the museum's pickled specimens of marine life, kept in glass jars full of alcohol, rapidly lost their form and color after death and therefore did not sell well to other museums. In a stroke of marketing genius, he set up a collaboration with the leading Czech glassmakers Leopold and Rudolf Blaschka, having them make exquisite lifelike glass replicas of octopuses, squids, jellyfishes, sea anemones, and many other kinds of sea creatures to offer to museums around the world. Today many of these rare and delicate pieces are still housed in the collections of major museums, and they are sought-after collector's items fetching high prices at auctions.

Back at the National Museum in Prague, Antonín Frič proved himself to be a real polymath, brilliant at everything he did. He completed major scientific works across the entire natural history spectrum, from fishes to birds, minerals, and geology. Unlike today, when most of us tend to specialize in studying one group of organisms—like fossil sharks of a certain age range—his work delved deeply into the entire spectrum of past animal life, including dinosaurs. His greatest love was working as a paleontologist, and his work on fossil sharks was unprecedented for his day. It stands as a shining example of the power of detailed paleontological descriptions accompanied by meticulous hand-colored drawings of superb specimens and adorned with restorations of key anatomical features. His papers are still widely used today as a primary source of information about the xenacanth sharks.

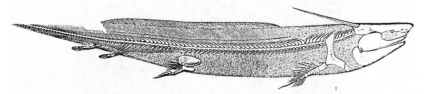

**One of Frič's excellent reconstructions of** *Triodus* **(then called** *Pleuracanthus*) FROM FRIČ 1895 PLATE 92.

Frič's work on the xenacanth sharks of Bohemia (the western part of Czechia) was important as the first attempt at sorting out these strange-looking sharks whose remains were largely known (up to then) only by their isolated neck spines and distinctive pronged teeth, sporting two large stout cusps with a smaller median cusp. He published the first detailed account of a set of three different xenacanth sharks (*Orthacanthus, Xenacanthus,* and *"Pleuracanthus"*), all known from relatively complete, well-preserved specimens from the Permian period.

Frič named many of his new Bohemian xenacanth species as varieties of *"Pleuracanthus,"* but the name was later found to be already taken by a South American beetle, so it had to be changed. Most of these sharks named by Frič are still valid species today but are now placed in the genus *Triodus.* As a result of Frič's work, many species of *Triodus* were found and studied from around the world, including some from the famous Permian Red Beds of Texas and the black Triassic shales of Sydney, Australia. The oldest *Triodus* start around 314 MYA and range right through to the end-Triassic (201 MYA); we will hear more about it in subsequent chapters.

Frič's *Orthacanthus* (meaning "straight spine") is the most common xenacanth shark known from the Late Carboniferous–Early Permian of

North America and Europe, mostly from teeth and fin spines with some Czech and German sites yielding complete specimens. It was another large shark reaching around 9 to 10 feet long. Its teeth bore powerful, razor-sharp twin cusps on each base. This shark had the largest teeth of any xenacanth. Such teeth are perfect for grasping and piercing prey like fishes, small amphibians, or other sharks. As we shall learn, the remains of their prey have been preserved inside them, so we actually know they fed on these creatures.

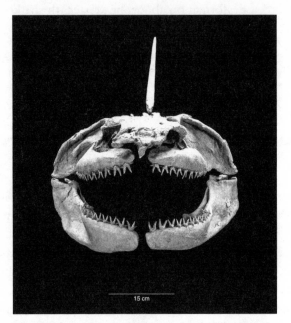

**The toothy grin of a large Texas xenacanth shark, *Orthacanthus*** IMAGE COURTESY OF THE SMITHSONIAN INSTITUTION, USNM PAL 617468, DEPARTMENT OF PALEOBIOLOGY, NATIONAL MUSEUM OF NATURAL HISTORY

*Lebachacanthus* was the giant of the xenacanth group, with several complete specimens more than 10 feet long and fragments of bigger beasts suggesting body lengths up to 12 feet long. It is found from deposits in western Germany in an area called the Saar-Nahe Basin, near the town of Lebach. The females were around 20 percent larger than the males, with distinctly longer dorsal spines.

# How xenacanths ruled the rivers

Let's put the xenacanths into a modern-day perspective. The commonest living sharks that enter freshwater rivers, like the bull shark (*Carcharhinus leucas*), also frequent shallow seaways. Bull sharks on average are around 6 to 8 feet long, so the largest xenacanths were much larger, taking the crown for being the largest predatory freshwater sharks that ever existed. Furthermore, once xenacanths left the sea, they never went back; they were henceforth restricted to just freshwater and maybe some brackish habitats. This is something we don't see any living shark doing, as no species is entirely restricted to just freshwater. Xenacanths were unique among all sharks, living and past, as freshwater apex predator specialists. Yet one xenacanth was probably even more specialized than its apex predator kin.

I think one of the most interesting of all the xenacanths, the one showing the range of their diet, was *Barbclabornia,* from the Early Permian of Texas, discovered by Gary Johnson, whom I fondly know as the "xenacanth king of Texas." Gary, now in his late eighties, has spent almost his entire working career studying xenacanth sharks, mostly from isolated teeth, and he sees *Barbclabornia* as one of the most extreme examples of the group. One specimen shows an incomplete large upper jaw cartilage, measuring over a foot long, with teeth attached. Based on other well-known xenacanth jaw shapes, the jaw might have been about 1.5 feet long. Using this measurement we can determine the maximum size of this shark at around 16 feet or so. This finding serves as a caution to those who calculate maximum size of sharks based on teeth alone, as here we have the enigma of tiny teeth (1/20 to 1/8 inch high) associated with an enormous upper jaw.

Although Gary suggested that *Barbclabornia* was a "grasper and swallower," a hunter of less active prey like crustaceans, small fishes, and amphibians, others have a different opinion. Michal Ginter suggested that it was probably a gigantic filter feeder, like a basking shark or whale shark. This idea adds a new dimension to xenacanth ecology, for if it is true, it implies that they weren't all simple "clutch, grasp, and swallow" hunters, but that some turned to feeding on plankton in freshwater to brackish habitats. There is no equivalent for this kind of behavior in freshwater sharks today, as the only filter-feeding sharks and rays live in the open oceans. The only filter-feeding animals that live on plankton and small organisms in freshwater habitats today are creatures lacking teeth, like the American paddlefish (*Polyodon*) and flamingos.

While nearly all xenacanth species thrived in freshwater, a few were also comfortable in brackish and shallow marine conditions, as shown by new research based on detailed isotopic studies done on fossil shark teeth by Jan Fischer of the Urwelt-Museum Hauff in Germany. He studied the strontium and oxygen isotopes captured in the formation of the enameloid layers of the sharks' teeth and found that freshwater species have more heavy isotopes deposited in such tissues than marine species.

Xenacanths are the most common type of shark fossil found in coal deposits formed in freshwater and brackish swamps and mires from the Carboniferous through to the Triassic (359–201 MYA). I went to Spain to search for xenacanth fossils and learn more about their biology from a true-life xenacanth toreador, Rodrigo Soler-Gijón.

## The voracious xenacanths of Spain

Rodrigo Soler-Gijón grew up in the very small village of Almarzin, near the coal mining town of Puertollano, in Spain. As a teenager, he used to escape the confines of the village and venture out on guerrilla missions to find fossils. After school, and after the workers had finished for the day, he would sneak into the quarries through holes in the wire fences. Once inside, carefully negotiating the slopes bristling with seams of black coal, he soon learned that the best fossils he could find were the teeth of ancient sharks called xenacanths. He knew where to look—in the layers above the coal, exactly as occurs in the Mazon Creek and Illinois Basin sites.

The Puertollano coal mines of western Spain are still active today, and I visited them with Rodrigo in June 2022, as part of a paleontology conference field trip. This time we had the blessing of the mine owners. They are now very proud of the paleontological heritage of their mines, established through the hard work of Rodrigo and his colleagues over the past three decades. They have even built a museum to house the paleontological treasures found in their mines.

Keeping a wary eye on a nearby brush fire, we scrambled like a row of highly educated mountain goats down the quarry track in single file

Rodrigo Soler-Gijón at a Puertollano mine, explaining
the geology of the site to us   AUTHOR PHOTO

behind Rodrigo. He would periodically stop to point at a rock layer
and tell us about the geology of the site. We did not need to go deep
into the coal mine, because, as we've learned with Mazon Creek, the
best fossil sharks are to be found in the top layers, immediately above
the coal. Luckily for us, this was not too far down from the top of the
quarry. The layers here are quite rich in fossils, but fossil shark teeth
and spines are still quite rare.

The most common thing we found that day were strange squashed
white blobs representing the mineralized poop of xenacanth sharks.
Scientists politely call these **coprolites** (I found a rather cute little
one at this site, which sits on my desk right now). Coprolites can tell
us a lot about ancient ecosystems when imaged using the latest high
technology methods. Most of the coprolites found at the quarries of
Puertollano can be attributed to xenacanth sharks, as they have the
characteristic spiral shape of shark feces. This is the result of a little-
known fact—sharks excrete their waste through spiral-shaped intestines, which create spiral-shaped poop.

Other fossils from the same coal mines show that xenacanth sharks
like *Orthacanthus* (the species here being about 3 feet long) were feeding in the ancient swamps on smaller xenacanths like *Triodus* (about
one foot long). Rodrigo documented one specimen of *Orthacanthus*

with several *Triodus* spines in its gullet, suggesting that it liked this tasty, if painful, entrée very much. Distinctive *Orthacanthus* coprolites from the Pennsylvanian of New Brunswick, Canada, studied by another Irish paleontologist, Aodhan O'Gogain, revealed a terrible secret in the closet of the *Orthacanthus* family dynasty: Tiny juvenile teeth of the same species were found, confirming that *Orthacanthus* fed on their own young. Aodhan's detailed geochemical work also showed that these sharks lived in a range of water salinities from shallow marine shelf conditions through to brackish tidal estuaries. He took his science outreach to the next level in 2019, when he helped make a TV documentary about these findings called *Cannibal Sharks*.

**Xenacanths preyed on their kind. Here a large *Orthacanthus* hunts a smaller *Triodus*** ARTWORK BY JULIUS CSOTONYI

So far in this chapter we have seen a range of different sharks that lived mostly in freshwater and near-shore lagoons and estuaries, and how we found the first evidence for the use of electroreception in sharks in the tiny Mazon Creek *Bandringa*. The serpentine xenacanth sharks diversified from marine forms in the Early Carboniferous to become the most specialized and largest freshwater sharks of all time. Today some sharks and rays occasionally venture into freshwater, but none are entirely freshwater specialists as xenacanths were—they stand alone in shark history. We will learn more about the xenacanths, and discover what fate awaits them, in the next few chapters.

It's time now to change gears and move back to the oceans to witness the rise of some of the largest yet most bizarre sharks the world has ever seen: sharks with giant wheels of teeth for cutting, slashing, and sawing their unfortunate prey. Let's take a swim in an ocean off the west coast of the old North American landmass about 270 MYA to see what these strange sharks were into.

CHAPTER 8

# RISE OF THE BUZZ-SAW SHARKS

*How Sharks with Wheels of Teeth*
*Dominated the Oceans*

---

### Pangaean sea, 270 million years ago

Named after the Perm Mountains of Russia, the Permian period started around 300 MYA. This period ended 252 MYA with the greatest mass extinction event Earth has ever endured. However, almost 20 million years before that dire event, the northern Pangaean sea is calm and cool, a climate created by the fluctuating glacial pulses of the last tens of millions of years. These cold snaps sucked up precious water into a great southern ice cap, draining water from the shallow seas and exposing the old coastlines as new land. It is our planet, but not as we'd recognize it. Ocean currents adjusted to a new world order—the single supercontinent of Pangaea locked together all our modern continents as one landmass, leaving the rest of the world covered by the expansive Panthalassic Ocean, with just one smaller body of water called the Paleo-Tethys Ocean (see map on p. 199). However, many smaller marine realms still existed, each defined by its latitude and proximity to land.

Here, around the continental slopes off the western shores of Pangaea, we find the sea abounding with life, with sharks thriving in the clear, cool waters rich in microscopic nutrients. The increase in oceanic oxygen and nutrients came from the flood of cold waters emanating from the melting northern ice caps, recharging the food chain with an abundance of shelled nautiloids and ammonoids, fishes, sharks, and many other sea creatures. These bountiful oceans are the product of the Permian Chert Event, starting around 290 MYA, when silica-rich sponges flourished and died, shedding trillions of their microscopic silica needles onto the seafloor. This resulted in the formation of large volumes of siliceous chert deposits along the coastal margins. One of the world's most spectacular sharks emerged at this time to surf the top of the food chain.

A giant male shark, a *Helicoprion*, swims swiftly through the temperate high-latitude Paleo-Tethys Ocean, which then lapped over the land that is now the Perm Mountains of Russia. The speed of the shark is astounding, akin to a piscine thoroughbred racehorse. Its sleek, tapered body, reaching around 30 feet in length, features a high-angle tail fin, reminiscent of those seen on today's fast-swimming marlin and mako sharks. About 300 feet ahead of the shark is a large school of nautiloids, squid-like creatures about 2 to 3 feet in length that live inside coiled shells, much like the living *Nautilus* in our seas today. The giant shark approaching them is characterized by a most unusual weapon at the front of its lower jaw—a single solid coil crowned by large, triangular, serrated teeth, aligned like the cogs on the gears of a killing machine. This distinguishing feature projects from the short, deep lower jaw. *Helicoprion* has evolved a totally new way of inflicting death on its prey—the bizarre and effective saw jaw.

The shark closes in on the school of nautiloids, building up speed with a powerful flick of its tail. As it closes the gap on a large nautiloid, it slowly opens its mouth, swinging the lethal lower saw jaw forward. It lashes and grabs the nautiloid head-on, quickly sinking its lower teeth into the fleshy head of the animal, neatly avoiding its hard shell. With a lightning flash it pulls its lower jaw back, scooping the soft and tasty parts of the nautiloid from its shell and ravenously swallowing the fleshy bundle in one mouthful. The shell floats slowly to the surface, buoyed by a pocket of air inside, as the shark turns around for another mouthful. It spies its favorite morsel nearby, a large squid, and chases after it. The squid, though, has seen it coming, sprays a cloud of black ink in the shark's face, and vanishes from the scene. *Helicoprion* is left to wander on in search of its next snack.

Sometime later, our male *Helicoprion* sights two more of his kind in the clear surface waters above. He rises slowly but warily to check out what is going on. He becomes excited as he spots a large female being courted by the competition—another large male. The competitor has gently grabbed the female by her heavily scaled pectoral fin, twisting his head sideways so his deadly tooth saw does not pierce her fin. This considerate lovemaking move allows some pressure on the fin, using the side of the saw for this special ritual.

*Helicoprion* **reconstruction**   ARTWORK BY JULIUS CSOTONYI

The approaching male seizes his moment and charges at the other male from below, crashing into him and dislodging him from this most intimate moment. Surprised, he accidentally bites down harder than expected on the female's fin, inflicting a small wound that stains the water with red blood. The two males face off and charge at each other with their saw jaws thrashing in the bloodstained water. They swerve and clash with their deadly saws, inflicting deep wounds on each other's faces. Finally, they lock their jaws together in a timeless macho battle of brute strength. They twist and writhe as their large, powerful bodies push together, until one gives up and rapidly swims away. The victor moves on, now pumped up with adrenaline and hormones, eagerly searching for the female. However, she has long since moved away from the violence and into deeper waters. Stunned and wounded by the unexpected bite to her fin, she is in no mood for amorous advances.

Much of this reconstructed scene is based on cutting-edge science from recent studies on *Helicoprion,* a most bizarre and intriguing animal. From detailed mechanical analysis of its jaws and the nature of the wear on its teeth, we have concluded that *Helicoprion* likely fed on nautiloids and squids. The mating scenes and male courtship battle at the end, though, are pure conjecture on my part, based on modern shark mating behavior.

The story of *Helicoprion* and its kin is important to understand how highly specialized predators emerged in response to changing environmental conditions. It is the story of how sharks with a unique way of feeding dominated the marine food chain for some 40 million years, then mysteriously went extinct well before the Great Dying at the end of the Permian period. *Helicoprion* is also the story of how scientists solve long-standing mysteries in biology, as for more than a century we knew it only by its peculiar coiled teeth. Let's go back to the beginning and take a look at some of the fascinating history behind the discovery of this truly awesome ancient shark.

## Early discoveries and the enigma of *Helicoprion*

The story of *Helicoprion* starts in Western Australia back in 1884, when a local farmer found what looked like a row of large shark-like teeth in a red sandstone outcropping along the banks of the Arthur River in the Pilbara region. The farmer passed the fossil on to the local clergyman, who corresponded about the fossil with the great Henry Bolingbroke Woodward, Keeper of Geology at the British Museum of Natural History in London. Woodward immediately expressed keen interest in seeing the fossil in person, so it was duly sent to London.

Henry Woodward wrote a paper about the fossil that established a new species of a then well-known Carboniferous shark, *Edestus,* as *Edestus davisii*—the species name honoring the farmer who found it. Woodward was stumped by the weird fossil. He did not see it as part of a jaw, but instead as a strange fin spine, based on similar finds in North America.

A few years later the story of these strange shark-like teeth shifts into another gear as the first large, complete tooth whorls of this beast were found in Russia. Alexander Karpinsky rose to become Russia's most

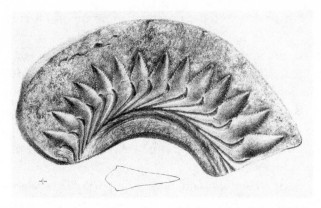

These strange sections of jaw with shark teeth found in Western Australia in 1888 were later identified as the first species of *Helicoprion* FROM HENRY WOODWARD, 1888

powerful scientist immediately following the 1917 revolution, yet he will always be remembered by paleontologists for his discovery of, and lifelong obsession with, the strange fossil shark *Helicoprion*.

Karpinsky, born in 1847, hung around the local mines as a kid and became fascinated by rocks and geology. He studied intensely at the St. Petersburg Mountain Corps and at thirty years of age was elected a professor in geology of the Mining Institute, where he gave lectures on a wide range of topics. In 1885, as director of the State Geological Committee, he began compiling the first detailed geological maps of Russia. His main research was devoted to determining the ages and nature of the country's sedimentary rock layers so that resources could be found. He began his study of fossils to determine the relative ages of the rocks and thus ensure the accuracy of his geological maps.

Alexander Karpinsky discovered and named *Helicoprion* and would be fascinated by its mysteries all his life   IMPERIAL ST. PETERSBURG ACADEMY OF SCIENCES, PUBLIC DOMAIN

Karpinsky's paleontological work

was exceptionally advanced. He worked on everything from ammonites and other fossil cephalopods to microfossils of all kinds, algal spores, and fossil fishes, including his study of a strange kind of tooth "wheel" (also called a whorl, or coil) coming out of the Perm Mountains. Each fossil was like a coil lined with up to 180 broad, sharp serrated teeth, some as large as 5 inches high, on its outer side. Nothing like it had ever been seen before. It was both puzzling and fascinating, instantly creating a biological mystery around how it could have functioned in a living creature.

In 1899 Karpinsky published his magnum opus on this new, strange fossil, which he appropriately named *Helicoprion*, meaning "saw spiral," and he created a new species, *Helicoprion bessanowi*. His 110-page monograph, enriched by 72 beautiful illustrations, presented detailed information on all the known fossil tooth whorls from Russia and compared them with others found in other countries. He recognized that the west Australian fossil teeth were not *Edestus*, as Woodward had suggested, but belonged to his *Helicoprion*, so he designated them as the original first species found (*Helicoprion davisii*).

Karpinsky then turned his attention to determining what kind of

One of the perfect complete tooth whorls found in the Perm Mountains of Russia. This is a cast of the holotype specimen of *Helicoprion*. In 1998 it was stolen    NATURAL HISTORY MUSEUM, LONDON / AUTHOR PHOTO

creature made these whorls—a shark, a fish, or something else? Based on their shape and his study of their cartilaginous tissue structure, he established that the teeth came from a kind of shark. However, he had no idea how such bizarre teeth could function for feeding if attached to a standard set of shark jaws. He suggested that the tooth whorl was probably used as a weapon to fight other large sharks or fishes.

Karpinsky was tormented by *Helicoprion* for many years. He couldn't quite reconcile these peculiar whorls as being teeth located inside the mouth of a shark, as no living shark has anything like this structure. His first interpretation was that they protruded from the shark's snout, much like the upcoiled trunk of a monstrous marine elephant, suggesting that it subdued its prey by slashing downward using its deadly snout. This fantastical idea was not unreasonable, because living saw-sharks and sawfishes with sharp teeth lining their snouts slash their prey, and the males battle each other using their snouts as weapons. He suggested that the *Helicoprion* tooth wheel might be encased in cartilage because he identified prismatic cartilage patches on the fossil, but because such a structure didn't show up in his early restorations, he wrestled with the issue for some time.

A worldwide debate ensued. The doyen of the Victorian fossil fish world in England at the time was Arthur Smith Woodward, who argued that *Helicoprion* represented the remains of a shark "with sharp, piercing teeth that were never shed, but became fused into whorls as the animal grew." This position was in stark contrast to that of our American fossil fish and medieval armor expert Bashford Dean, who studied the similar tooth whorls of another form, *Edestus*. Dean saw the fossil as a spine embedded in the musculature of the shark's back. Two other American paleontologists, Charles Rochester Eastman and Oliver P. Hay, fully supported the whorls as being fin spines in the middle of the body, with the serrated teeth always curled around just above the skin so the shark didn't slice grooves into its own back.

In 1910, the German paleontologist Friedrich John suggested that the whorl was a coil of teeth emerging from the front of the lower jaw, spiraling forward and downward. It was a brilliant insight. If not for the scale of the tooth whorl being enormous in his reconstruction (roughly the same size as the shark's head), he would have been spot-on with what we think is the correct interpretation today. Back then, as

**Karpinsky's first strange reconstructions of *Helicoprion* (1899)**
FROM KARPINSKY, 1899

evidence to prove which of all the interpretations was correct was still lacking, speculation continued.

With many different views about *Helicoprion* now emerging from scientists all around the world, Karpinsky felt he needed to weigh in on the interpretations of his fabulous sea monster once again. In 1911 he wrote a major review of *Helicoprion* and the Edestidae, the family that contained all the sharks with similar strange tooth whorls. He used the hard evidence from the fossils while also formulating new ideas from his observations of living sharks. He studied the nature of the prismatic cartilage and the symmetry of the symphyseal teeth (special small teeth situated where the lower jaws meet) of living horn sharks (*Heterodontus*) to make comparisons with *Helicoprion*. He discarded his original idea of the funky nose whorl (so socially awkward) and argued that it was more likely to be located somehow inside the head. Whether it would have been on the upper or lower jaw he couldn't decide, but his writings lean toward its being a lower-jaw structure, based on his vivid comparisons with the lower-jaw symphyseal teeth in some living sharks.

Karpinsky was totally intrigued by *Helicoprion,* and he kept coming back to it all his life with new studies. In 1922 he erected another new species, *Helicoprion ivanovi,* based on large whorls coming from much older layers at the top of the Carboniferous period, found near Moscow. This material was recently restudied and referred to a new genus of shark named in his honor, *Karpinskiprion* (more about this one below).

It's an understatement to say that Karpinsky had a stellar scientific career. He was elected president of the Russian Academy of Sciences in

1917—Russia's tumultuous year of political and social upheaval. He immediately became responsible for leading Russian science through the uncertainty of the Revolution, setting its agenda for the prosperity of the new regime. When he died in 1936, his ashes were interred within the Kremlin Wall Necropolis, the national cemetery for the Soviet Union. He is the oldest person, by birth date, to be interred there.

While he seemed to be close to cracking the mystery of the tooth whorls, Karpinsky could take his work only so far because at that time he had only the teeth of *Helicoprion*. There were no body remains, no skulls, nor jaws with tooth whorls still attached. Nor any synchrotrons or CT scanners. From the time of Karpinsky's last seminal paper on *Helicoprion* in 1922, its mystery would remain unsolved for almost another century.

Periodically, other breakthroughs advanced our knowledge of the biology of these strangest of all sharks. The next big contribution to our understanding of *Helicoprion* came in 1966 from the Danish paleontologist Svend Bendix-Almgreen. I met Svend in 1982 when I visited the Geological Museum in Copenhagen to work in the fossil collections. He was a quiet, friendly man who kindly assisted me in accessing what I wanted to see. Svend studied a range of specimens of *Helicoprion ferrieri* from the Phosphoria Formation of Idaho. He homed in on one particular fossil, known locally as Idaho No. 5, that he had borrowed from the Idaho Museum of Natural History. Earlier researchers had identified this specimen as being more than just the tooth whorl, for it also contained some cartilages of the head.

Svend showed conclusively that the tooth whorl sat above the lower jaw symphysis, resting on the lower jaw cartilages, so most of it in life was hidden by soft tissue, all except the oldest part of the whorl, exposing the largest teeth in the mouth. He also showed that it bore additional small rhombic flat teeth to grind up its prey after it had been hooked by the jagged whorl. It got even more interesting. This unique specimen from Idaho also showed that the whorl grew from a primary small tooth in the middle, the smallest first tooth, with the oldest teeth at full size, emerging outward from the mouth. This interpretation was anatomically the best at that time, but still not quite 100 percent correct. Svend's work nonetheless pointed to the key specimen required for future researchers to take it even further.

Scientific interest in *Helicoprion* grew. Everyone wanted to solve the mystery of how they used their strange tooth whorls. What the heck were these sharks eating? And how did they eat? It took a young Canadian scientist based in Idaho to finally solve this mystery that had puzzled paleontologists around the globe for almost 150 years.

## *Helicoprion's* secrets finally revealed

Leif Tapanila works as a buzz-saw shark wrangler. The son of parents who were a teacher and an artist, he grew up in Salisbury, Ontario, Canada, and in college he took geology units that led him to study trace fossils in hard material, such as animal borings, for his doctoral dissertation. Sharks were far from his area of interest in those early days. In 2012, as an up-and-coming young paleontologist working at Idaho State University, Leif became interested in the numerous *Helicoprion* fossils found in the phosphate mines of Idaho. He loved a challenge, so he set his mind to trying to solve the long-standing mystery of how the heck this coily-toothed shark could eat.

A complete large tooth whorl of *Helicoprion* from Idaho. This specimen was recovered by the FBI during a raid, as it came from an illegal site   PHOTO BY LEIF TAPANILA

Once he settled into his new position, Leif quickly became aware of the many specimens of *Helicoprion* from the Phosphoria Formation of that state. The best-known *Helicoprion* tooth whorls in the world have come from this formation in Idaho, a deposit rich in siliceous chert, representing open ocean conditions. This is also the location where the largest *Helicoprion* specimens were found.

One of his brightest students, Jesse Pruitt, probed him one day with questions about how *Helicoprion* fed. Leif was baffled. He couldn't answer them. Together they set about doing a series of studies to figure out all the possible motions that a shark like *Helicoprion* could perform with its strange tooth whorl. Leif realized that studying the whorls in isolation was not enough to solve the mystery; they needed more of the head and skull to see how and where the tooth whorls fitted into the overall cranial anatomy. Looking through the literature, they came across the superb specimen that Svend Bendix-Almgreen had described earlier. It had associated cartilages of the head preserved with its whorls in a rock.

The specimen, Idaho No. 5 (also mentioned above), had been borrowed by Svend from the Idaho Museum of Natural History back in 1961, and after a few years it got lost in the collections of the Natural History Museum in Denmark. It was rediscovered in 2017 and finally entered in the database of that museum. An astute curator at the museum, Ane Elise Schrøder, noticed the writing on it and notes accompanying it, so she immediately suspected it wasn't one of their specimens but an old loan. In early 2023, Leif decided to go to Copenhagen to research specimens there, so he contacted the museum. They told him about the specimen's being found, resulting in its swift return to Idaho in July 2023. It will be studied in detail in the near future. However, we are jumping ahead of our story—let's go back to around 2013.

In 2013, even without Idaho No. 5, Leif and Jesse were able to pick up where Svend left off by identifying another *Helicoprion* specimen (Idaho No. 4) with possible skull and jaw parts preserved. Using a high-powered CT scanner on this outstanding specimen, they revealed its jaws and part of the braincase, preserved in life position inside the rock. This was the first time any fossil of this weird shark had confirmed the tooth whorl in its natural position. They discovered that the coil of teeth developed from the lower jaw, and that there were tiny flat, plate-like teeth lining the upper jaw. While the snout of *Helicoprion* is missing from the specimens, the shape of the rest of the cranium supports the idea that it bore an elongated snout. Next they focused on the functional aspects of the coiled tooth whorl of *Edestus,* a form closely related to *Helicoprion.* This examination revealed the long-mysterious mechanism driving how these coily-whorled sharks fed. Finally!

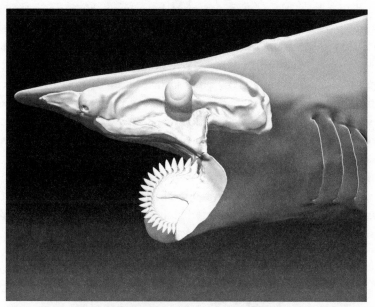

**How the tooth whorl fitted into the head of *Helicoprion*** SKETCH BY JESSE PRUITT

Leif called in support from the shark jaw mechanics expert Cheryl Wilga of the University of Alaska and the French CT expert Alan Pradel. Together they discovered that while all these bizarre buzz-saw sharks had developed their strange tooth whorls at the place where the lower jaws meet, some of them had also grown tooth whorls elsewhere.

*Edestus,* for example, differed from *Helicoprion* in that it bore a second whorl of deadly serrated teeth on its upper jaw. New CT scans of its skull and jaw cartilages revealed that it had a long **quadrate cartilage**, which connects the articulation of the lower jaw to the upper jaw, as in the quadrate bone of a snake. This meant that *Edestus* could open the jaws very wide, just as a python does when swallowing an alligator whole. It could then retract the lower jaw tooth whorl quickly backward to snag, slice, and swallow its prey. Such a swift action could work with hard-shelled prey, like ammonoids, pulling the hooked soft body out of its hard shell while simultaneously slicing it up for easy swallowing.

The largest *Helicoprion* whorl found so far in the Phosphoria Formation is almost two feet in diameter. When I spoke to Leif, I eagerly asked him how big *Helicoprion* must have been. He replied that any way you scale up the *Helicoprion* tooth whorl to potential body size, based on other shark teeth and jaw ratios, it was monstrously huge, growing to a maximum estimated size he thinks was between 32 and 39 feet long. *Helicoprion* was the biggest shark in the history of shark evolution since they first appeared 465 MYA. No future shark would surpass its size for another 150 million years after it died—making it the largest shark in the oceans over a span of 440 million years, until megalodon arrived on the scene!

Let's get back to what this monster might have chowed down upon. With their wealth of stunning material and modern scanning technology, Leif and his team suggested that both *Helicoprion* and *Edestus* probably ate the large squid-like creatures that were abundant at the time in the cooler Panthalassan ocean. Using the tooth whorls, the

These *Edestus* fossil jaws from the Carboniferous of Iowa look like nature's jagged scissors   AUTHOR PHOTO OF USNM V 7255, DEPARTMENT OF PALEOBIOLOGY, NATIONAL MUSEUM OF NATURAL HISTORY, SMITHSONIAN INSTITUTION

*Helicoprion* could concentrate their biting power into one narrow plane of force to catch the squid and pull it into its mouth in one swift scissor-like action.

Of the more than ten thousand *Helicoprion* teeth that Leif has examined, he has seen breakage or wear on hardly any of them, save for maybe three or four. This points to *Helicoprion* as more likely to have been feeding on squids rather than cracking the shell-covered ammonites and nautiloids. Or it was processing shelled nautiloids and ammonites in a highly specialized way, extracting the soft body of the animal from its shell using the serrated tooth whorl without damaging the shell or wearing down its teeth. Either way, it was a true maestro at what it did.

While this theory is the best so far for resolving the mystery of *Helicoprion*'s eating habits, with science you never know what new informative specimens could yet emerge from the vast territories in which their remains have been found (Australia, Russia, Greenland, Japan, the United States), and from the return of missing specimens like Idaho No. 5 with promising preservation.

## Other significant buzz-saw sharks

At around the same time Leif and Jesse were drawing their conclusions, a mild-mannered American physicist was turning his attention to another unsolved mystery of the buzz-saw shark world—how the peculiar double tooth whorls of the much older Carboniferous shark *Edestus* might have functioned. What did it eat, and how did it use its magnificent dual tooth whorls?

We have heard of *Edestus* earlier in this chapter, as it was discovered and named well before *Helicoprion,* by the American paleontologist Joseph Leidy in 1856, from older rocks in the Carboniferous period. With all the new work unraveling the secrets of how *Helicoprion* fed, attention soon began to turn to solving the equally challenging mystery of *Edestus*. Unlike those of *Helicoprion,* the tooth whorls of *Edestus* were only slightly curved, not coiling back upon themselves. Its teeth were still very large (4 inches high) and sharply serrated for sawing flesh. Like *Helicoprion,* it, too, was huge, perhaps up to 20 feet long.

Wayne Itano of the Museum of Natural History at the University of Colorado has said "the tooth whorls of *Edestus* are perhaps the most enigmatic dental structures of any known vertebrate." Wayne is a gentle, quiet-spoken man who has led a curious double life, with one half of it spent achieving incredible work on weird fossil sharks, and the other . . . He studied physics, following the hard sciences after his father, Harvey, who was a student of the legendary Professor Linus Pauling, a two-time Nobel Prize winner, and in 1979 he became the first Japanese American to be elected to the U.S. National Academy of Sciences. That same year, Wayne started his long career working at Boulder's National Institute of Standards and Technology. While gradually becoming more and more interested in fossil sharks, he landed an honorary position as Curator Adjoint at the Museum of Natural History in Boulder. Wayne was especially fond of the more peculiar fossil shark specimens, like *Edestus,* so he began investigating how it might have used its strange tooth whorls.

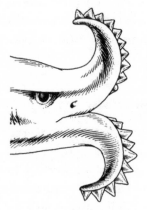

**Early restoration of the head of *Edestus*, which had two tooth whorls, upper and lower** ARTWORK BY R. GARY RAHAM

In one study, Wayne focused on the nature of the tooth wear on a new specimen of *Edestus minori* from San Saba County in Texas. The specimen is a large symphyseal tooth from the whorl that shows no wear on its sides, but the crown of the tooth is cut off and worn smooth. This is totally unlike *Helicoprion,* in which the teeth are not worn. He saw this as evidence that the teeth were used during the life of the shark, abrading the crown. The tip, he concluded, possibly broke off during the forceful striking of a prey object. Continued contact with similar hard-surfaced prey, like scaly fishes or other sharks with tough sandpaper-like skin, could have ground the biting tooth down to a smooth, rounded surface.

He suggested that *Edestus* might have used its deadly whorls in the vertical plane to make the most effective use of the serrated edges. This could have been done by moving the front part of the body up and down with the jaws in a fixed position. The prey could have been slashed to pieces with a strong downward motion of its head. Such an unusual mode of predation would be atypical for all sharks, yet not very dissimilar to how living sawfishes slice their prey, by slashing sideways with the sharp tooth-like denticles lining their snout.

In a later study, Wayne reported a high number of fine linear scratches on the sides of *Edestus* teeth, and wear on the tips of several teeth. This reinforced his earlier idea that the shark killed or disabled its prey in a special way—using its jaws in a scissor-like action to cut prey trapped between the large saw-edged teeth. Grisly, but effective.

*Helicoprion* and *Edestus* were both extremely specialized sharks, despite being separated by some 50 million years. It's important to remember that not all these buzz-saw sharks fed in the same manner, since they didn't all have the same types of teeth. Others, like *Agassizodus*, was the opposite to *Helicoprion*, as it had one deadly upper jaw tooth whorl with the sharp cutting edge aligned across the front of the lower jaw, not parallel with the long axis of the shark's body. It also had an area of flat pavement teeth in the lower jaw. *Agassizodus* must have used its powerful upper jaw whorl to hook its prey and then crush the prey with its flat pavement teeth. Having the whorl on the upper jaw might have been advantageous for raking the seafloor to find creatures with hard shells, like clams, and then scooping them into its mouth using its flat lower jaws.

So, while we have bucketloads of tooth whorls, how do we know what the rest of *Helicoprion* and *Edestus* might have looked like? A closely related form named *Fadenia* provides the best clue to how we can reconstruct the full body shape of *Helicoprion*. *Fadenia* is known from a relatively complete body fossil from the Permian marine rocks of East Greenland that was briefly studied by Svend Bendix-Almgreen. No one has yet followed up on a detailed study of this intriguing specimen, probably because so few specialists exist who are interested in such fossils. Svend passed on his notes depicting how he thought it looked to Rainer Zangerl, who included this rough sketch of the body outline of *Fadenia* in his 1981 *Handbook* on Paleozoic sharks. The sketch showed the smooth, streamlined fusiform shape, like a torpedo, with a high-angle tail and strong dorsal fins, while entirely lacking the paired pelvic fins or an anal fin. It basically had the body of a shark athlete, something that evolved for very fast swimming.

Associated streamlined scales found with the Russian specimens also support a fast-swimming lifestyle for these sharks. Such a combination is highly unusual and somewhat contradictory—to have a body built for speed but a mouth like *Fadenia*'s built for eating slow-moving prey like clams, crabs, and snails. Unless, that is, it also fed on ammonites, tasty cephalopods that lived inside large floating coiled shells. If that was the case, the ability to swim fast to catch the ammonites, then to crush its hard shell with its battery of teeth, would be a winning combination.

Another extinct giant within the *Helicoprion* clan was the appropriately named *Sarcoprion,* whose full name means "gluttonous flesh saw," suggesting that it was a ravenous and effective killer. *Sarcoprion* had a long snout with a short row of lower-jaw symphyseal teeth curling back from the tip of the jaw into the mouth, with a single row of smaller upper jaw symphyseal teeth ready to meet them when it closed its mouth. Estimates of maximum size place it at around 20 feet, so it was clearly capable of catching large prey of various kinds—large fishes, other sharks, squids, and other shelled cephalopods.

A new addition to this coily-toothed clan was named in 2023 as a new genus of giant buzz-saw shark closely related to *Helicoprion,* called *Karpinskiprion,* by Oleg Lebedev from the Moscow Paleontological Institute and colleagues. It was based on a study of new specimens of the species first studied by Karpinsky and named as his *"Helicoprion" ivanovi. Karpinskiprion* had large highly coiled tooth whorls similar to *Helicoprion* but in a slightly different alignment, with the main axis of the teeth pointing backward. The teeth showed lifetime wear facets on them, something we don't see on any *Helicoprion* specimens. It was using its teeth in a different way from *Helicoprion.* The growth of the tooth whorl in *Karpinskiprion* shows that the tooth crowns grew steadily through the life of the shark, with subtle changes to the saw blade wear patterns as it grew. These suggest that it was eating large fish as it matured.

*Karpinskiprion* was also enormous, probably reaching over 20 feet in length. The growth rate of *Karpinskiprion*'s tooth whorl suggests a 10-to-20-time size increase throughout its life span. Significantly, *Kar-*

*pinskiprion* also shows that large buzz-saw sharks with coiled teeth were around well before *Helicoprion* ruled the seas; its remains came from much older layers dated at between 307 and 299 MYA, right at the end of the Carboniferous period.

**Karpinskiprion appeared in the seas around 300 MYA, well before its slightly bigger cousin *Helicoprion***
AUTHOR PHOTO, COURTESY OF THE NATURAL HISTORY MUSEUM, LONDON

The last questions surrounding *Helicoprion* and its toothy kin are where they fit in the shark family tree, and what they have taught us about shark evolution in general. From what we have seen so far, it's clear that *Helicoprion* and its kin are a very important part of the shark story. Their place is special not just because they represent the rise of the first mega-predatory sharks, successfully seizing the top position in the ocean's food chains around the globe. They also highlight how certain sharks became extremely specialized to take advantage of just one kind of abundant prey. This highly specialized lifestyle also led to their demise. Ammonites suffered a great collapse at the extinction event at the end of the Permian period. Most (but not all) of these grandiose sharks probably went extinct as a result of the loss of their main food source.

Mike Coates from the University of Chicago has confirmed the earlier hypothesis that *Helicoprion* and its kin diverged from a side branch onto the line leading to modern holocephalans. He explained to me that the main features that push *Helicoprion* and its closest relatives into holocephalan land are their braincase and teeth. Living chimaerid embryos have a special kind of jaw articulation to the braincase that is shared with *Helicoprion* and its kin. Other features are the brain shape and tooth whorl development—basically, the big tooth whorls of *Helicoprion* are built in the same way that chimaerid crushing plates develop. Unlike true sharks (selachians), there are no independent rows of separate teeth, a defining feature of all modern sharks.

While spectacular fossils like the big tooth whorls of *Helicoprion* fire up our imagination about past worlds, they are also valuable artifacts sought after by collectors. Important museum specimens have been known to go missing when they become the focus of nefarious doings.

## *Helicoprion* goes missing

One of the best *Helicoprion* tooth whorls from Russia was stolen from the Paleontological Musuem of St.Petersburg State University in 1998. The remarkable story of how it was miraculously recovered and returned to its rightful home is worth retelling, as it shows that even the best-planned heists can be thwarted by clever paleodetectives. It all started when a fabulous specimen of a *Helicoprion* tooth whorl from Russia (the same one figured on p. 171) was offered to a U.S. fossil dealer based in Florida. To find out more about *Helicoprion*, the dealer opened a well-known guidebook on fossils written by my friend David Ward (whom you'll hear more about later in the book). The dealer recognized that the fossil pictured in the book was the same one being offered to him for sale. He contacted David immediately to learn that the photo in the book was of a replica of the holotype specimen of *Helicoprion bessanowi* in the Natural History Museum in London. The holotype is the sacrosanct specimen used to define the characteristics of a new species, so it is incredibly valuable as a scientific resource. They are the treasures of any museum collection and are never, ever put up for sale.

David immediately contacted his colleague based at the St. Petersburg Paleontological Museum in Russia, where the specimen was usually kept locked in a vault in a high-security room. His colleague ran to check on the specimen and confirmed it was missing. What happened next sounds like something out of a John le Carré novel.

David contacted the U.S. dealer and learned he had bought the *Helicoprion* specimen, so he told him to immediately cancel the check, which the dealer was able to do as it hadn't yet been cashed. In the next few weeks, the specimen arrived in Florida, posted from a location in Finland. Finally, as an assurance that they had the correct specimen, David arranged for the dealer to take it to a local fossil shark tooth expert, Gordon Hubbell, in Miami. They needed to confirm it was indeed the real holotype and not a possible fake, as fakes are all too common in the black market fossil trade. Gordon confirmed it was the real deal. David then rang the director of the St. Petersburg museum to inform him that they had found the missing holotype fossil of *Helicoprion* and had safely stored it until ready for return.

The U.S. dealer agreed to carry the fossil to California, where he handed it to another of David Ward's friends, who then transported it to Paris. The following year, this courier handed it in person to David. *Helicoprion* then spent a few months over the winter in a hatbox in a corner of David's London study. David set about planning how to carry the specimen by hand back to Russia. Despite his noble intentions with this Russian national treasure, carrying any illegal specimen back into any country can be a risky business. David called his contacts, including the museum director in Russia, to hatch a plan. He requested the director to organize documentation for the official retrospective loan of the specimen. Then, armed with copies of this document in both Russian and English, with the fossil carefully packed in his hand luggage, David boarded a flight for St. Petersburg.

After sweating it out in the customs line, he finally made it outside, where he was promptly met by the museum director and his old colleague. They drove David to the museum, and the holotype specimen was officially handed back to the museum in a small ceremony. The happy moment was celebrated with special German brandy, coffee, and cakes. David received a special medal and certificate of honor from the museum for his role in returning the priceless national treasure.

The thief who stole the specimen from the museum's collection was never found, although a few months later a well-known German fossil dealer was arrested in St. Petersburg with a Mercedes van packed full of illegally sourced Russian fossils. When questioned about the *Helicoprion* specimen, he denied any part in its theft.

It is unfortunate that thefts like this can occur from major museums, but even more alarming is the proliferation of faked fossils on the market. These can be of a few types. Some are specimens made up from several broken parts of individuals spliced together and "touched up" to make it look like a stunning complete fossil of some exotic new type. Others are faked material made to look like a real fossil. Examples abound, but some are so good that even museums fall for them, despite having experts inspect them. In one example, a Chinese dinosaur nicknamed *Archaeoraptor* offered for sale turned out to be made up of parts of several different dinosaurs. A CT scan revealed the truth in the end. Fake fossil sharks' teeth are often sold on the internet. While some are honestly labeled as casts or resin replicas, others are not. If you see a large megalodon tooth at a very low price, it is likely a replica, so be cautious before you buy anything like this online. Similarly, there are many *Helicoprion* tooth whorls for sale on the internet, but most are replicas.

These bizarre and evocative fossils of *Helicoprion* understandably get a lot of attention. But what other kinds of sharks lived alongside them in these Permian oceans? It's time to leave *Helicoprion* and its kin and see what other sharks were competing with the buzz-saw clan. Were there other gigantic killer sharks in these oceans that posed a threat to a beast as large as *Helicoprion*?

## *Helicoprion*'s competition

*Helicoprion* may have been large, but as we've just seen, it was a very specialized predator targeting certain kinds of mollusks. Other gigantic predatory sharks lived at this time, probably capable of attacking and killing medium-sized *Helicoprion*. These include late-surviving species of the voracious ctenacanthid group, such as *Kaibabvenator*, whose large teeth were found from rocky outcrops of the Kaibab Formation in the Grand Canyon National Park in Arizona, representing a late Permian shallow seaway about fifty miles off the coastline. Its teeth are more than an inch wide, and it's been estimated to have been 16 to 18 feet long. *Kaibabvenator* was a rather beefy predatory ctenacanth whose large, chunky teeth were discovered and named by the national parks paleontologist John-Paul Hodnett. It differs from all other ctenacanths in that it has serrated teeth, suggesting a shift from the typical piercing and grabbing of prey to a tearing dentition more suited to tackling large prey. Wear on the large tooth tips even suggests they were hitting their prey very hard, as when white sharks attack seals. Who knows what they were eating, but whatever it was, it was very big—possibly other large sharks like *Helicoprion* and its kin.

**Teeth of the giant predatory shark *Kaibabvenator* measure just over an inch wide. They were found in and around the Grand Canyon National Park in the United States. This shark could be up to 17 feet long** PHOTOS BY JOHN-PAUL HODNETT

Other sharks from the Grand Canyon area include the delicate toothed *Nanoskelme,* whose jaws were adapted for grasping and swallowing small fishes and squids. The Carboniferous giant *Saivodus* survived to this time and is found at this site, but it is now diminished to a much smaller shark no larger than 6 to 10 feet long. We can see from this diversity within the one site that there was predatory partitioning as the ctenacanths of the Kaibab Formation occupied different size ranges while living in the same habitats. We also see this predatory partitioning going on among the many reef sharks around today.

Several sites around the world have yielded many other species of Permian sharks and holocephalans. Fossils of the latter group are particularly common, represented by a great variety of rounded to flat crushing tooth plates, like *Tanaodus,* with flat petalodont teeth also found in abundance. We also have *Orodus* with its humped crushing tooth plates, indicating that many large shark-like forms, between 10 and 15 feet long, were feeding on the seafloors on the rich assemblages of marine invertebrates. None of these would have worried *Helicoprion.*

Outside the oceans, medium- to large-sized xenacanths like *Orthacanthus* continued to dominate the freshwater estuaries and rivers of the world. They were accompanied by a gradual invasion of hybodontiforms, like *Wodnika,* in the shallow marine habitats. Leonard Compagno, a guru of early shark studies, suggested that the generalized littoral zone—areas close to the shore—best characterize the primitive feeding ecology for sharks, and it's not surprising that this habitat is where most sharks were living in the Late Permian.

We will now wrap up our tour of Paleozoic sharks with a remarkable fossil shark discovery from Germany that tells us about the role of sharks in food chains of the day.

## A Permian turducken

Aside from inferring from their tooth shapes what these Permian sharks ate, what direct evidence do we have of shark predation at this time? I guess you could say it's a case of "You are what you last ate." We will answer this question by looking at an awesome fossil shark from this time found in Germany whose stomach contents are well preserved.

Jürgen Kriwet is one of the world's leading researchers on Permian and Mesozoic sharks. He grew up in southern Germany wanting to become a veterinarian. However, when he was nine years old, his mother bought him a book titled *The Super Shark*, showing how sharks were the perfect predators for their environments, and from then on, sharks had him hooked. Today he's a professor of paleontology at the University of Vienna (see bio on p. 419).

Back in 2007, Jürgen was based at the Natural History Museum in Berlin, where he studied a most extraordinary fossil specimen that provided a rare insight into the hunting abilities of the xenacanth sharks. The fossil was found at Lebach, in southwest Germany, from an ancient freshwater lake called Lake Humberg, which was at its peak around 290 MYA. Inside the concretion he found the remains of the xenacanth shark *Triodus sessilis*, which, if complete, would have measured about 1.5 feet long. Inside its gut he found the remains of its last few meals—the skeletons of two ancient crocodile-like amphibians called *Cheliderpeton* and *Archegosaurus*. Both were consumed not yet fully grown; they were still larval forms. An even more surprising find was the remains of a juvenile stem shark, *Acanthodes bronni*, inside the gut of the consumed *Cheliderpeton*.

Here we have a unique example of a Permian turducken. *Triodus sessilis* is the only known fossil shark to contain two amphibians in its stomach. Both were swallowed tail first, indicating that they were actively chased before being consumed whole. From this superb fossil we can reconstruct three levels of a food chain. The remains of its last meals were well preserved, not strongly etched by stomach acids, which would have indicated a much longer time of digestion. The success of the *Triodus sessilis*—a predator able to consume two amphibians within a short time frame—testifies that sharks had now well and truly adapted to hunting in freshwater environments. In fact, they seem to have become rather good at it.

**Permian lake food chain as shown by a xenacanth shark that ate an amphibian that ate a stem shark**   ARTWORK BY BRIAN CHOO

By analyzing the other fossils found in the ancient Humberg lake, Jürgen and his team were able to construct the entire food chain of this freshwater ecosystem. At the top sits the even larger, more fearsome xenacanth called *Lebachacanthus*. This beast grew to around 10 feet and could have taken down larger amphibians, such as the mature *Archegosaurus* (reaching 6 feet long), the smaller *Cheliderpeton* (about 2 feet long), or other xenacanth sharks like the 8-to-9-foot-long *Orthacanthus* and *Xenacanthus*. These predators fed on ray-finned fishes, lungfishes, and stem sharks like *Acanthodes*. Below this level we have a crustacean called *Uronectes,* along with smaller microscopic plankton and ostracods (tiny bivalved crustaceans).

What all this means is that we have a unique ecosystem operating in this Permian lake in that sharks are the main apex predators. From the Jurassic onward, this would change, as large bony fishes and reptiles, like early crocodiles, become the dominant apex predators in lakes and rivers. But for a brief time, the xenacanth sharks ruled the roost in the ancient Permian rivers and lakes.

## The Paleozoic shark wrap-up

So far in our story we have witnessed the first 230-million-year history of sharks. They first appear as enigmatic scales about 465 MYA, and we know little about them until proper sharks' teeth begin appearing in rocks around the world around 419 MYA. While we have few species of sharks from the Ordovician period, the mass extinction event at its close didn't affect sharks at all; they sailed right through. These nebulous first 50 million years of sharks are dominated by early "proto-sharks" known only from scales and spines, as well as the group of archaic "stem sharks" (or acanthodians), most bearing fin spines in front of all their fins, and often bearing strange jaws with or without teeth.

In the Devonian period, the shark story accelerates as we begin to find many different types of sharks' teeth in rocks all around the world, along with the first articulated or near-complete body fossils of sharks that begin to look something vaguely like a modern shark. This shows that sharks' first superpower had kicked in—their ability to produce many teeth and shed them regularly to replace worn or broken teeth with sharp new ones.

*Doliodus* from Canada, living around 409 MYA, shows a nearly complete body with braincase and teeth as they were positioned in life. While it retained fin spines in front of its pectoral fins (making it a stem shark), it was still shaped very much like a true shark. It demonstrated that some of these early stem sharks were a lot closer to the line leading to modern sharks than others. It also showed the start of another shark superpower—a superb sense of smell. Its broadly spaced nasal capsules suggest that it hunted like living sharks by swaying its head from side to side to home in on its prey.

In the Middle to Late Devonian, sharks really take off. We have hundreds of different species, but most of them are known only from their teeth. Sharks begin invading large freshwater river and lake system habitats, and some forms, like *Portalodus*, emerge as the first giant predatory sharks, reaching up to about 10 feet long.

Sharks also developed another remarkable superpower by the end of the Middle Devonian—the ability to develop radical new tooth shapes and new tooth tissues. This allowed them to break free from being one-trick ponies that hunted live prey with sharp-pointed teeth. The appearance of rounded, arched teeth reinforced with dentine pillars in the Late Devonian heralds the dawn of the holocephalan line that branched away from the main shark line around 380 MYA. Holocephalans differ from regular sharks not only by their crushing teeth, but also by their strange and wonderful new body designs. In the last few million years of the Devonian (359–357 MYA), sharks became very diverse in our oceans with common forms like *Cladoselache* around 4 to 6 feet long, and giant predators like *Ctenacanthus* growing to around 16 to 20 feet.

In the past 70 million years leading up to the end of the Devonian, sharks had been kept under the dominance of the armored placoderm fishes. Placoderms were more diverse, larger, and more abundant than sharks up until around 357 MYA. Finally, near the close of the Devonian, sharks win the placoderm wars. Not only do they begin to outnumber placoderms in ocean environments, but some, like *Ctenacanthus,* were much bigger than the largest known predatory placoderms such as *Dunkleosteus.* Sharks finally reign at the top of the marine food chain, but only in restricted areas.

The end-Devonian extinctions—a deadly double punch of two chaotic events—bump out placoderms, and sharks immediately take over both the marine and freshwater habitats of the world in the Carboniferous period, starting around 359 MYA. The early part of this period sees the biggest speciation event of shark and holocephalans so far known; they become the most abundant fishes in our seas in some places, with sites like Bear Gulch home to more than eighty species of sharks and kin all living in one small area. Meanwhile, serpentine-shaped xenacanth sharks with deadly neck spines take over the rivers, swamps, and lakes of the world. Sharks like *Bandringa* show us that sharks are now using another of their superpowers: electroreception for sensing the weak electric fields of living creatures they cannot see or smell.

Some sharks now grow to enormous sizes in the open oceans, with *Saivodus* perhaps peaking at around 24 to 28 feet long. The shallow seas of the world are now dominated by hundreds of bizarre species of holocephalans, like the petalodonts and cochliodonts, all with highly specialized tooth plates adapted for feeding on clams, snails, crustaceans, and other hard prey on the seafloors. These are the holocephalan glory days; they will slowly decline in diversity from here on.

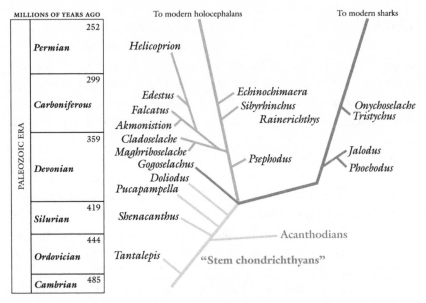

A simplified tree of the evolution of sharks up to the end of the Paleozoic era   DIAGRAM BY AUTHOR

The first buzz-saw sharks, like *Edestus,* also appear in the Early Carboniferous. They were an offshoot line of the great holocephalan species radiation that occurred at this time. Forms like *Edestus* would give rise to the gruesome dynasty of gargantuan saw-toothed sharks that peaked in the Early Permian with *Helicoprion,* gigantic in size (33 to 39 feet), but with each species highly specialized to feed on specific prey like squids or shell-covered cephalopods.

Perhaps the most important group of sharks to appear at the start of the Carboniferous period and carry on through to the Mesozoic era are the hybodontiforms like *Tristychius,* with its strange suction feeding method. They are the one group of sharks we have encountered so far in our story that are most closely related to the modern living sharks; we will hear much more about them in the next chapter.

At the end of the Permian, the world was hit by a massive evolutionary punch to the face and life was knocked out cold. What caused this mass extinction event, and how long would it take for our oceans, rivers, and lands to recover from the sudden loss of so many species? How did sharks manage to survive not just another devastating planetary event, but the worst one of all time? Brace yourself as we tackle these questions and see how sharks managed to steal second base in the evolutionary game of the eon in the next chapter.

# PART 3
# SHARKS UNDER PRESSURE

*Ostenoselache*

CHAPTER 9

# SHARKS AND
# THE GREAT DYING

*Earth's Biggest Extinction Event Shapes
Shark Evolution*

---

## The Great Dying

Let's pause for a moment to revisit the evolutionary story of life so far. At the moment you picked up this book, life had been around on this planet for around 3.6 billion years, evolving slowly from simple single-cell organisms into the complex multicellular animals and plants we know and love today. Throughout this unthinkably vast span of time, there have been several major disruptions to the rhythmic expansion of life, often abruptly curbing its diversity and decreasing the abundance of some branches of its complex family tree.

Of the five major mass extinction events that have occurred in the past 500 million years, this next one was *the big one,* and absolutely the most devastating for all life on Earth. This event, which comes next in our story, is commonly known in geological circles as the Great Dying, and it was as cataclysmic as it sounds. Close to 87 percent of all species went extinct, with estimates varying from 81 percent to 97 percent of all marine species being wiped out.

The event started around 251.9 MYA, a date defining the end of the Permian period and the beginning of the Triassic. (It's also the boundary separating the end of one long geological era—the Paleozoic, 541–252 MYA—from the start of a new one—the Mesozoic, 252–66 MYA.)

It was triggered by enormous and prolonged volcanic eruptions emanating from one place on Earth that ripped deep gouges in our planet's crust, allowing molten lavas to billow forth for thousands of years. These lava flows created the Siberian Traps, whose name refers to the vast area of the eruption site in northern Russia and the Swedish word *trappa,* or "stairs"—because cooled lava rocks erode to form stepped landscapes. The traps emitted around one million cubic miles of lava, covering over 3 million square miles of land (an area roughly the size of Australia) and filling the skies with deadly clouds of carbon dioxide and other aerosols that hindered plants from photosynthesizing.

On land we see the loss of diverse landscapes dominated by vast forests of tree-like seed ferns and conifers, but not only from the changes to the atmosphere. Wildfires raged across the globe, and higher temperatures stressed all living organisms. The volcanic flows pulsed in violent episodes of eruptions, bringing about dramatic extinctions of reptiles, insects, plants, and many other species on land. The number of all known land-dwelling animals, amphibians, and reptiles decreased by two-thirds. Even insects suffered; some scientists say this event was the only known mass extinction to impact this class of creatures.

In the oceans, though, the killer took on a more insidious nature. Stored methane was suddenly released from the ocean floors, causing a rise in acidity levels and compounding the high levels of carbon dioxide degassing from the lava flows. Sea surface temperatures rose by between 10° and 14°F. These increased sea temperatures drove certain marine microorganisms to increase the recycling of surface water nutrients, like phosphate. This process sucked all the oxygen from certain areas of the oceans, allowing the spread of seawater layers rich in hydrogen sulfide to expand into the richly biodiverse shallow continental shelves, rapidly causing the bottom-dwelling organisms to suffocate. All these factors combined to create a supercolossal cluster bomb of an environmental crisis. The Great Dying wiped out nearly all life in the seas over a roughly sixty-thousand-year period.

The Great Dying was, any way you measure it, the biggest extinction event ever. Although the massive asteroid impact that would wipe out the dinosaurs almost 200 million years later is much better known, it would not cause the disappearance of as many species, nor would it take as long afterward for life on Earth to fully recover—around 8 or 9 million years. In recent years, scientists have increasingly turned their attention to studying this end-Permian extinction, as it shows exactly what can happen when carbon dioxide levels reach dangerous levels in our atmosphere. The Great Dying is a tragic foreshadowing of where we are heading today if we do not do something soon to dramatically reduce our current levels of greenhouse gas emissions.

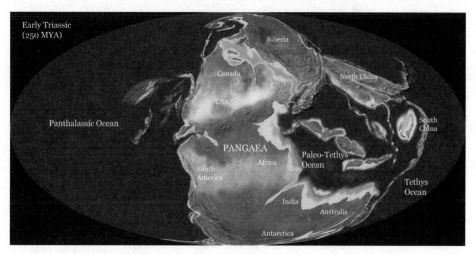

**The world during the Early Triassic period**   RON BLAKEY, DEEP TIME MAPS
(DEEPTIMEMAPS.COM)

The aftermath of the Great Dying was the dawn of a new era, starting with the Triassic period (252–201 MYA). The Triassic is the only time in Earth's long history that is bracketed by two global mass extinction events: the Great Dying (252 MYA) and the end-Triassic extinction event some fifty million years later (201 MYA). This latter event is ranked third out of the five extinction events in terms of its severity. These two periods of utter chaos on Earth acted as massive filters, sorting out the strongest and most adaptable forms of life to survive the cosmic double punch.

The Triassic period was nonetheless a landmark time for planet Earth. For the first 25 million years or so of the Triassic, life was in a biotic recovery phase—trying to get back to pre-extinction levels of diversity and abundance. The Early Triassic air had very low oxygen levels, between 12 and 18.5 percent, according to different studies, compared to 21 percent in today's atmosphere. By the end of the Triassic, oxygen had risen to around 19 percent. Global temperatures sat about 5°F above today's averages. Earth still had just one large supercontinent, Pangaea, which would endure throughout most of the Triassic.

On land we see a revolutionary change in the fauna during the Triassic. The first dinosaurs, mammals, crocodiles, tortoises, lizards, and flying pterosaurs appear about 25 million years after the Great Dying. Expansive new forests grew, dominated by ferns, pines, and new kinds of coniferous trees, with the first ginkgoes appearing. All this growth on land had repercussions in the oceans, as the rapid pace of reptile evolution and fierce competition on land drove some of these beasts back into the water. This return to the oceans resulted in many new kinds of aquatic reptiles entering the marine environments, challenging many of the shark groups at the top of the food chains for the same resources.

How did sharks make it through the crisis? What factors played in their favor? And how would they cope with the new reptilian invaders of the oceans?

## Sharks make it through the crisis

The Great Dying had major consequences for sharks. Several kinds of sharks and their kin went extinct at this boundary, but not as many as we might have expected. Species living in shallow-water environments that depended on seafloor resources like clams and snails were hit the hardest. The stem shark group, the acanthodians, which had hung on since the end of the Devonian period for around 110 million years in greatly reduced numbers, finally succumbed at this time. All the largest predatory sharks of the Permian went extinct, including most of the giant ctenacanths like *Kaibabvenator,* the *Phoebodus* group, and nearly all the tooth-plated or weird-toothed holocephalans we discussed in chapters 6 and 7. A few of the smaller buzz-saw sharks made it through, but the giant *Helicoprion* succumbed well before the event got started. The Early Triassic seas straight after the event did maintain a few shark species, and all that survived were small, under 6 feet long. We will see soon which of these old stragglers made it through.

Some of the oldest lineages of sharks, like symmoriiforms, which first appeared in the Late Devonian, were long thought to have gone extinct at the Great Dying, but they have recently been found to have survived the crisis. Even more impressive is that some of these groups of sharks then sailed right through another extinction event at the end of the Triassic period, to eke out a meager living well into the Early Cretaceous, finally succumbing about 140 MYA. That is an impressive 100 million years after the Great Dying. How did they do it?

These archaic sharks survived by using their old trick—moving out into the deep oceans. This reflects a trend, one of sharks' proven survival skills: When the going gets tough, the tough go deep. The sharks that moved to deep water were the ones that had sailed through previous extinction events—possibly at the end-Ordovician event (the mongolepids), but certainly at the end-Devonian extinction event (such as phoebodontids, symmoriids, and jalodontids). It was this same trick that enabled similar groups of sharks to survive the devastating end-Permian extinction. These species had remained undetected by most of the geologists because they had taken their sampling from shallow marine rocks, not from deep ocean basins. These species remain very rare after the extinction event. They were clearly the refugees, escaping the worst of the Great Dying by moving away from the chaotic zones that were hit the hardest.

There was an unprecedented diversity of several kinds of buzz-saw sharks, all living together in the Early Triassic seas between 252 and 247 MYA, immediately after the extinction event. These well-preserved articulated fossil sharks were found near Wapiti Lake in the high mountains of British Columbia, Canada. The fossils formed after burial in temperate latitudes, in deep-water continental shelf sediments on the western margin of northern Pangaea. Among the finds announced in 2008 were a skull of *Caseodus,* a moderate-sized shark, around 4 to 6 feet long, which was previously known only from much older Carboniferous sites, about 100 million years earlier. Also present were *Fadenia* and *Paredestus,* which were both up to around 4 feet long. All these closely related buzz-saw sharks had clearly won the battle at the end of the Permian, but it was a Pyrrhic victory, as they lost the Triassic war— the group evidently went extinct shortly after this time, as there is no younger fossil record of them.

The Wapiti Lake site has a hybodont called *Wapitiodus,* about 3 feet long, known from a beautiful and almost complete skeleton showing the head, fins, and stout body outlined with stunning clarity. Its teeth are wide and pointed in the front half of its jaws, with very elongated, crenulated crushing teeth in the back half. Unassuming little hybodonts like this one soon became the dominant sharks of the seas all through the Triassic and first half of the Jurassic periods. How did they do this, and why were they so darn successful?

## The hybodont invasion

By accident, in late 2022, I came across a gruesome cabinet in the fossil fish collections of the Natural History Museum in London. The cabinet is full of trays containing rows of mummified heads. I was not expecting such a confronting site—a stiff parliament of individuals all frozen in stone, their faces forever caught in expressions of stunned amusement. These remarkable specimens, collected from Bexhill, in Sussex, all belong to *Hybodus basanus,* one of the commonest species of the hybodont shark group. Most of these specimens were found in the nineteenth century. The skulls are all preserved in 3-D form, with jaws and holes for the eyes clearly visible, some bearing that strange selachian grin, like a devil smiling from deep time. During that visit, I had the entire fossil shark collections area to myself. I pulled one specimen out to study and simply couldn't restrain myself. I held it aloft in my right hand and softly whispered: "Alas, poor Yorick, I knew him, Woodward." You may be wondering what such a long-dead fossil had to tell me, but the real answer lies not in what we hear but what we can read from their petrified skulls.

While these fossils represent sharks that lived in later times, most of the successful hybodonts, including *Hybodus,* first appeared in the Early Triassic. Imagine a short, stumpy shark sporting sharp-pointed toothlike spines above its eyes. Most *Hybodus* species were around 5 or 6 feet long, with the largest one maxing out at nearly 10 feet. *Hybodus* takes its name from the Greek for "humped tooth," yet many hybodonts had solid cone-shaped teeth, while others had combinations of flat, wide, and pointed teeth. Their teeth are perhaps their most variable feature, even in a single species.

Hybodonts were the main shark group that sailed through two mass extinction events—the Great Dying at the end of the Permian and another at the end of the Triassic—without even blinking a nictitating membrane (that's a shark's eyelid). They were nearly all gnarly, robust little sharks with stout heads resembling a fist, and they rapidly ad-

*Hybodus* appeared in the Early Triassic and survived until the end of the dinosaur age (186 million years)    ARTWORK BY JULIUS CSOTONYI

justed to a chaotic underwater landscape with greatly diminished resources.

Hybodontids thrived after the Great Dying for the same reason that birds, small reptiles, and mammals survived the devasting end-Cretaceous event that led to the extinction of large dinosaurs. It's because they were small, and unfussy eaters. They needed fewer resources and were adapted for opportunistic feeding on a wide variety of prey. We know that one hybodont had the remains of some two hundred squid-like belemnites in its stomach, so they must have had a good hunting strategy to catch these fast-moving squid-like creatures. Hybodontids had developed powerful jaw muscles and a variety of tooth shapes. Some of them featured cusps lined with wavy ridges of enamel-like tissue for retaining prey within the mouth. They were essentially the rodents of the Mesozoic seas, happy to eat anything they could fit into their mouth.

Some hybodontids had teeth that were highly advanced relative to earlier forms, and in some ways similar to what we humans have today. As you can tell when you run your tongue around your mouth, we mammals have various kinds of teeth in our jaws that serve different purposes. Our front teeth (incisors) are for biting and cutting through food, and our pointy canines serve to grip and pierce food when we bite. They also make for great threat displays when we are angry, more so in our hairy primate ancestors. Our premolars are for shearing, and our molars serve to grind food down into smaller particles. All these functions serve to reduce food to digestible fragments through biting and chewing. This action eases the burden on our gut and is an important part of the process of extracting the optimal nutrition from our food.

Some hybodonts, like *Acrodus* and *Polyacrodus,* had jaws lined with slightly different kinds of teeth. The teeth of these hybodonts can vary from the front (pointed clasping, clutching teeth) to the back (flat or arched grinding, crushing teeth). While dentitions vary enormously between the different hybodont families, most have small sharp-pointed biting teeth at the front to clutch their prey, and broad, arched, or strongly humped crushing and grinding teeth lining the rest of the jaws. This gives them the capability to either grab and swallow prey as regular predatory sharks do, or to grab and grind hard prey such as clams, crabs, shrimps, or snails. Just as the rats of today have gnawing incisors at the front that grow throughout their lives, hybodonts had a similar power (albeit with a slight difference) to grow and replace their teeth throughout their lives. Like rodents, they maintained an ever-sharp row of front teeth.

Hybodont sharks have mouths full of domed crushing teeth, like those of *Acrodus* (left). Others, like *Hybodus* (right), had broad teeth with sharper cusps for gripping prey as well as crushing them   AUTHOR PHOTOS, COURTESY OF THE NATURAL HISTORY MUSEUM, LONDON

The hybodont sharks are also important because they are regarded as the closest group of extinct sharks to the group containing all living sharks, the **neoselachians**. Recent CT studies of hybodont skulls by John Maisey of the American Museum of Natural History showed that they were capable of detecting low-frequency sounds coming from specific directions underwater. This is based on their similar anatomy to the inner ears of living sharks with this ability: Sharks today can hear low-frequency sounds much lower than we can (at 10 Hertz, whereas our limit is at 20 Hertz). This ability is very useful for the shark because such sounds travel great distances underwater, where low-frequency sound is perfect for detecting wounded or struggling prey. This special hearing ability first appeared in these sharks in the oldest hybodontiforms, like our suction-feeding friend *Tristychius,* more than 100 million years earlier.

John Maisey used his anatomical findings in detailed analyses of hybodont shark evolution and applied sophisticated CT techniques to study the braincase anatomy of several living sharks for the first time. This research showed conclusively that all modern sharks (neoselachians, including elasmobranchs—sharks and rays) must have evolved from ancient hybodont stock.

Further new evidence from their tooth tissues supports these findings. Compared to modern sharks, the hybodont's teeth show that the **bundled enameloids** forming the outer layer of the teeth are less well organized. To make a neoselachian tooth, we can start with one from a hybodont with slightly better organized bundled enameloid, then just add another outer enameloid layer. This is why most scientists who study the evolution of sharks regard the hybodonts as the closest relatives of the neoselachians.

Why should all this detailed tooth histology really matter? Shark teeth structure has been studied in great detail because the vast majority of sharks are known only from teeth. The microstructure of the teeth empowers sharks with amazing properties. Neoselachians' teeth were built of three strong enameloid layers that made them stronger and more flexible than those of their ancestors. As teeth are the business end of any shark, such subtle differences are the make-or-break factor in evolutionary success, and this is exactly the case for hybodonts.

Hybodonts were abundant throughout the first half of the Mesozoic era, spanning the entire globe, and we will hear more about them at their peak time of diversity, the Jurassic period, in the next chapter. In the Triassic, they lived alongside the first neoselachians and the rise of marine reptiles, so our next sections deal with how they coped with the increasing competition in the seas from new groups of bony fishes and reptiles, and how the neoselachians emerged out of all this with their award-winning new stronger tooth structure.

## Arthur Smith Woodward and his fossil sharks

Some of the best early work on these hybodonts was carried out by a remarkable paleontologist based at the British Museum, Arthur Smith Woodward, who rose to acclaim as the world's greatest fossil fish expert of his day. His contribution to our understanding of fossil sharks was immense, to say the least. Over the course of his forty-two-year stint at the museum, Woodward worked on many groups of fossil sharks and fishes, churning out papers and books at an unbelievable pace. His 742 publications span seventy years, from his first paper at age fourteen (1878) through to one issued four years after his death in 1948 (it's rumored he was still working away in his coffin). He described and named an incredible 321 new species of fossil fishes, including around eighty new species of fossil sharks, rays, and holocephalans. Over forty years of public service at the museum (1882–1924), he apparently took only half a day off work, because he broke his arm.

Arthur was born in Macclesfield, Cheshire, UK, in 1864. The family was comfortably well off, but not wealthy. The eldest of four children, he was a very curious child who started winning prizes from his first year in school. He also loved lecturing to others, and at age eight would give speeches to his family while perched aloft on a soapbox in the kitchen. Arthur became totally fascinated with fossils at age twelve when he and his father were on holiday in northern Wales. He then became hooked on collecting fossils and learning everything he could about geology. In high school he struck it lucky when his math teacher also happened to be a gung-ho geologist who started a course on the topic. At age fifteen Arthur gave his first public lecture on geology, entitled "The World Before the Deluge." By the end of his schooling, he had won prizes in many subjects and been given special awards for his various fossil projects.

In 1882 his dearest wish came true when he entered the newly built British Museum, better known even then as the Natural History Museum, its official title today, as an assistant working with the fossil collections. The museum had acquired two extraordinary collections of fossil fishes soon after he arrived, so he began cataloging and research-

**Arthur Smith Woodward was one of the world's leading authorities on fossil sharks** PHOTO, CIRCA 1900, COURTESY OF THE NATURAL HISTORY MUSEUM, LONDON

ing these rich collections, a pursuit he continued for the rest of his career. He also studied part time for a university degree and joined various learned societies. He was a master networker and collaborator who would later hold memberships in some fifty scientific societies over four continents.

In 1894, Arthur married Maud Seeley. As the daughter of Professor Harry Glover Seeley, a famous paleontologist, she was probably already accustomed to the rugged fossicking life. They traveled together frequently for Arthur's research projects, collecting fossils or examining collections in foreign countries. In 1901, Arthur became Keeper of Geology at the museum and received an honorary doctorate from the University of Glasgow. He received many medals, honors, fellowships, and awards, even a knighthood. He enjoyed an unrivaled career, unmatched by any other geological or paleontological scientist of his time. Yet he always found the time to carry out groundbreaking research, particularly on fossil sharks.

Arthur worked on many different fossil sharks, rays, and holocephalans of all ages, including the famous Jurassic and Cretaceous hybodonts found at Lyme Regis and sites near Bexhill. Perhaps his most significant contribution was his monster catalog *Fossil Fishes in the British Museum*. This multivolume set is still used throughout the world as a standard reference to these and many other fossil shark specimens held in the collections. The catalog was not just a list of specimens, but a beautifully illustrated work embellished with his own insightful comments and short notes on many of the key specimens. Arthur also worked on the complete fossil neoselachian sharks from the Jurassic of Solnhofen, Germany, and the Cretaceous of Lebanon. His work ranged right through almost the entire geological time charts across the world, from Devonian to recent species, including his paper describing the last of the xenacanth sharks from the Triassic of Australia.

Arthur made a seminal contribution to the study of hybodont sharks. There are some twenty different species of *Hybodus*, based on teeth (seventeen species) or fin spines (three species). He sorted out the earlier classification problems with *Hybodus* and determined that certain species known only from teeth and fin spines were not valid species. His detailed investigations on the anatomy of certain well-preserved British hybodont fossils is just one of his many lasting legacies to our current understanding of shark evolution.

## Challenges facing sharks of the new era

Sharks didn't have it easy in the oceans of the Triassic, as more trouble was lurking around the corner. By the late Triassic period, around 210 MYA, that threat would increasingly become a reality. The seas were becoming home to a growing band of new predators that would challenge sharks for millions of years to come. Large, voracious reptiles were taking over the seas.

Reptiles had ruled the land for the past 200 million years, but fierce competition on land—no doubt owing partly to the dramatic rise of the first dinosaurs—drove some to explore the ocean as a source of sustenance. By the end of the Triassic, the vast majority of sharks were still small (under 3 feet) with maybe a couple species about 10 feet long. Suddenly they found themselves dwelling in the shadows of massive air-breathing marine reptiles nearly 70 feet long.

## Sharks become the underdogs

This aquatic reptile invasion took on a myriad of forms. Some reptiles had long necks; others sported short, stumpy bodies. One weird-looking group, the **placodonts**, were seal-sized, barrel-chested beasts with short snouts and mouths full of flat crushing teeth. Some had shells on their back and started to look like strange marine turtles, but they were not at all closely related to them. They fed on a variety of crunchy marine clams, snails, and crustaceans, so they clearly challenged the hybodontids for the same food resources.

The **ichthyosaurs**, meaning "fish lizards," are marine reptiles that resembled long-snouted dolphins. All their limbs developed as flippers, making them fast swimmers. They also bore shark-like dorsal fins for stability when darting through the water. We know they were fully aquatic reptiles, as they gave birth to live young; exquisite fossils from Holzmaden, Germany, still have babies inside the mother's uterus (laying hard-shelled eggs underwater would be fraught with problems). Ichthyosaurs represented the largest and fiercest competition for sharks throughout the last half of the Triassic period. Let's meet some of them to see what sharks were up against.

Remarkably, it took only 2.5 million years from their first appearance as cute little shore-hugging lizard-like critters, like the 1.5-foot-long *Cartorhynchus* from China (249 MYA), before ichthyosaurs transformed into truly gigantic and formidable marine tyrants. This rate of evolution is five times faster than what we see much later in the evolution of whales, from tiny four-legged hoofed creatures to gargantuan aquatic behemoths.

What was driving this turbocharged development? One theory is that the seas became filled with new groups of tasty squid-like creatures. One group was the **ceratitid ammonites** (squid-like mollusks that lived in coiled shells); the other was the **belemnites**. These critters resembled a modern squid, up to 2 feet long, with a special thick, pointed, calcified part of the shell, called the guard, inside their long, straight body. These guards readily preserve as fossils, so they tell us how abundant they were in ancient times. Other groups appearing at this time include the vampire squids. These cephalopods fueled the feeding frenzy in the oceans, and everything that could swim and had teeth suddenly had a taste for calamari.

Heralding the dawn of a new age of marine reptile dominance starting around 246 MYA, we find the ichthyosaur *Cymbospondylus*, which reached around 58 feet in length and weighed up to 48 tons. Its remains have been found at the Fossil Hill site in Nevada, as well as at sites in central Europe and Spitsbergen. Only 4 million years later, we see even bigger forms like *Shonisaurus* reaching proportions almost equivalent to those of the massive blue whales of today. While articulated skeletons show sizes around 69 feet, isolated limb bones from the same group scale them up to a staggering maximum length of 85 feet. We know from their stomach contents that these giant ichthyosaurs were eating mostly cephalopods like squids and ammonoids. Some, like *Himalayasaurus,* bore huge teeth, so they were clearly predatory and chasing after large prey like other large reptiles, sharks, and other fishes. Their huge body mass and feeding capability would certainly have impacted sharks, as cephalopods were likely the regular food source for the vast majority of sharks living at this time.

The giant shashtasaurid ichthyosaurs peaked at close to 85 feet long, and competed with sharks for squid-like prey   ARTWORK BY BRIAN CHOO

The other new reptilian kids on the oceanic block were the **plesiosaurs**. These long-necked paddle-limbed marine reptiles had a small, rounded head, instead of the dolphin-like beaky snout of the ichthyosaurs. They were well adapted to eating small fishes, by virtue of their curved, spindly teeth and small heads. This would have enabled them to sneak within striking range before the fishes knew what was coming at them. Plesiosaurs like *Bobosaurus* from northern Italy entered the seas in the latest Triassic with small forms, reaching around 10 feet long. They were quite rare in the Triassic seas and would rise to prominence in the next period, the Jurassic.

How did sharks respond to this competition toward the end of the Triassic? A surviving jalodontid shark from the Late Triassic of England, *Keuperodus,* shows that the group made a move from their ancestral habitat of deep-water open ocean into freshwater river and delta environments. These were small sharks, probably under 2 feet long. By moving out of the crowded and more threatening deeper waters, they got the hell out of Dodge, just as they'd done to escape the Great Dying. Strange as it may seem, this was the only time in the entire history of sharks when most sharks in our oceans were small, benign species, and the absence of large, voracious predatory forms was the norm.

While all these marine reptile groups became established at the top of the food chains of the Middle to Late Triassic seas, another small, almost inconspicuous group of sharks called neoselachians was emerging. They would eventually diversify and then take over the niches occupied by the now abundant hybodont sharks. It is exciting to learn about the neoselachians, as they are the basic stock that gave rise to all modern sharks and rays. They tell us how modern sharks became successful primarily because of subtle changes in their teeth.

## Rise of the neoselachians

As they did in the Devonian, sharks are now about to reactivate their superpower of making new types of teeth with improved new structures that will give them a big push forward in their evolutionary trajectory. Modern sharks and rays (elasmobranchs) are all survivors from a new diversification of sharks that started in the Triassic period when the first neoselachians appeared. Conclusions about these first fossil neoselachians are based entirely on fossil teeth, and it's in the structure of these teeth that we find the first major improvement in sharks' teeth since the first Devonian holocephalans invented their powerfully hard crushing teeth (and by doing so diverged from the line leading to modern sharks). Sharks are about to cash in big-time on their dental plan yet again.

The oldest neoselachian teeth were at first thought to belong to hybodonts because their teeth look exactly alike. Back in 1836, the Swiss paleontologist Louis Agassiz (see bio on p. 432) described the Late Triassic *Hybodus minor* based on teeth and fin spines found along the Somerset coast sites of England. It took another 157 years before a detailed study of these teeth by the British researcher Chris Duffin (see bio on p. 423) showed it to have a distinctive three layers of enameloid. This is one more layer than is found in hybodonts or any earlier sharks—

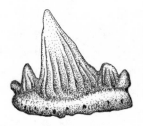

Tooth of *Rhomphaiodon,* the
oldest known member of
the modern shark group, the
neoselachians. It is 0.13 inch
wide    ARTWORK BY AUTHOR

a feature that defines the neoselachian shark group. He renamed it
*Rhomphaiodon.*

The tooth is up to one-third of an inch high, so *Rhomphaiodon* was
a small shark, perhaps up to 3 feet in length. It is very similar in shape
to the teeth of some hybodonts, so it probably lived its life in much the
same way as they did, feeding on a wide range of prey from crabs and
shellfish to small fishes, with the added benefit of being equipped with
slightly stronger teeth. These stronger neoselachian teeth signal a shift
back to an earlier lifestyle that sharks were always good at—being
predators—but now with new added tearing, slicing, and dicing teeth.

Some new work on the Early Triassic shark faunas from Oman led
by Martha Koot from Plymouth University, UK, challenged the pre-
vailing idea that neoselachians were rare in the Triassic period. The
marine faunas from Oman contained a surprisingly higher number of
the allegedly rare neoselachians, outnumbering the common hybodonts. The faunas were dominated by a particular neoselachian
group called **synechodontiforms**, after the common genus *Synechodus.* These sharks had very sharp teeth made up of a central flattened
cusp forming a curved triangular blade, with dagger-like cusps flanking
it. The teeth were tiny (a tenth of an inch), yet they still signify a major
change in the direction of shark evolution: that neoselachians had
started to increase in numbers.

In the Middle Triassic (247–237 MYA), more neoselachian sharks
emerged, sowing the seeds for some of the first of the living shark family groups to appear. One of these new kinds of small predatory sharks
developed an entirely new way to feed on the flesh of larger fishes and
marine reptiles. *Pseudodalatias* got its name from its resemblance to
the living kite shark, *Dalatias,* because its teeth are very similar. *Pseudodalatias* tooth fossils are easily recognized, as they have a distinctive
set of fused triangular teeth attached to a root base that forms the en-

tire lower dentition as one unit. The teeth are curved and deadly, exactly like those of the living cookiecutter shark (*Isistius brasiliensis*), and the resemblance didn't end there. *Pseudodalatias* most likely fed in the same way, cutting plugs of flesh out of its ambushed prey. Typically, it might have attached itself with its teeth to a large fish or marine reptile passing by. It would then proceed to cut out a plug of flesh by rotating itself around the wound, digging in with the jagged lower jaw and working it like a can opener.

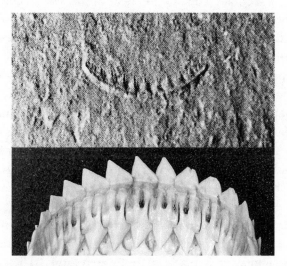

*Pseudodalatias,* an early cookiecutter-type parasitic shark (above); kitefin shark jaws (below) NATURAL HISTORY MUSEUM, LONDON

Just as the neoselachians were taking off, their world was about to get a massive shake-up, when life in the seas would face another round of utter chaos.

## Another mass extinction shakes up sharks

The end-Triassic extinction event was caused by another massive pulse of volcanic eruptions. Known as the Central Atlantic Magmatic Province, this cataclysmic sequence of eruptions saw voluminous amounts of lava spewing forth, raising global greenhouse gases to super-high levels. This caused extremely elevated global temperatures, widespread global fires, and warming seas. And when I say warming, the oceans not only became so hot that they cooked the clams but they became so acidic that they also dissolved their shells. On land we witness the demise of many archaic reptilian groups (phytosaurs, procolophonids, tanystropheids, aetosaurs, and others). We also see the decline of most of the world's giant freshwater amphibians, like *Paracyclotosaurus*—the crocodile-like forms called **temnospondyls**.

About as many groups of marine invertebrates were wiped out as had been eliminated in the Great Dying, including certain families of ammonites and corals. The development of the world's coral reefs was hindered or interrupted for tens of millions of years. This was no doubt also brought about by increased acidification of the oceans, an inescapable phenomenon that devastates all creatures relying on calcium carbonate deposition for their shells or hard frameworks. Conodonts, a previously abundant group of worm-like marine protovertebrates that had populated the world's oceans for the previous 300 million years, also died out at this time. All told, at the end of the Triassic, some 76 percent of all species on land and in the seas were wiped out.

Even with so many prey items now off the menu, the oceans were taken over by more kinds of monster marine reptiles, so sharks needed all their evolutionary advantages and talents for survival. Sharks were now good at surviving global mass extinctions; this was their fourth one. It had little effect on the sharks living in the oceans and seaways but had a significant impact on freshwater forms. The reason that ocean sharks survived this event so well is simple: Most of them were small, and both major shark groups (hybodonts and neoselachians) were generalist feeders. The most vulnerable species to extinction are those that are highly dependent upon just a few major food sources. If those food sources suffer a rapid decline, then the species dependent on them can also go extinct. There were likely very few sharks in the entire Triassic that were specialists feeding on just a few species of prey. That is why they survived.

But in the freshwater, the xenacanths, the major group of sharks that had ruled the rivers, lakes, and estuaries for the past 100 million years, were lost to this extinction. Other groups of sharks that didn't make it through this boundary include forms that existed much earlier in the Triassic, like the straggler buzz-saw sharks, which succumbed around 247 MYA. Overall, sharks and bony fishes suffered little during this extinction event, compared to the hammering suffered by the shallow-water marine invertebrate communities.

By the time of the end-Triassic extinction, the many marine reptile groups that had quickly diversified throughout the period had been filtered down into a few major groups, leaving mainly the ichthyosaurs and plesiosaurs. From the start of the Jurassic (201 MYA), they rapidly began diversifying to become the top predators of the marine food chains. The strange stumpy-toothed placodonts did not make it through, nor did the two families of gigantic ichthyosaurs like *Shon-isaurus*.

You could say sharks thrived under pressure, for the next chapter of their evolutionary history tracks how sharks went on to even greater evolutionary success. The coming Jurassic is a landmark time in the history of shark evolution, because it's when most of the modern shark groups, including some genera alive today, start appearing and diversifying in our oceans. Yet Jurassic sharks would still not have it easy.

## CHAPTER 10

# THE JURASSIC RISE OF MODERN SHARKS

## Little Sharks Hold Their Own

---

### Solnhofen Archipelago, 150 million years ago

The temperate archipelago seas are bursting with life. Small patches of pink and pale green coral reefs stand out between the rounded domes of gray-green sponge mounds. Flower-like sea lilies wave their arms in the currents to sift out tiny plankton. Small worms, crabs, and snails feed on and around the reefs, with tiny shrimps darting in and out of the dead coral canyons and bigger lobsters with wide flat shells hiding in small caves at the base of the reef base—it's like some sort of fantasy underwater Garden of Eden.

Not far away on the seafloor, an angel shark, *Pseudorhina*, waits hidden, perfectly camouflaged under the clean white silts, waiting for its next meal to come along. A small fish swims slowly above it, completely unaware of the hidden predator who senses its presence from the fish's weak electric fields. In a lightning move, it darts up, opens its wide mouth lined with hundreds of tiny triangular teeth, and sucks the helpless fish into its gullet. A billowing white cloud of fine mud now marks the crime scene.

Schools of long gar-like fishes shimmer in the dappled sunlight as they quickly swim after a large school of smaller fish in the surface waters. From the sky above, a leather-winged *Pterodactylus* swoops down and grabs one of the small fishes in its toothy long beak, then makes a beeline for the safety of the skies. Out of the midwater haze emerges a 15-foot-long marine crocodile, *Dakosaurus*. It is fully equipped for the seagoing life with its streamlined body, four limbs developed as powerful paddles, and long muscular tail ending in large, forked flukes to push it swiftly through the water. Its head is deep rather than flat like a modern crocodile's, and its jaws are lined with strong, sharp conical teeth. It has salt glands in its palate that allow it to drink seawater and excrete the salt, so it never has to go onto land.

A small bullhead shark, *Heterodontus*, a near mirror image of the living form, cruises slowly along the side of the reef, probing its fat face into every dark hole it can find. Suddenly a spot inside its brain tingles as it senses the electric field of a lobster that thinks it is safe tucked behind a tangle of dead coral. The shark goes to work applying its strong plier-like jaws to bite away the dead coral, spitting out mouthfuls of broken pieces as it pushes deeper into the lobster's hidey-hole. With one last effort it thrusts its head into the far reaches of the hole and plucks out the struggling crustacean with its pointed front teeth, moving it quickly to the back of the mouth, where its domed teeth easily split open the shell like a nutcracker and it begins to feast on the tasty flesh.

*Dakosaurus* spots this drama going on below and swoops down to grab the bullhead shark in its long jaws and swallow it in one big gulp. Suddenly it opens its mouth again and coughs out the shark, as it doesn't like the intense pain caused by the little shark's saber-like dorsal fin spines as they pierce its palate. The bullhead shark, only slightly rattled, settles back down to the seafloor, content to rest and digest its tasty meal. *Dakosaurus* then spots a large tuna-like fish about 6 feet long, *Caturus*, and begins chasing it. The pair quickly disappear into the distance.

Swimming out of the gloom at a fast pace is *Sphenodus*, the top shark predator in these seas. This one is an old, large male, about 10 feet long, and its wide head bears a very large mouth bristling with slender, razor-sharp, blade-edged teeth. It is on the hunt for a big bony fish, *Thrissops*, resembling a modern tarpon, which unknowingly swims toward the big shark poised below, still, waiting for the precise moment to launch its attack. With an explosive thrust of its tail it bolts upward to the surface and seizes the fish in its mouth, then bites hard to break the fish's body in two, swallowing one half, then seconds later turning around to clean up the other part. Satisfied with this meal, it swims away slowly.

About three hundred feet away, a school of squid-like *Belemnotheutis* is being hunted by a shoal of large predatory gar-like fishes. Another large shark, about 6 feet long, a hybodont called *Asteracanthus,* sits on the seafloor among the sponge mounds and beds of seaweed. It is not interested in the fishes; it has the group of *Belemnotheutis* firmly in its sights. It emerges from its weedy shelter and in a fast burst shoots toward them and seizes a mouthful of the squid-like delicacies, crunching down hard with its powerful jaws and crushing teeth. Its teeth are perfect for this particular meal, as the *Belemnotheutis* have a hard, calcified rod supporting their bodies that would shatter the teeth of lesser sharks.

Meanwhile, back on the coral reef, a group of catsharks, *Bavariscyllium,* each about a foot long, are becoming active as the evening settles in, while a guitar-shaped ray, *Spathiobatis,* starts moving slowly along the seafloor in search of buried creatures hiding below the surface. The catsharks dash in and out of the reef, busily on the hunt for small crustaceans and little fishes hiding between the branching corals.

On the surface of this blue-green warm sea, a strange decaying carcass floats along and drops slowly to rest on the clean, limy seafloor. It is soon investigated by a curious large shark, a streamlined form with a flat head adorned with a set of fleshy tendrils surrounding its mouth, much like today's wobbegong shark. The carcass belongs to a little dinosaur, *Compsognathus,* a member of the ruling group of land animals at that time. Without hesitation, the shark, a *Palaeocarcharias,* takes what's left of the rotting dinosaur in one gulp and swims away as the sun begins setting on this beautiful scene.

## The incredible sharks of Solnhofen

This scene represents a major milestone in our story. It is the first time we encounter a diverse fauna of sharks living in one place where most of them belong to groups that are still alive today, with just a few archaic forms, like the hybodonts, still hanging around. We are lucky to be able to study these ancient Solnhofen sharks as complete specimens rather than isolated teeth, because these fossil sites in southern Germany are the best anywhere to study sharks of this age (153–150 MYA).

Let's find out why these sharks are so well preserved, and then learn what these fossil sharks can tell us about the rise of modern sharks.

The Solnhofen Limestone of southern Germany has since Roman times yielded stone for making tiles or constructing buildings. The layers preserve a truly beautiful pageant of ancient life that once lived on or around a sparkling island archipelago at the edge of the Tethyan seas. These sites are another set of *lagerstätten,* for they store great numbers of extraordinarily complete fossil organisms. The Solnhofen sites in the northeast represent an ancient archipelago with complex lagoon environments containing stagnant salt-rich bottom waters. The Nusplingen site about eighty miles away is slightly older and represents a restricted lagoon setting with water probably less than 330 feet deep. This last site has yielded more than four hundred different species of animals and plants, making it the richest Jurassic fossil site in the world.

The preservation of the fossils began when the lagoons became extremely salty with waters periodically starved of oxygen, helping preserve the carcasses that fell into the lake and were buried by the fine limy sediment. These shallow lagoons sometimes dried up, so at times the sticky mud would be exposed to trap insect and plant debris. The flow of new sediments then buried the remains, preserving them whole and preventing them from being scavenged by other organisms. The Solnhofen site is most famous for having given us the world's oldest bird, *Archaeopteryx,* essentially a dinosaur adorned with feathers, along with many species of pterosaurs, marine reptiles like *Dakosaurus* and ichthyosaurs, dinosaurs like *Compsognathus,* nearly a hundred different species of bony fishes (like *Caturus* and *Thrissops* in the scene), marine invertebrates like lobsters, crabs, shrimps, snails, crinoids, and many other species, as well as some terrestrial insects and plants that washed into the sea. These lagoons were packed full of tasty creatures that sharks liked to eat, such as shrimps, lobsters, and a huge number of small bony fishes. More than 750 species of plants and animals have now been recorded from all the sites in the Altmühltal Formation (previously known as the Solnhofener Plattenkalke).

Until recently, only a few species of sharks from the Solnhofen sites had been studied in detail, but a recent review led by Eduardo Villalobos-Segura of the University of Vienna, aided by a crack team of eminent shark paleontologists including Jürgen Kriwet, have revealed that some thirty-two species in twenty genera of sharks, rays, and chimaerids lived here, with the majority of the sharks being neoselachians, representing modern groups. Some of the earliest studied key specimens, like the chimaerid *Ischyodus egertoni*, were unfortunately lost during the bombing of the Bavarian State Collection of Palaeontology and Geology in Munich during the last week of World War II. Fortunately, more recent collecting has turned up new, complete examples of all the lost species, as well as revealing many new species that are still waiting to be formally studied.

Let's take a quick look at just a few of the most significant sharks from this amazing site, particularly the ones featured in our scene above.

Solnhofen has given us the oldest known complete angel shark, *Pseudorhina*. Its body shows the same shape seen in today's angel sharks like *Squatina*. These are fearsome ambush predators that lie in wait on the seabed, snapping upward in a lightning-fast action to grab passing prey by rapidly opening their mouth, using powerful suction to draw the unsuspecting prey into their gullet. *Pseudorhina* grew to about 3 feet and has large, rounded pectoral fins that meet the elongated round pelvic fins to give it a kite-shaped body form. The body and head are very flattened. Its teeth are very small, and unlike the modern angel sharks'; they are more like those of a wobbegong shark.

TOP: an ancient bullhead shark, *Heterodontus*. BOTTOM: the oldest possible lamniform shark, *Palaeocarcharias*. RIGHT: an early complete ancestral angel shark, *Pseudorhina*. All from the Late Jurassic Solnhofen site in Germany IMAGES SUPPLIED BY JÜRGEN KRIWET

Today's bullhead sharks are mostly small sharks living in warm waters above 70°F in temperate to tropical shallow seas. At Solnhofen we see the earliest fossil record of the living *Heterodontus*, looking almost exactly the way it does today. *Heterodontus zitteli* differs from modern **heterodontiform** species in very minor respects, like the shape of some of its teeth. Like the modern bullhead sharks, these fossil Teutonic bullheads were well adapted to eat a wide range of prey, hunting along the bottom of the reefs and seafloors for lobsters, crabs, shrimps, clams, and other invertebrates.

While many of the Solnhofen sharks fit surprisingly neatly into families of sharks alive today, some others are more problematic. Complete specimens of the predatory *Sphenodus* are also found here. John Maisey has informed me that their distinct pattern of lower gill arch bones suggests that they are both **galeomorph** sharks—the overarching group comprising four major living shark groups (orectolobiforms, heterodontiforms, lamniforms, and carcharhiniforms). A large incomplete specimen shows that *Sphenodus* possibly reached 10 feet in length. Its streamlined body shape suggests that it was a fast-swimming, active predator, one of the top apex shark predators in this tropical ecosystem. It likely preyed on bony fishes, squids, other sharks, and marine reptiles.

A closely related form to *Sphenodus* also found at Solnhofen, *Paraorthacodus*, tells us something special about the mating habits of these sharks. It was a chunky-looking shark with a wide head and one large dorsal fin near the tail fin, similar to the modern sixgill sharks. Interestingly, the specimens are well enough preserved to detect subtle but distinct differences in the teeth of the males and females. Male *Paraorthacodus* have an extra pair of smaller cusps on the side of the main tooth, a feature lacking on female teeth. This tells us that the male sharks were probably biting the females on the pectoral fins during mating. We find similar sexual distinctions in the teeth of some living sharks for exactly this reason. The male must grab on to the female to get purchase while copulating. Not only are the teeth different, but the skin of the female shark is often thicker than the male's to protect her fin while the male is biting her.

Three different hybodonts are found at Solnhofen, but we will look at the one depicted in the scene, *Asteracanthus*, as it tells us a lot about the group. While several specimens of this shark were previously known, an exceptionally fine new one was recently studied by Sebastian Stumpf of the University of Vienna and his colleagues. It shows the jaws and all the teeth arranged in neat files and rows in perfect life position. The specimen has an orange-colored body and fin outline contrasting clearly with the light gray limestone. It's reminiscent of a tragic ghost shark, twisted in death yet frozen forever as an exquisite natural artwork, a true expression of life's deep-time beauty. This female is almost 6 feet in length, and she's missing only the tip of her tail fin. The teeth of this *Asteracanthus* display both a grasping style of dentition at the front of the mouth and lower, wider teeth filling out the rest of the jaws. The multiple rows of teeth were capable of biting all at the same time, something modern sharks do not generally do, much to the relief of surfers. The squat shape of the body and fins reinforces that it was a sluggish swimmer, adapted for cruising just above the seafloor. Large, isolated teeth from other sites suggest *Asteracanthus* might have grown up to nine and a half feet long. We know that one German *Hybodus* specimen had a bellyful of belemnite fossils, so I've used this information to reconstruct its closely related kin here

Nearly complete hybodont *Asteracanthus* from the Late Jurassic Solnhofen site in Germany, with a restoration of the shark below. Scale bar is approximately 20 inches IMAGES COURTESY OF JÜRGEN KRIWET, WITH ARTWORK BY FABRICIO DE ROSSI

doing a similar thing. One *Asteracanthus* found in Britain also showed an association of its teeth preserved within a fossil reptile carcass, suggesting it might also have been a scavenger.

One of the most interesting sharks from the site depicted in our scene is *Palaeocarcharias,* one of the oldest possible **lamniform** sharks, as shown by its special osteodont tooth structure. Lamniforms include living white sharks, thresher sharks, makos, and some other forms. *Palaeocarcharias* was first identified from a single tooth from England in older rocks dated at about 167 MYA. The complete remains of it are quite rare at Solnhofen. At around 3 feet long, *Palaeocarcharias* had a streamlined body shape with large, rounded pectoral fins. Its broad head was adorned with fleshy barbels and lobes coming off the mouth area, like the living wobbegongs—a feature not seen in any other known lamniforms. Its body was perfectly suited for a predatory lifestyle, its teeth adapted for clutching slippery fishes or squids. Solnhofen incidentally also has the oldest complete member of the wobbegong shark group, the **orectolobiforms**, represented by a small bottom-dwelling flat shark called *Phorcynis.*

The **carcharhiniform** sharks today are a very diverse group of 291 species in ten families, including the ground sharks, reef sharks, some catsharks, hound sharks, hammerheads, and tiger sharks. Carcharhinids from the site include the oldest complete catshark (family Scyliorhinidae) known in the fossil record, the little *Bavariscyllium,* which was about 8 or 9 inches long. It probably lived close to the bottom,

*Bavariscyllium* **was one of the oldest carcharhinid sharks in the world, a little catshark that lived on the Solnhofen archipelago. This one measures just over 8 inches long** IMAGE COURTESY OF JÜRGEN KRIWET

searching the nooks and crannies of the reefs and lagoons for small fishes and shrimps, as do living catsharks.

The Solnhofen neoselachians also include the earliest well-preserved sixgill sharks like *Notidanoides,* a form very close to the living *Hexanchus.*

The last group of elasmobranchs known from Solnhofen are the fossil rays. I have saved these for the end of this chapter, as they deserve special treatment. Without doubt the biggest event in shark evolution for the entire 186-million-year Mesozoic era occurred in the Jurassic, when certain flattened sharks appeared on the scene—the first rays; but more about that remarkable evolutionary transition later.

The Solnhofen fossils are a rare window into a diverse fauna of sharks living in the closing few million years of the Jurassic period. For the first time, we find the seas dominated by elasmobranch sharks, forms that are directly on the line to, or include, some early living sharks. We find that most were specialist predators, rather than generalists like the few hybodonts living in the same seas.

Now we need to find out the missing middle part of the shark story: How do we get to a Late Jurassic sea (150 MYA) full of modern-type elasmobranch sharks from a Late Triassic sea (202 MYA) dominated by archaic hybodont sharks? We need to jump back to the beginning of the Jurassic to take stock of what kinds of strange and unusual sharks were around, and the competition they were up against immediately after the seas were battered by the chaotic end-Triassic extinction event. Then we will hear about another environmental crisis taking place in the oceans around the middle part of the Jurassic period that would drive a monumental shift in shark faunas and change the oceans forever.

## Hybodonts at the top of their game

We saw in the last chapter that hybodont sharks were the most abundant sharks of the Triassic period. They continued to dominate in the Early to Middle Jurassic seas and rivers as the neoselachian sharks slowly gained ground. Let's take a brief look at the continued success of the hybodonts and hear an interesting snippet of their history involving a well-known amateur paleontologist named Mary Anning.

The area around Lyme Regis, in Dorset, England, is famous for its range of Jurassic and Cretaceous fossils, including sharks. It was here that the renowned fossil hunter Mary Anning found the first well-preserved fossils of marine reptiles, the ichthyosaurs and plesiosaurs, back in the first decades of the nineteenth century.

Mary was born into a struggling family and was raised in Lyme Regis. Her parents had had many children, but only Mary and her brother Joseph survived to adulthood. When they were young children, their father, Richard, would take them out on walks along the coast to search for fossils. Beautiful ammonite shells could be found by cracking rocks on the beach. Occasionally, fossil fishes or bones of ancient reptiles could be collected from the cliffs. Fortunately for Mary and her family, there was a market for such unique natural specimens. After her father passed away, the family lost their main source of income, so Mary began seriously collecting fossils to sell to museums and collectors. With a keen eye and her early years of experience, she was able to help her family maintain a living. Many of her prize specimens were sold to the newly formed British Museum, including excellent examples of the Jurassic fossil sharks *Hybodus* from the lower parts of the coastal cliffs.

Some people regard Mary as just a very talented fossil collector and preparator. She is discounted by some people as not being a "real paleontologist" because she had no formal education in science. I think that anyone who can make a major contribution to paleontology, through collecting new significant fossils or doing research into fossils, can rightly call themselves a paleontologist, even without a university degree. There are several examples in this book of untrained fossil shark researchers who have made enormous scientific contributions to our understanding of shark evolution. Anyone who can publish new paleontological research work in a respected scientific journal can certainly call themselves a paleontologist. And Mary did. In 1839 she published her only scientific article, a sole-authored paper in the *Magazine of Natural History,* an important contribution to our knowledge of the fossil shark *Hybodus*. In honor of her tremendous contributions to the science of paleontology through the many significant new fossils she found, in 1843 she was honored with a new fossil hybodont shark called *Acrodus anningae.*

*Acrodus* (meaning "high tooth") was one of the most abundant hybodonts, ranging from the Early Triassic to the Early Cretaceous. It bore bulbous crenulated teeth at the front of its mouth coupled with very elongated, grinding teeth adorning the rest of its jaws. A closely related form called *Polyacrodus* (meaning "many high teeth") was common in the Jurassic of England and Europe. Some of its back crushing teeth are enormous, as seen by specimens in the Natural History Museum measuring around 4 inches long, suggesting that the creature spanned 8 to 10 feet long, making it one of the biggest of all hybodonts.

Hybodonts reached their peak of diversity in the Middle Jurassic, about 170 MYA, when thirty different species thrived in a range of shallow marine and freshwater habitats. These Jurassic hybodonts had not been idle. They refined their feeding style, increasing their ability as opportunistic feeders that could eat a wide variety of prey items rather than narrowing their options and becoming too specialized. They achieved this by evolving teeth that were more highly crowned, sharper, and pointier along with combinations of differently shaped side and back teeth. Such complex combinations facilitated a more predatory diet than their Triassic forms.

## Jurassic sharks and the rise of marine reptiles

While hybodonts were thriving at this time, a new reptilian menace began appearing in the Jurassic seas to create even greater competition for sharks. From the start of the Jurassic, these reptiles rapidly began diversifying to become the prevalent top predators of the marine food chains. Early Jurassic ichthyosaurs were now mostly smaller fish-eating forms around 6 to 10 feet long, with a few notable exceptions, such as *Temnodontosaurus* at up to 33 feet long. It inhabited the deeper, open marine parts of the oceans. The first long-necked plesiosaurs were of moderate size, most under 17 feet long, yet they became very abundant as they diversified in the Early Jurassic seas. Imagine a long-necked serpent-like beast with a small head like a lizard, a stout body, a short, pointed tail, and four large flippers, propelling itself gracefully through the water pursuing a school of fish. These animals were not much of a threat to sharks, except that they may have occasionally preyed on their young. In the Middle Jurassic, plesiosaurs evolved into another group, called **pliosaurs**, characterized by their shorter necks and much larger and longer crocodile-like heads armed with banana-sized teeth. By the end of the Jurassic, these beasts would reach significant sizes, like *Lio-*

*pleurodon,* which at 22 feet long was the apex predator of the Middle to Late Jurassic northern Pangaean seas.

Pliosaurs like *Liopleurodon* were formidable predators. The long-necked plesiosaurs like *Plesiosaurus* ate small fishes    ARTWORK BY NOBU TAMURA

With the continued rise of these formidable marine reptiles in the Jurassic seas, you might think sharks would surely undergo a corresponding decline in abundance and diversity. Yet, surprisingly, the opposite occurred. The two main groups of sharks around then, hybodonts and neoselachians, were well enough equipped to deal with any challenge. The neoselachians, armed with their newly developed strong and variable teeth, would rise to the challenge by rapidly evolving into many new lineages, most of which would be successful, survive another mass extinction event, and carry on to modern times.

How do we know which sharks survived this time without being threatened by the giant marine reptiles? The answer lies in finding out exactly what the reptiles were eating.

A study looking at four medium to large Jurassic marine reptiles (an ichthyosaur, two regular long-necked plesiosaurs, and a large short-necked pliosaur) found in Wyoming revealed they were far less interested in sharks than in other kinds of food. All of them had stomach contents well preserved, with squid-like belemnites making up the primary food source of all four species. One of the plesiosaurs in the study, *Tatanectes,* included a specimen with hybodont shark teeth and denticles in its gut, suggesting they occasionally preyed or scavenged upon these sharks.

## Neoselachians pushing the limits

We saw in the last chapter that the neoselachian sharks were fairly rare throughout the Triassic. They began increasing steadily by the end of that period, but most of these creatures from the Early Jurassic are known only by teeth. However, one site in northern Italy revealed a complete specimen of one such shark, showing a totally unexpected body shape not mirrored by any living shark. Let's head to Italy to learn what this unbelievable fossil shark find tells us about how some sharks began hunting in strange new ways.

*Ostenoselache* was a small shark named after the town of Osteno in Lombardy, where it was discovered along the picturesque shores of Lake Lugano. Fossils from this site are well preserved, representing complete bodies of sharks and fishes that lived in an enclosed marine bay. Their carcasses sank to the bottom and were buried and rapidly fossilized sometime around 198 MYA. They have been collected there since 1964, and today some forty-seven examples of this strange shark are known in varying states of completeness.

*Ostenoselache* has a very long, pointed snout, like a shark trying to be a swordfish (see restoration on p. 195). It was a small neoselachian, a tad under a foot long, with an elongated body, a whip-like flat tail, and a very long anal fin, similar to the ancient xenacanths. It is so well preserved that some specimens even show soft tissues, including the stomach and the coiled spiral-shaped intestine. Its mouth was filled with tiny spindle-like teeth around one-twentieth of an inch or so in length, adapted for clutching its prey.

The most intriguing thing about *Ostenoselache* is that its body shape closely resembles that of the electric knifefishes of South America, called **gymnotiforms**. Its elongated body shape and long anal fin are seen today only in these electric fishes, so Chris Duffin, who studied it, suggested it also had similar capability to generate electric pulses to either stun its prey or locate prey in dim waters. While the Osteno fossil deposits are thought to be marine, they also record periodic influxes of a large amount of terrestrial plant material. It is quite possible that

*Ostenoselache* lived primarily in freshwater and its carcass was washed into the marine embayment where it was finally buried.

The fossil remains of some large predatory crustaceans found at the same site show the remains of *Ostenoselache* vertebrae in their tiny gut cavities. Perhaps *Ostenoselache* were regularly washed into the marine embayment, where they died and the crustaceans feasted upon their decaying corpses. However, there is one other possible interpretation: *Ostenoselache* might have been an active marine shark with electrolocator properties, used to locate prey or detect predators in murky waters rather than hunting with strong electrical discharges. In such cases it could have generated weak electric pulses in dim waters to locate its prey at special times when salinity levels were extremely low in the bay.

We may never know the answers to how *Ostenoselache* lived and hunted, but it remains a truly intriguing shark, a weird Jurassic experiment, perhaps the only shark ever to use electrical discharges for hunting or stunning prey. Living electric skates can also do this, but they would not appear in our seas for quite some time yet.

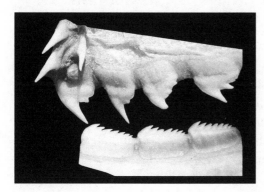

Some sharks have differently shaped teeth in the upper and lower jaws to help them cut chunks of flesh off their prey, like this sixgill shark, *Hexanchus* PHOTO BY ROSS ROBERSTON, SHOREFISHES OF THE TROPICAL EASTERN PACIFIC ONLINE INFORMATION SYSTEM

While *Ostenoselache* represents an odd neoselachian that was a dead-end line with no living relatives, other neoselachians appeared in the Early Jurassic that include the first member of a still-living genus of shark—the sixgill shark *Hexanchus*. Hexanchiform sharks today include the most primitive of all living sharks, such as the frilled sharks and the sixgill and sevengill sharks called **hexanchids**. Today *Hexanchus* is known by three living species including the widespread bluntnose sixgill, *Hexanchus griseus*, which commonly inhabits the seas between 700 and 3,500 feet deep. Growing up to 16 feet, it hunts squids, bony fishes, and other sharks and rays, while the biggest individuals are known to eat seals and feed on whale carcasses.

*Hexanchus* first appeared in the shallow continental seas that once covered the now mountainous Switzerland, about 199 MYA, based on very distinctive teeth named *Hexanchus arzoensis.* The teeth of all sharks in the hexanchid family show long, serrated jagged lower teeth matched against shorter, hooked upper teeth, allowing these sharks to tackle a huge range of prey, as discussed earlier. Because *Hexanchus* teeth have changed little since the Jurassic, we can assume that the extinct species fed on similar prey, with larger, more feisty forms possibly taking on the larger marine reptiles.

So how did the neoselachians eventually take over the seas from the predominant hybodonts and other archaic forms? A massive environmental crisis in the middle of the Jurassic gave them just the opportunity they were looking for.

## A Jurassic crisis changes the game for sharks

The Toarcian crisis struck the Jurassic oceans as a devastating two-phase event. The first phase occurred at around 182.7 MYA, when oceans in the northern part of the newly formed Tethys Ocean and around Spain cooled and shrank, drying up the shallow shelf seaways. This resulted in many localized extinctions of common marine organisms like microscopic foraminifera (shelly single-celled protozoans), brachiopods and clams, ostracod crustaceans, ammonites, and belemnites. Immediately after this initial crisis, the first kinds of living bullhead sharks appear. These are small sharks, under 5 feet long, today represented by nine living species of *Heterodontus* found only in the Pacific and Indian oceans. Fossil teeth of the extinct form, *Paracestracion,* found in Belgium, mark the origin of this iconic group at this time.

A second, more deadly phase of the crisis occurred 174 MYA, known as the Toarcian oceanic anoxic event. Around 5 percent of all marine animal families were lost, with some localized areas having around 50 percent of their marine invertebrate species going extinct. The event was probably triggered by massive volcanic eruptions that caused a huge release of methane hydrate, a greenhouse gas, from the seafloor, possibly released from coal deposits heated by rising magma.

The fatal effect of all these greenhouse gases accumulating in the atmosphere and oceans was the sudden warming of sea temperatures by around 13°F, causing widespread depositing of sediments rich in organic matter that further sucked oxygen from the seawater. This caused a drastic reduction in the populations of some abundant marine species like the belemnites as well as localized extinctions of several bottom-dwelling invertebrates like clams and brachiopods. Many other marine species suffered declines in abundance too. All this caused a significant reduction in the biomass of food resources for sharks. The drastic decline in belemnites, one of their main food sources, would certainly have affected the hybodont sharks.

Charlie Underwood, from Birbeck College in London, has studied Jurassic sharks and what these Toarcian events meant for shark evolution. I sat down with him one day at the Natural History Museum in London to have a chat about it over a coffee. Charlie is a quiet-spoken but wonderfully enthusiastic scientist. He told me that immediately after the Toarcian crisis, the neoselachian sharks clearly underwent a turning point, when lots of new lines of sharks began rapidly diversifying into new groups of species. I can see now that the proof of this idea is the twenty different neoselachians found at Solenhofen, and the many, many more species known just by their teeth. Yet the turning point for neoselachian sharks wasn't just the dramatic increase in their diversity, it was also the overall rise in their abundance. This larger variety of species could now occupy new ecological niches. Why? Because there was more to eat. Their rise coincides with the rise of the dominant bony fish group, the **teleosteans**, and with the change in abundance of marine invertebrate groups, both supplying new sources of food following the major upheaval by the Toarcian crisis. Many of the new Toarcian neoselachians lived in open ocean environments. We do not yet have the full story regarding what the sharks living in shallow-water habitats were doing at this time.

Let's now turn to the grand finale of the Jurassic, the most significant event in shark evolution since a radical group with a new type of tooth split off the main line of sharks to become the holocephalans about 380 MYA. In the Jurassic, another group of sharks broke away from the mainstream line of neoselachians that represent the modern shark families and invented a completely new body plan that enabled them to swim in a new, more efficient way. This new group would ultimately become the most successful group of "sharks" ever—in fact they are now so different from what sharks look like today that we call them by another name: the rays.

## Dawn's early rays

Some years ago, when I was taking a vacation with my three young kids, we arrived at the scenic waters of Hamelin Bay in the south of Western Australia. We went for a walk along the clean sandy beach, taking in the vista of the deep blue Indian Ocean shimmering in the spring sun. Then we noticed something odd. A man was standing in the shallows, in about a foot of water, holding a fish in his hand just above the surface. I could see a large black shadow moving slowly through the water toward him; he saw it too, yet he didn't seem worried. Once the shadow reached his feet, a large black thing reared itself out of the water and we saw that it was an immense stingray. It gently plucked the fish from the man's hand and swam away.

We watched in total silence, awestruck by this amazing sight. I'd never seen a tame stingray before and was simply gobsmacked by just how gentle and poised the big ray was, elegantly arching itself up out of the water like a ballet dancer. It was a Smooth Ray, *Dasyatis brevicaudata*, which grows to about three feet across the wings. We then noticed a number of them scattered around the bay, mostly lying still, with a few slowly milling around in the water, waiting for something. Eventually more people came down to the beach carrying fish and began feeding the eager rays. Each one gently approached their human and delicately took the morsel on offer then swam away to somewhere more private to quietly savor its meal.

We waded into the water and were astounded to see that the rays were not afraid of us, but accepted us, unperturbed by our curiosity. Today the friendly stingrays of Hamelin Bay are internationally famous. The site has become one of the top tourist attractions for the state. However, the interaction between rays and humans has not always been strictly friendly here. One of them, nicknamed Stumpy by the locals, used to be known as the "friendliest ray in the bay." In 2011, Stumpy was caught and killed by fishermen (legally) and brutally hacked to pieces in front of families that were horrified by the scene. The locals petitioned their minister for action and succeeded. Ever since that gruesome incident, the rays here have been protected by law with fines of up to $3,400 for harming one of them.

**People interacting with the friendly smooth rays at Hamelin Bay, in Western Australia**   PHOTO BY TREVOR PADDENBURG, CAPE TO CAPE TOURS

Rays are truly extraordinary creatures that evolved from regular "sharky sharks" back in the Jurassic period, so consider them all as highly modified sharks, which is why we classify them with sharks in the group called elasmobranchs. Rays are called batoids in scientific terms; they principally differ from regular sharks in having their gill slits underneath the head rather than on the side. Unlike regular sharks, most rays fly through the water by flapping their greatly expanded wings, which are simply enlarged pectoral fins. They thus have no need of a powerful tail, so in most species that has been reduced to just a stick-like rudder. Many rays bear deadly venomous stingers (which are modified dorsal fin spines) to protect themselves, while others emit powerful bolts of electricity to stun their prey or deter attackers. Stingray venom is exceedingly painful and can be fatal. The famous Australian wildlife presenter Steve Irwin was killed in 2006 by a stingray he encountered while searching for tiger sharks to film. He found a large 8-foot short-tailed ray and followed it to investigate for another project. The ray must have thought he was a tiger shark—their main predator—so it struck him with its barb. It pierced his heart and thoracic cavity, and he died of this traumatic injury, rather than the effects of the ray's venom, soon after.

One way of better understanding how rays evolved is to consider the parallel story of another great transformation in evolution, that of birds evolving from small running predatory theropod dinosaurs driven by their legs. As these dinosaurs gradually lengthened their arms and developed more expansive feathery coverings, they morphed into a new body plan, becoming the first birds. Birds fly through air using their wings, which are just modified arms, structures that themselves evolved from the pectoral fins of fishes.

While this dramatic transformation was taking place on land, sharks underwent the same dramatic change underwater. Long-bodied sharks that propel themselves in water with their powerful tails slowly transformed to flattened rays that fly through water by flapping or undulating their wings (modified pectoral fins).

Rays also had to undergo major dental work to form a new kind of harder dentition—batteries of broad, flat crushing tooth plates to better serve them for devouring hard-shelled prey living on or below the seafloor. The rays first emerged rapidly in the Jurassic and would soon become more successful in terms of diversity than all other sharks—today there are more than 600 species of rays compared to around 536 species of sharks. The largest ray is the gentle filter-feeding manta ray,

**The manta ray is the largest ray ever, with a 23-foot wing-span** ANONYMOUS SOURCE

which grows to 23 feet in width and can weigh up to 6,600 pounds. Rays also include an offshoot group looking like long-bodied flat sharks with long sword-like snouts armed with sharp tooth-like denticles—the sawfishes, which can grow to just over 20 feet long and weigh more than 1,300 pounds.

The first well-preserved complete fossil rays come from the Soln-hofen sites in Germany. Only two species are known from here, *Astero-dermus* and *Spathobatis*, although both are represented by perfect specimens. They superficially look similar, although *Asterodermus*, at around 2 feet long, has just one dorsal fin, while the much larger *Spathobatis*, reaching almost 5 feet, has two prominent dorsal fins. *Spa-thiobatis* also has a short sharp point protruding from its head. The teeth are well preserved on both forms and help confirm their identifi-cation as rays as they are flat, small teeth with distinctive features only seen on living rays. Distinct fossil egg capsules of rays matching those of living forms are also found at Solnhofen, so we know that right from the start they were reproducing similar to modern rays. What all this tells us is that at this time rays could fill a specialized niche as specialist bottom feeders and were thriving by the end of the Jurassic.

Rays later evolved some ingeniously effective forms of defense, from deadly venomous spines in their tails to the ability to deliver powerful electric shocks, being the only group of elasmobranchs today that de-veloped such weaponry. The electric rays can shock a prey item with a powerful jolt of electricity, stunning it long enough to eat it. More ef-fectively, they can deliver deadly shocks to attackers, powerful enough to deter them from coming back for a second try. The discovery of electric organs in rays by John Walsh was published in the *Royal Society of London* journals way back in 1773, and some science historians think that the early descriptions of these shocking organs were read by Ital-ian physicist Alessandro Volta, so may well have inspired him to invent the voltaic pile battery.

The last years of the Jurassic was a time of intense competition for sharks, and was also an opportunity, as the bony fishes kept diversifying and expanding their range of body shapes and sizes. The largest bony fish that ever lived appeared at this time, the 54-foot-long *Leedsichthys*, a gentle filter feeder. Such tasty prey was the choice of large marine reptiles like the gigantic pliosaurs with their ghastly long heads and huge sharp teeth. Sharks throughout the entire Jurassic were smaller in

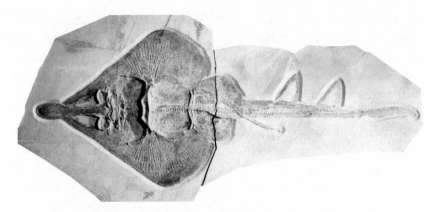

*Spathiobatis,* an early ray, from Solenhofen   IMAGE COURTESY OF JÜRGEN KRIWET

size, with none growing larger than around 10 feet long. The emerging new guild of modern sharks—the elasmobranchs, including extant shark and ray families—had now appeared on the scene from earlier neoselachian stock. Would they be up to the fight when even more fearsome predators from the saurian world would soon invade the seas? Soon all this would change as a new arms race began in the Cretaceous period, driving sharks to reach new heights of diversity and ferocity.

In the next chapter we will meet some of the most spectacular new sharks emerging in the Cretaceous period—the longest geological period, spanning 79 million years (almost 20 percent of total shark history)—when many more of the modern shark families appeared. With competition at its peak, sharks had to find new niches to occupy. While some opted for deadly razor-sharp teeth and larger body sizes, becoming as ferocious as the modern-day white sharks, others became bottom-feeding or filter-feeding specialists, eventually taking this lifestyle to extremes—one giant shark of this age reached well over 30 feet long and possibly lived on giant clams up to 6 feet wide. The result of all these evolutionary experiments would be ultimately tested by the astronomical mass extinction event heading toward Earth at the end of the Cretaceous.

## CHAPTER II

# SHARKS GO LARGE

### Sharks Versus Giant Marine Reptiles

---

### Cretaceous seaway, Italy, 90 million years ago

The giant sharks seem to come out of nowhere. The 26-foot-long *Cretodus* pushed swiftly through the water using its blunt snout, its wide mouth opened to show arcades of curiously chunky yet sharp dagger-like teeth. The shark resembled a gnarly, gigantic white shark that had been punched in the nose. Not far behind it, a 23-foot *Cretoxyrhina*, also known as the Ginsu shark for its knife-sharp teeth, appeared, sleek and terrifying, like the Super Sports model of the *Cretodus*, its tapered body looking like a mako on steroids. Both sharks had detected blood in the water and felt the vibrations of a struggle going on, so both had closed in to see if a meal could be had.

The disturbance was caused by an encounter between an 8-foot mosasaur named *Romeosaurus* and a 15-foot long-nosed sawfish, *Onchosaurus*, just above the seafloor. As the mosasaur darted in with jaws open, thinking it could take on the much larger ray-like fish, it turned its head sideways to wrap its jaws lined with curved conical teeth around the tail of the sawfish. Suddenly the sawfish twisted its huge head around in a flash, wielding its deadly long rostrum packed with sharp triangular teeth, slashing across the underside of the mosasaur's head, deeply cutting its throat. Billows of red blood filled the water as the mosasaur writhed in pain. Unperturbed, the massive *Onchosaurus* lifted its broadsword-shaped head, then pushed itself forward and upward through the water column to slowly swim away.

The mosasaur's pain stopped a second or two later when the gigantic *Cretoxyrhina* rushed in from above and grabbed its upper torso in its powerful jaws. It snapped the mosasaur's upper body in two, then ravenously gulped down the head and shoulder region in one gulp. Another *Cretoxyrhina*, around 16 feet long, swooped in to snap its jaws on the other bleeding half of the reptilian carcass, then thrashed around to dismember the carcass. More blood billowed out as the second *Cretoxyrhina* finally swallowed a sizeable chunk of the dead reptile, while smaller shreds of the mosasaur were scooped up by silvery bony fishes darting in between the frenzied sharks.

As feeding took place, another large shark entered the scene from below. It was a massive beast, some 30 feet long, named *Ptychodus*, that slowly cruised just above the seabed, scooping up 3-foot inoceramid clams with its enormous rounded bullish head. Its wide jaws were armed with hundreds of bulbous crushing tooth plates, each with a strongly crenulated surface for gripping and crushing—the perfect set of choppers for grinding up hard clams. As it crunched them up, its spiracles, two open holes behind the eyes, pump out clouds of fine muddy sediment and bits of powdered shell. *Ptychodus* coughs out larger pieces of broken shell, then swallows the gluey mixture of clam meat mashed with tiny shell fragments, but it doesn't seem to mind, as its large gut will bathe the shell pieces in strong gastric acids to eventually dissolve them.

The huge *Cretodus* has spied far better prey, so it leaves the *Cretoxyrhina* pack and the *Ptychodus* to their meals. Ahead in the surface waters above it recognizes the familiar outline of its favorite food, a large turtle, almost 7 feet long. The turtle glides along with slow, rhythmic strokes of her large paddles, relishing the sunlight in the upper few feet of the warm water. She is not interested in food right now, for she is on a long migratory journey to her breeding island. The shark closes in on the turtle at a moderate pace, knowing instinctively that its prey has no chance of spotting it below her. Then, with a quick thrust of its large tail, it accelerates in a direct line to its prey, gaining speed for the final attack. Millions of years of evolving stout, powerful teeth and jaws with incredible muscle power enable it to make a deadly snap strike on the turtle shell from behind, crushing it in one powerful blow. As it releases the turtle from its maw, it leaves a tooth embedded in the shell. The wounded turtle panics as blood clouds around it. *Cretodus* then circles to launch another attack from the front, enveloping the remains of the wounded turtle in one large gulp. The whole broken mass of turtle, shell, and gore is swallowed and pushed down the throat of the shark to its stomach. The shark at one instant flinches as it feels a sharp pain in its gullet. While the *Cretodus* has eaten many turtles in this way before, something doesn't feel right this time.

*Cretodus* swam off and slowly began digesting its large meal, but a week or so later it died of an infection caused by a jagged part of the broken turtle shell that tore a wound in its throat. After it died, its body floated to the surface, where it was scavenged upon by seabirds and numerous small sharks and bony fishes before rupturing, the remains of its carcass finally dropping slowly to settle on the seafloor near a pile of shells of dead ammonites and sea urchins. It was eventually buried by the rain of fine reddish limy sediment falling down the water column.

## The amazing fossil sharks of the *lastame*, Verona

This scene gives us a quick snapshot of a small shark community living in the Middle Cretaceous open seas of the northern Tethys Ocean 92 MYA. The tale is based on a set of truly awesome fossils that you can see in a small village museum high up in the Lessini Mountains, about twenty miles north of the city of Verona in northern Italy. I took a trip there in November 2022 to see these shark fossils and learn about the geology of the site, driving up through the mountains of this forested part of the Veneto Province in the foothills of the Italian Alps. Stunning views across the plains took my breath away as we kept climbing into the lightly forested hills. Small villages dotted the hillscapes, adorned with vineyards and occasional quarries. These were the major industries up here. Some of Italy's finest buildings feature the beautiful pinkish-white nodular limestone known locally as the *lastame*, within the *Scaglia Rossa*. In Milan, the next day, I was outside the main shopping mall in the city center when I spotted a big ammonite fossil in the pavement, then suddenly realized it was a slab of this same beautiful stone.

The quarries near the town of Sant'Anna d'Alfaedo have produced

**View of the Italian Alps with the *lastame* rock quarries that yield the giant fossil sharks at right** AUTHOR PHOTO

this quality building stone for centuries. Occasionally when quarry workers split the large slabs, extraordinary fossils are found. The quarrymen keep the fossils aside and often spend their free time preparing them before they are lodged in the town paleontology museum (I think every town should have one). I turned up on a Sunday morning outside the museum's regular open season. The people that run the museum were pleased to open it to show me their prized fossils. Inside I immediately clocked the giant slabs of reddish limestone, each with a complete or nearly complete giant fossil shark, or remains of mosasaurs and other marine fossils. The sea here 92 MYA was rich in ammonites, horn-shaped clams (rudists), giant oyster-like clams (inoceramids), and sea urchins, in particular. Most of the large vertebrate animals were preserved either as whole or as partial skeletons, the result of when complete carcasses are first scavenged, then fall to the seafloor to be buried.

The star sharks in our story are the gigantic *Cretodus crassidens*, based on a stunning, nearly complete fossil missing its tail but still almost 15 feet long, and *Cretoxyrhina mantelli*, based on two skeletons, each spanning between 20 and 21 feet in length. *Cretodus* and *Cretoxyrhina* were the largest predatory sharks at this time. The *Cretodus* jaws measure about three feet wide, with hundreds of well-preserved sharp teeth still in their life positions. *Cretodus* grew to up to 26 feet long. Annual growth rings in its backbones tell us it died at twenty-three years old. It might well have reached 35 feet long if normal growth had ensued, although no isolated giant vertebral discs have yet been found to substantiate this projected upper size estimate.

The remains of a 6.5-foot protostegid turtle are nestled within the *Cretodus* with one of its teeth stuck in the shell, the smoking gun proving this shark was its killer. Protostegids are extinct sea turtles which include the largest known marine turtles ever, like *Archelon,* which grew to about 15 feet long. Their shells were not as hard as those of regular turtles, being more like those of today's leatherback turtles. Our *Cretodus* had the entire turtle in its stomach with added evidence of digested bones etched by stomach acids, proving the shark was alive for a short time after eating it.

The author with the 15-foot-long *Cretodus* specimen with turtle remains in its belly at Sant'Anna d'Alfaedo Museum of Palaeontology and Prehistory in Italy, 2022   PHOTO BY HEATHER ROBINSON

Jacopo Amalfitano, a keen young *paleontologo* from Italy, worked on these sharks with other local paleontologists and colleagues from Padua University in northern Italy. He got interested in fossil sharks from collecting fossils in the northern Italian mountain regions, where he lives. I contacted him to ask if he had any idea what might have killed the giant *Cretodus.* He said that there was no obvious evidence of how it died, so I admit that the postmortem shell injury scenario is my own creation.

Both *Onchosaurus* and *Ptychodus* remains are also on display in this museum. *Onchosaurus* was an extinct kind of sawfish (a **sclerorhynchid**), like the modern *Pristis* species with a long snout covered in tooth-like denticles. These deadly little knives along the sides of the snout were replaced with larger ones as the fish grew. It was about 16 feet long. Remains of the other strange shark, *Ptychodus*, from this site are quite rare, and there is only one slab of rock bearing its teeth. I'll tell you a lot more about this beast later.

My story above focused on two impressive predatory sharks living around 92 MYA, but the Cretaceous period started 145 MYA. What was going on in the world, especially in the seas, during the first 45 million years of this period, and what kinds of sharks were living then?

## Living the Cretaceous life

The Early Cretaceous spans 45 million years (145–100 MYA). The world at this time was arid and warm, with humid climates dominating the mid-latitudes. While temperatures were generally much warmer than today, they would continue to rise steadily and peak in the early part of the Late Cretaceous when rising sea levels inundated the interior depressions of large continents, bestowing North America with a vast inland sea called the Western Interior Seaway. It was a continuous body of water that divided the continent into two halves from the Gulf

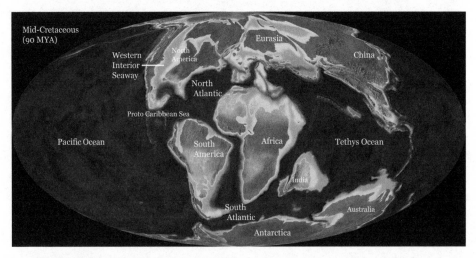

**The world during the Cretaceous period** RON BLAKEY, DEEP TIME MAPS (DEEPTIMEMAPS.COM)

of Mexico through to Alaska. Elsewhere the southern part of Pangaea, Gondwana, had split away, and it too was slowly beginning to break up as India rifted northward away from Antarctica and Australia. The narrow southern Atlantic opened up as the Tethys Ocean expanded; later it would become the Indian Ocean. The Panthalassic Ocean has now morphed into the modern Pacific Ocean.

Suddenly, around 133 MYA, ocean temperatures drop by around 5° to 7°F and ice caps begin to form around the South Pole area. This short but dramatic climatic change is called the Weissert event. As seas cooled again, nutrients upwelled, and life really took off, as oceans were now packed full of marine life, including large ammonites, plus abundant reefs built by horn-shaped rudist clams and crinoids (sea lilies) covering the seafloors or drifting on the surface attached to logs. The sea abounded with many kinds of snails, clams, shrimps, crabs, lobsters, lampshells, sea mosses, and solitary corals. Fishes were highly diverse at this time, dominated by many families of teleosts, including the first guild of large predatory bony fishes called **ichthyodectids**. These include fishes like *Cooyoo* with long, pointed teeth reaching maybe 4 feet in length, but many forms up to three times larger would emerge later in this period.

Sharks had a lot to deal with throughout this time. Several new kinds of gargantuan predatory reptiles would ascend to the top of the food chain, creating even fiercer competition for sharks. Let's take quick stock of what exactly the competitors were for sharks at this time.

The long-snouted fish-eating ichthyosaurs took over as the reptilian dolphins of the seas, reaching around 20 feet long. Cretaceous ichthyosaurs were basically all the same shape, designed for eating squids and fishes. They were not much of a threat to sharks, apart from competing for similar food resources. Living alongside them were long-necked plesiosaurs, some up to 40 feet long, also feeding on small fishes and squids. Like the ichthyosaurs, they were no direct threat to sharks, as they had very small heads with spindly teeth for catching small prey. However, any shark dependent on fishes as its primary food would surely be up against competition from these reptiles.

Giant seafaring crocodiles like *Machimosaurus,* reaching 23 feet, would have preyed on larger fishes, sharks, and other marine reptiles. Luckily for sharks, and for us, they died out by the end of the Early Cretaceous (100 MYA). Later in the Cretaceous seas, the short-necked but massive-headed pliosaurs, up to around 40 feet long, and giant lizard-like mosasaurs that reached almost 60 feet long, were the oceans' most powerful apex predators of the day. More about those beasts and how sharks interacted with them later in our story.

**More stiff competition for sharks emerged as enormous crocodiles like the 23-foot-long *Machimosaurus* took to the Early Cretaceous seas**
IMAGE COURTESY OF NOBU TEMURA

New lines of sharks arose in the Early Cretaceous that would later give us the four modern culprits guilty of causing most human fatalities: white shark, tiger shark, bull shark, and oceanic whitetip shark. These predatory sharks of the lamniform and carcharhiniform groups were all master creations spawned out of the fierce competition for food during the dinosaur age.

Recalling the smaller sharks of the Triassic and Jurassic, when most species were around 2 to 3 feet and none over 10 feet in length, something radical happened in the Cretaceous that enabled some sharks to suddenly grow much larger once again. All the previous groups of sharks and rays that were around in the late Jurassic continued through to the Cretaceous, yet the elasmobranchs kept diversifying, so several new families of unusual sharks appeared. The dominion of the archaic forms, like hybodonts, greatly diminished; they fade away gradually until the end of the period, when they quietly go extinct. The presence of many early lamniform sharks was the real success story of this age.

Let's meet some of the new sharks appearing in the Cretaceous seas that represent the first members of several families of living sharks. This really is the dawn of the modern shark age.

# New Cretaceous sharks appear

The sharks living in the first 40 million years of the Cretaceous were all small forms under 6 feet or so long. Here, though, it's not size that matters, but what you do to earn a living in the face of strong competition. Many of the sharks appearing at this time likely arose because of increases in marine nutrient levels after the Weissert event cooling, as well as by taking advantage of new currents that formed when major continents rifted apart to create new seaways.

Today much of the world's oceanic food chains begin in Antarctica with the algae forming under the floating ice sheets, which start the entire food chain that the vast masses of krill feed upon. The algae are food for small marine invertebrates which feed small fishes that are prey for larger fishes, penguins, seals, seabirds, orcas, and the massive baleen whales. It's not hard to see how the formation of polar ice caps would have triggered a similar massive boost to the food chains in the Early Cretaceous seas. Recent research has confirmed the existence of a small northern polar ice cap as well as a southern one at this time.

Several modern shark families appeared in our oceans during this part of the Early Cretaceous, alongside other new lineages of ancient sharks that wouldn't make it through the next mass extinction event. These newbies include the first member of the sawsharks (**pristiophorids**). Sawsharks are flattened sharks with extended snouts armed with thin, sharp, tooth-like denticles (rostral denticles) along their sides. They are mostly under 6 feet long and live in shallow seas to upper continental slopes of the South Atlantic, West Indian, and West Pacific oceans. Unlike their much larger ray cousins, the sawfishes (**pristids**), they are distinguished by having their gill slits open along the side of the head, with mouths armed with small conical teeth up front for grasping prey and batteries of flat grinding teeth along the jaws for crushing crabs and other crustaceans.

The oldest known sawshark appeared around 100 MYA at the very end of the Early Cretaceous, based on numerous undescribed teeth and rostral denticles found in Western Australia by Mikael Siversson of the Western Australian Museum. Complete specimens of the living genus *Pristiophorus* date back to about 85 MYA from the sites in Lebanon, which we will discuss later.

Sawsharks like this *Pristiophorus* first appeared over 100 million years ago   FROM FISHES OF AUSTRALIA, CSIRO, AUSTRALIA

The bramble sharks (**echinorhinids**) are so named because they have thick thorn-like skin denticles. Their oldest fossils come from the inland sea deposits of France, dated at around 135 MYA. Their teeth are like a set of flat serrated cusps pointing away from each other, not un-like the teeth of the Devonian *Maiseyodus*. Today, there are two living *Echinorhinus* species, the bramble shark and the prickly shark. Both are sluggish deep-water forms with the prickly shark growing to around 14 feet long. The oldest fossil teeth of this genus are around 0.4 inch in length, suggesting this ancestral form was a small shark up to 5 feet long that lived in deep-water habitats.

We find the first members of the collared carpet sharks (**parascyl-lids**), long-tailed carpet sharks (**hemiscyllids**), and nurse sharks (**gin-glymostomatids**) also appearing at this time in shallow seaways. These are all families within the broader diverse group called orectolobi-forms—a group of around forty-five living species containing wob-begongs, carpet sharks, nurse sharks, and zebra sharks, as well as the largest living shark, the whale shark. They all have small barbels com-ing off the mouth area for sensing food, although in some forms, like the whale shark, these structures can be hard to spot, as they are greatly reduced.

The collared catsharks have the amazing ability to change their body color to match the seabed to blend in, to hide from predators or am-bush prey. They are mostly around 2 feet long and inhabit various coastal to deep-water continental slopes down to around 550 feet. Some are nocturnal and hide in caves during the day, then come out and feed at night on small invertebrates and fishes. The oldest member of this family is *Pararhincodon*, known from tiny fossil teeth found in England, dated at around 113 MYA.

Epaulette sharks are a wonderful example of the long-tailed catshark

**The epaulette shark can walk over reefs using its fins like arms and legs**  PHOTO BY NIGEL MARSH

group, showing how this group evolved a peculiar way of moving. They evolved specialized fins that enable them to walk around exposed coral reefs using their paired pectoral and pelvic fins like arms and legs. The oldest fossil of this family are 120 MYA teeth from England identified as belonging to the living genus *Chiloscyllium*, represented today by various small species of bamboo sharks. *Chiloscyllium* diversified by the late Cretaceous when we have 4 distinct species known from various sites around the world.

The nurse sharks live in rocky and coral reefs; they hunt along the bottom, sucking up small crabs and fishes. They include some large forms up to 13 feet long. While the fossil nurse shark *Cantioscyllium* is known from a nearly complete skeleton found in Kent, England, which was studied by Arthur Smith Woodward in 1889, the oldest fossil teeth of this genus were found in rocks from the mountains of Aragon in Spain, dated at around 128 MYA.

The flattened angel sharks (**squatinids**) are represnted by the living *Squatina,* that appears around 135 MYA, based on distinctive teeth found in Poland. Angel sharks are a homogeneous group of sharks known from over 22 species in the single genus *Squatina*. They get their name from their large wing-like pectoral fins and broad pelvic fins that meet, flanking a very flattened body shape. They are small ambush predators, mostly under 5 feet in length, that often hide in the sandy or muddy seafloor, partly buried, and seize their unwary prey with a lightning-fast deadly strike, sucking the prey inside the mouth as soon as it opens.

The carcharhiniform groups that first appear in the Cretaceous include members of the hound sharks (triakids), a diverse group today of forty-five species that are all active predators with broad triangular teeth perfect for catching octopus, squids, and bony fishes. The oldest record is based on just one tiny isolated tooth from 130 MYA, found in England, while the living catshark *Scyliorhinus* first appears around 137 MYA. Catsharks include some fifty living species. They are poor swimmers that hunt only at night, feeding on small invertebrates and fish around reefs and rocky areas. They diversified rapidly to become much more common in the Late Cretaceous seas.

The living broadnose sevengill shark (*Notorynchus*) also appeared in the Early Cretaceous of Europe. It might have been around 13 feet long, based on lower jaw teeth 50 percent larger than those of the living species. The living sharpnose sevengill shark (*Heptranchias*) pops up later in the Late Cretaceous (75 MYA), known from relatively small teeth, similar in size to those of the living species.

Several new groups of rays appeared in the Cretaceous period. The ancient groups include the spectacular sclerorhynchids, large forms with sawfish-like long snouts armed with hooked, narrow, teeth-like denticles. Just four families of rays out of a dozen that appeared in the Cretaceous period survived to our time. These were the eagle rays (**myliobatids**), whiptail stingrays (**dasyatids**), thornback guitarfish (**platyrhinoids**), guitarfishes (**rhinobatids**), and the biggest clade, the skates (**rajids**).

At the opposite end of the monster shark spectrum we find the remains of the smallest shark that ever lived at this time. To hear about it, let's go digging in Spain.

## The smallest shark of all time

In 2022 I had the opportunity to visit one of the most renowned fossil sites of the Early Cretaceous age, the Spanish heritage protected site at Las Hoyas, not far from the UNESCO World Heritage town of Cuenca in central Spain. It was a hot summer day and we arrived at the quarry late morning with my good friend Dr. Jesús Marugán-Lobón from the Madrid Autonomous University, who was leading our group from a paleontological conference on a dig at the site. Jesús has been collaborating with me over the years on various projects, so I was excited to finally see the site with him there to explain its history and geology, its fossil diversity.

The Las Hoyas site was formed when creatures that died nearby were washed into large ponds, then sank to the bottom to be buried by very fine lime-rich muds. These ponds were situated in a lowland area surrounded by verdant wetlands formed between 129 and 126 MYA. It is renowned for its excellent preservation of plants and animals of many kinds, including exquisite dinosaurs, birds with feathers, mammals with fur, delightful fossil plants, and extraordinary fossil fishes and sharks.

Jesús directed us to the place where fossil sharks had previously been found, and my wife, Heather, and I began excavating the layers slowly with hammers and fine chisels. We found some nice fossil fern leaves, and the occasional fish scale, but we soon realized it takes a lot of human-powered excavations over many, many years to find the top-shelf goods here.

The Las Hoyas shark was the smallest shark that ever lived, probably under 8 inches long when fully grown. Being small helped it survive at a time when everything underwater was out to get you  PHOTO BY JESÚS MARUGÁN-LOBÓN, MADRID

The sharks from Las Hoyas are particularly fascinating because they are all so tiny, with complete juvenile sharks only 1.3 inches in length, which is even smaller than the embryos of the smallest living sharks, like the dwarf lantern sharks (*Etmopterus perryi*, reaching between 8 and 9 inches long). The Las Hoyas shark was identified by its teeth as being a hybodont called *Lonchidion*. It might be a miniaturized species that evolved its small body size for survival. The tiny *Lonchidion* suggests it was a rapidly maturing shark with a short life span, a pattern seen in many living dogfishes and rays. A closely related form called *Lissodus* shares similar narrow, arched tooth plates with *Lonchidion*. It has a body ranging to just under one foot long, with fin spines that are four times larger than those of the tiny Las Hoyas sharks, suggesting that an adult Las Hoyas shark, if it followed the same growth pattern, could have reached 8 inches or so, on par with today's smallest sharks. Being small helped *Lonchidion* become a very successful shark, whose known range spans 160 million years, from the Triassic to the end-Cretaceous, as it is known by more than a dozen different species.

At the end of the splendid afternoon digging at the Las Hoyas site we thanked Jesús for a great day. When I next saw him he excitedly told me that during the major excavation at the site that summer the team had found two excellent new complete tiny sharks at precisely the spot where we were digging that day. They now call that spot "shark gully."

Now we can leave the smallest sharks to look at the rise of the group, the lamniforms, that would one day produce the largest and most frightening shark of all time: megalodon.

## Rise of the deadly lamniforms

The next stage in our story concerns the biggest shark success story of the Cretaceous, the lamniforms, the group containing today's white sharks, makos, goblin sharks, and threshers. Lamniform sharks have distinctly cylindrical bodies and large mouths that extend behind the eyes. Their fossil remains are characterized by their heavily calcified vertebral discs and powerful teeth built of a special dense type of dentine (osteodentine) tissue. The oldest definite lamniforms are the teeth of *Protolamna* found in Poland, dated at around 135 MYA, which belong in an extinct family allied to one of the sharks featured in our opening scene, *Cretoxyrhina*. The tiny, sharply pointed teeth testify this first lamniform shark was a small predator around 2 to 3 feet long.

The oldest living lamniform families are the goblin and sand tiger sharks that appeared around 125 MYA. The living goblin shark, *Mitsukurina*, is known from one living species, the only surviving member of this archaic shark family that includes many extinct species. The goblin shark gets its name from its gruesome appearance featuring many spindly teeth poking out of its mouth in all directions. These sharks, which can grow up to 20 feet in length, inhabit the deep oceans. They have a bizarre head featuring an elongated flat paddle-shaped snout full of electroreceptors, enabling it to detect prey like squids or small fishes in pitch-dark waters. It rapidly projects its jaws while opening them, a bit like the Alien in the sci-fi movies, to seize prey with its many jagged teeth.

The sand tiger shark (*Carcharias taurus*) is a common species around the world that can often be seen in aquaria. It's a slow, powerful swimmer that hunts fishes in shallow seas at night. It is the sole survivor of the **carchariid** family, which contains many fossil species that lived in the Cretaceous period. The oldest fossil sand tiger shark, *Hispidaspis,* appears around 130 MYA. The sand tiger has one of the slowest reproductive rates of all sharks—they give birth to only one pup each breeding season, the victor after eating its siblings inside the womb. The males look after the female mothers after birth, increasing chances of the pup surviving longer. Their fossil teeth are similar to modern species, suggesting they haven't changed much at all over time. The closely related deep-sea sand tiger sharks (*Odontaspis*) are similar in appearance but prefer deeper seafloor habitats. Like the sand tiger, it also has long spindly teeth with side cusps perfect for grasping fishes or squids.

My personal favorite fossil sand tiger shark was one first discovered in Australia by Mikael Siversson of the Western Australian Museum (you'll hear more about him soon). It was a small shark he recognized from the older scientific literature as one previously named as a species of *Odontaspis* from the United States. Armed with a newly collected set of specimens, Mikael separated it from the U.S. species and named a new genus after me, *Johnlongia.* The teeth are tiny, around one-third of an inch high. He recognized that it first appeared in Queensland's inland sea about 110 MYA, and since his detailed work on it from Western Australia, it is now known from sites in the United States, Mexico, Europe, and Russia. Other new families of archaic sharks also appeared at this time but would not survive beyond the end of the Cretaceous.

*Cretalamna* appeared about 102 MYA and persisted throughout the Cretaceous, later giving rise to the *Otodus* lineage that would lead ultimately to the gigantic megalodon. The oldest teeth of *Cretalamna* are small, suggesting it was around 4 to 5 feet long, although it would gradually increase its body size throughout the Cretaceous, a trend seen in several other lamniform sharks. One famous fossil find from Japan shows that some eighty-seven *Cretalamna* teeth were found with the carcass of a long-necked plesiosaur *Futabasaurus,* which was almost 30 feet long. Some of the teeth were stuck in its bones, so we have good evidence that it was either scavenging or hunting large plesiosaurs about 85 MYA.

The "crow sharks" (because they were thought to be carrion feeders), are a peculiar group of lamniforms called the **anacoracids** that appeared at the end of the Early Cretaceous. They are best known by *Squalicorax,* which spanned the last half of the Cretaceous between 110 and 66 MYA. It was a very successful shark that diversified into at least thirty-four species found around the globe. This group has distinctive hooked serrated teeth somewhat resembling those of a living tiger shark. It is thought they occupied the same niche as opportunistic predators capable of killing a wide range of fishes, sharks and marine reptiles as well as being scavengers on larger animals. Fossils from Kansas show *Squalicorax* teeth embedded in bones of mosasaurs and dinosaurs, the latter having been washed into the Western Interior Seaway as carcasses.

The largest species, *Squalicorax pristodontus,* lived at end of the Cretaceous, and grew to around 16 feet long, making it one of the largest sharks in the oceans anywhere at this time, well after *Cretoxyrhina* and *Cretodus* go extinct. In 2014 a complete fossil *Squalicorax* just over 13 feet long was found in a drawer at the Canadian Fossil Discovery Centre, in Manitoba, forgotten since 1975. Other complete *Squalicorax* specimens showing nearly the complete body outline have been recently found in Lebanon, but have not yet been studied. *Squalicorax* had a broad head with a wide, toothy mouth, as befitting their predatory and scavenging nature. The crow sharks died out at the end of the Cretaceous.

The last lamniform group to appear at this time was another extinct large group, the **cardabiodontids**, after the genus *Cardabiodon,* first discovered by my good friend the Swedish shark paleontologist Mikael Siversson. Mikael's story and his superb fossil shark discoveries, which put Australian fossil lamniforms on the map, intersected with my own career, and I dare say we are both lucky to be here after a motorcycle adventure we once embarked on.

## Mikael Siversson, the fossil shark Terminator

In 1994, Mikael won a postdoctoral scholarship from the Swedish Natural Sciences Research Council and decided he wanted to work at the Western Australian Museum, where I was the curator of vertebrate paleontology, so I was appointed as his supervisor. We hit it off immediately, often going out on the town together after work, spending more time drinking beer and talking fossil sharks than getting into any real trouble. We bought large, powerful motorcycles on the same day. At the time, we also shared a love of physically challenging sports. While I trained at karate each week, Mikael was into weightlifting. In later years he would achieve a world record in the unusual competitive sport of grip strength.

Once we were both nearly killed by a pack of emus. It's an interesting story that shows paleontologists are still humans who sometimes do crazy things. We were traveling on our motorcycles from Perth to the outback town of Kalgoorlie to attend a party. Cruising along at the speed limit between the major towns, we eventually reached a long, desolate straight road stretching 116 miles with no towns or gas stations. Mikael pulled up beside me and asked if I could take his backpack for a while.

"You see," he began in earnest, "I would like to really test out my

**Mikael Siversson collecting fossil sharks in outback Western Australia** WESTERN AUSTRALIAN MUSEUM

bike, push it a little." I nodded, shouldered his backpack (as my own gear was lashed to the back of my bike), and set off down the lonely road sitting on the "outback speed limit." Five minutes later I heard the rising drone of his bike's engine roaring up behind me when unexpectedly a large flock of emus, Australia's ostrich-like flightless birds, scrambled onto the road directly in front of us. I slammed on the brakes as Mikael belted straight through the pack of confused birds, feathers and creatures scattering in all directions. Later, when I caught up with him, I asked him how he managed to avoid hitting the flock of emus on the road, and he simply looked at me and said in his deadpan voice, "What emus?" To this day I think we were both extremely lucky not to have been wiped out by those darn phantom birds.

Mikael not only looks a bit like Arnold Schwarzenegger, but he also works like the Terminator in the field, using his physical strength and laser-focused attitude to find new fossil sharks. He put his great strength to good use collecting large bags of sediment from the site, then processed the material back in the lab, screen-washing it to break it down. He would then carefully pick out the tiny sharks' teeth, working long hours peering down the microscope. Through this methodical approach he was able to extract large numbers of fossil sharks' teeth and discover many new species from sites throughout Western Australia, a large state about four times the size of Texas. He identified more than fifty different sharks from just one Middle Cretaceous site near Kalbarri, making it one of the richest fossil sites in the world for fossil sharks of this age.

Mikael's most spectacular find was the giant shark *Cardabiodon*, which he discovered near Kalbarri in Western Australia in 1996. He and his ex-wife spotted a number of fossil teeth and a few large vertebral discs lying on the ground, eroding from a shallow creek bank. By carefully digging they found more and more teeth and excavated a series of additional large vertebral discs, so he knew at once that he had one large shark. It was a new genus dated around 97 MYA, which he named after the locality and station owner (Rick) at Cardabia Station— *Cardabiodon ricki*, meaning "Rick's Cardabia tooth."

*Cardabiodon ricki* is known from almost the entire shark's dentition, consisting of more than one hundred teeth, plus fifteen large vertebral discs. These fossils suggested that its full size was near 18 feet long,

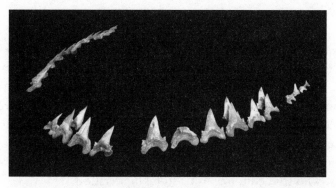

Restored jaws of *Cardabiodon ricki,* the first large shark to appear in over 150 million years. It was around 18 feet long PHOTO BY MIKAEL SIVERSSON, WESTERN AUSTRALIAN MUSEUM

making it the largest shark of its time. Because its vertebral discs were so well preserved, Mikael and his colleagues were able to reveal information about its birth size—pups born at 1.3 to 2 feet long—and how fast it grew—just under one foot each year—from the growth rings. These rings showed it was between twelve and thirteen years old at death, so not fully mature. Some large living lamniform sharks can take a long time to grow, such as white sharks that attain sexual maturity between twenty-one and thirty-three years of age, so this fellow would no doubt have been much larger when mature. Another new undescribed cardabiodontid shark was identified by Mikael that once swam in the vast Queensland inland sea around 105 MYA. Its fossilized vertebrae suggest it was bigger than the Western Australian species, at least 26 feet long, making it the largest shark in the Early Cretaceous seas.

Mikael has contributed many interesting ideas about the rise of large superpredatory lamniform sharks, including how they exerted competitive pressure on both ichthyosaurs and pliosaurs, contributing to the extinction of both groups. He thinks that some of the estimates for pliosaur maximum size are exaggerated, and that the largest cardabiodontid sharks were easily capable of bringing them down as prey just as white sharks can bring down much larger baleen whales.

If they were so large and powerful, why did cardabiodontids suddenly go extinct?

# A tale of Ginsu sharks

The giant cardabiodontid sharks were likely driven to extinction by the arrival of a new, more efficient lamniform shark on the scene, one we met at the start of this chapter. *Cretoxyrhina,* nicknamed the Ginsu shark for its knife-like teeth, likely caused the demise of the cardabiodont sharks by outcompeting them. We know a lot about this giant Cretaceous predator from its complete fossil remains. Its backstory begins more than a hundred years ago with a bright young American scientist who studied it for his doctoral dissertation.

A superb specimen of *Cretoxyrhina mantelli* was found from the Late Cretaceous Smoky Hill Chalk deposits of Kansas in 1891 by the famous American dinosaur hunter Charles H. Sternberg. It was a complete shark about 20 feet long, containing around 250 teeth, calcified vertebral discs, parts of the braincase, jaws, scales, and even some fin cartilages. Sternberg was in the business of selling fossils to museums, and this one was bought by the Ludwig Maximilian University in Munich and transported to Germany.

A young American student named Charles (Chas) Rochester Eastman, from Cedar Rapids, Iowa, traveled to Germany to study this fossil for his doctoral degree at the university in Munich. His groundbreaking monograph describing the anatomy of *Cretoxyrhina* came out in 1895. The specimen was another Rosetta Stone shark, so significant that Eastman's study sank more than thirty scientific names of other fossil sharks that were based on isolated teeth thought to represent separate species, whereas in fact they all belonged to different parts of this one specimen, then named *"Oxyrhina" mantelli.* A Russian paleontologist later studied it and changed the name to *Cretoxyrhina.*

*Cretoxyrhina* lived in the Late Cretaceous seas between 100 and 78 MYA. The British paleontologist, geologist, and obstetrician Gideon Mantell first found eight teeth of what he named *Oxyrhina* in East Sussex in 1820, and these were later referred to the species named in his honor as *Cretoxyrhina mantelli.* Based on its teeth and vertebral discs, which compare closely to those of living lamnid sharks like the white shark, we estimate that *Cretoxyrhina* grew to at least 26 feet long. Its large tail was similar in shape to that of a white shark. One study recently calculated that it could likely swim in short bursts at speeds up to 43 mph, which is 8 mph faster than the fastest recorded white shark,

The giant *Cretoxyrhina*, at nearly 26 feet long, was the largest shark of its time and could easily have taken out a mosasaur ARTWORK BY JULIUS CSOTONYI

and faster than any mako or thresher, which can accelerate in short dashes up to 31 mph, making *Cretoxyrhina* the fastest predatory shark that has ever lived.

*Cretoxyrhina* had to be fast, as it lived alongside the gigantic **mosasaurs**, seagoing lizards as long as 60 feet that reached their peak in the last 20 million years of the Cretaceous period. The teeth of *Cretoxyrhina* are found all around the world, so we know it had a wide habitat range, living in subtropical seas down to high-latitude oceans with waters as cold as 41°F, which suggests it was warm-blooded, like several living lamniforms.

Chas Eastman landed a job at Harvard University and continued his paleontology work, publishing more than a hundred scientific papers, along with major bibliographic volumes on fish fossils. His work expanded our knowledge of buzz-saw sharks, the Solenhofen sharks from Germany, the great Cretaceous lamniforms, the complete Eocene sharks and fishes of Bolca, Italy, and Cenozoic sharks from several U.S. and other localities around the world.

His career took a dive in October 1900 when he was indicted for the murder of his brother-in-law. The charge resulted from an incident involving them shooting rifles at targets and then getting into an argument late one night. Eastman spent some months in prison awaiting trial, spending his time behind bars doing some great paleontological research. He eventually won the case and was released. Sadly, his life ended in 1918 when he drowned after falling off a boardwalk into the sea on Long Island. Also tragic was the loss, some twenty-five years after his death, of his *Cretoxyrhina* fossil from Kansas that he spent several years working on in Germany. The German museum housing the fossil was totally destroyed by bombing during World War II. Eastman will always be remembered for his magnificent work on *Cretoxyrhina*.

*Cretoxyrhina* teeth have been found in the Late Cretaceous to about 78 MYA. It survived by being large and fast, coupled with the fact that for most of its reign it was bigger than most mosasaurs, except maybe *Tylosaurus*. The largest mosasaurs arrived on the scene from about 80 MYA, close to when *Cretoxyrhina* went extinct.

However, another, even larger shark swam these oceans at the same time as *Cretoxyrhina* and *Cretodus*. It was not a fearsome predator, but a gentle giant we have briefly encountered in Italy, called *Ptychodus*. It demonstrates how some Cretaceous sharks were boldly branching into new lifestyles, and the successful ones could grow to gargantuan sizes.

## *Ptychodus,* the crushing monster

The largest shark of the entire age of dinosaurs, the Mesozoic era, was not something we would necessarily fear if we encountered it in a time-traveling deep-sea dive. The giant *Ptychodus* was a bizarre shark whose jaws were lined with hundreds of mostly wrinkly, dome-shaped crushing teeth. Shawn Hamm from Wichita, Kansas, studied *Ptychodus* for his master's thesis at the University of Texas, and he has since published more than twenty papers on this mysterious giant shark. He thinks it was likely an active swimming predator reaching between 26 and 42 feet in length that fed on large swimming ammonites, and clams like *Inoceramus,* some of which were more than 6 feet across.

The evidence that it preyed mainly on giant *Inoceramus* clams is

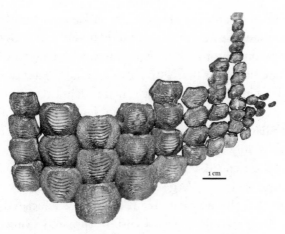

**The strange crenulated humped tooth plates of**
*Ptychodus*   PHOTO BY SHAWN HAMM, USA

circumstantial, based on *Ptychodus*-tooth-shaped marks on some fossil clams and the remains of clamshell fragments thought to be derived from either *Ptychodus* poop (coprolites) or coughed-up remains (regurgitates). It may well have had the ability to feed on large shells and excrete the small parts as fine sand, as the wrasses living on coral reefs do today, and if so, it could have played a role in creating new habitats in the seaways where it lived.

We met *Ptychodus* briefly at the beginning of this chapter as one of the *Lastame* fossils from northern Italy. That specimen comprised a single rare slab containing many associated teeth from one individual, as most of the world's specimens of this shark are just isolated teeth. Rare specimens of *Ptychodus* from the Niobrara Chalk in Kansas provide much more information about its body shape and lifestyle. A seminal study was done by Kenshu Shimada of DePaul University in Illinois some years ago. First, a little background on Kenshu, who is another quite extraordinary shark paleontologist whose work pops up several times throughout this book.

Kenshu became hooked on fossil sharks as a thirteen-year-old when he collected Miocene age teeth from sites around Japan and found an *Otodus megalodon* tooth, which he claims changed his life forever. It wasn't a giant tooth, but he found it by cracking open hard lumps of conglomerate. He won third prize in the national high school science competition in Japan and got to meet Emperor Hirohito. Although he wanted to pursue a career in paleontology in Japan, his grades wouldn't allow him to enter the university system there. He eventually got a foot in the door by moving to the United States, where he enrolled for his undergraduate degree at the University of Kansas. Later he completed a PhD at the University of Illinois on the dental evolution of fossil lamnid sharks. Today he is based at DePaul University in Illinois, working on fossil sharks found at many sites in the United States and around the globe, including *Ptychodus*.

In late 2008, Kenshu and some colleagues found a special fossil in Jewell County, Kansas—a large section of a *Ptychodus* jaw. Once prepared, it revealed about twenty teeth and many associated scales and sections of the right upper jaw. Kenshu estimated from this chunk of jaw that the maximum body length of *Ptychodus* was somewhere around 36 feet.

The mystery of *Ptychodus* was finally solved in April 2024 when several complete fossilized bodies of this shark, found in Mexico, revealed it as an active fast-swimming lamniform with a powerful tail fin, that likely hunted ammonites and turtles rather than clams. It occupied a

**The ammonite and giant clam eater *Ptychodus* was the largest Cretaceous shark known, spanning between 33 and 39 feet long** ARTWORK BY BRIAN ENGH

totally new niche for lamniforms, a lifestyle they would never return to again. *Ptychodus* went extinct around 75 MYA, about the same time as the giant *Cretoxyrhina,* at a time when larger and more specialized mosasaurs began increasing in numbers in the oceans. A simultaneous increase in the abundance of rays, which also competed for hard-shelled prey, might well have sounded the death knell for *Ptychodus.*

Some other large chondrichthyans around at the time that would have been competing with *Ptychodus* in the clam-chomping business include the gigantic extinct chimaerids such as *Edaphodon.* The largest tooth plates of *Edaphodon* suggest total body sizes between 18 and 20 feet. The giant *Edaphodon* sailed through the end-Cretaceous extinction event and thrived in the seas around the world until it went extinct around 3 MYA.

So far, the picture is emerging that most of our knowledge of Cretaceous sharks has come from the efforts of dedicated scientists who go out and bulk-sample the sedimentary rocks, then wash the sediments away and sieve out fossil sharks' teeth. Previously, collectors had focused on large specimens that one could spot and pick up in the field. This new approach means that in the last four decades we have found vast numbers of new specimens representing diverse new fossil shark faunas from various sites around the globe.

## Spectacular fossil sharks from Lebanon and Mexico

One of the pioneers of this screen-washing technique who has made an unfathomable contribution to our knowledge of fossil sharks of the Mesozoic and Cenozoic eras is a quiet-spoken Frenchman named Henri Cappetta. Henri stands with Louis Agassiz and Arthur Smith Woodward as one of the three greatest paleontologists who advanced our knowledge of fossil sharks. Over a career spanning fifty-five years (1966–2021) he wrote or co-authored more than four hundred articles as well as two foundational works, the *Handbooks of Paleoichthyology* on Mesozoic and Cenozoic fossil sharks, which deal in intricate detail with their teeth, anatomy, and evolutionary trends. His sole-authored *Handbook* published in 2012 has 512 pages filled with hundreds of excellent photos and drawings of fossil sharks of the past 250 million years. Henri has written entries on nearly nine hundred different fossil shark genera, and his oeuvre includes references to several thousand species. His work has been my bible as I was writing this book (for more, see his bio on p. 426).

**The late Henri Cappetta in Morocco in 2009, sieving for shark tooth fossils**   PHOTO BY SYLVAIN ADNET

Henri's study of the complete fossil sharks and rays from the Late Cretaceous sites in Lebanon were seminal studies at the time, but in recent years so many new sharks have been coming out of these sites at such a fast rate that they haven't had time to be formally studied. Lebanon is without doubt ground zero for the study of well-preserved Late Cretaceous sharks—it is the Solnhofen of the Cretaceous.

The spectacular complete fossil sharks of Lebanon offer a unique window into the diversity and biology of a large number of sharks and rays living about 20 million years before the last great extinction event that killed off the dinosaurs and giant marine reptiles.

The superbly preserved fossil fishes and sharks from the mountains near Byblos, Lebanon, have been known since the time of Herodotus. Here we find spectacular complete fossil sharks that look as though they died only yesterday and were compressed flat, like flowers between the pages of a book. Some have bands of pigment across them; some have remnants of soft tissues or their last meals. The fossils were thought to have formed when deadly blooms of algae proliferated following heavy rains leading to lack of oxygen and low salinity, causing the death by asphyxia

Examples of near perfect complete fossil sharks from the late Cretaceous of Lebanon. TOP, a predatory *Cretalamna;* BOTTOM, a fossil example of the living catshark *Scyliorhinus* IMAGES COURTESY OF FRIEDRICH PFEIL, MUNICH

and poisoning of the sharks and other creatures living in the bays. The most significant thing about these fossils is not just their perfect preservation but the stunning diversity of shark and ray species that have been found.

Without going into exhaustive lists of the species present, suffice it to say that around sixty shark and ray species are now known from two sites of slightly different ages. The most dramatic difference they show is a great increase in numbers of rays compared to the marine fauna at the start of the Cretaceous. A 2021 review of the Lebanese fossil sharks records a stunning thirty-eight species of fossil sharks, of which twenty-one are completely new records of shark and ray species awaiting further scientific study.

Living genera of sharks are represented by sixgill sharks, piked dogfish sharks, a sleeper shark, an angel shark, and a sawshark, as well as a variety of catsharks and small reef sharks. The bullhead shark occurs here, along with a range of small orectolobiform sharks. Lamniforms include the living deep-sea sand tiger shark alongside other extinct forms. We also find various goblin sharks and the crow shark, plus *Cretoxyrhina* and *Cretodus.* There are some ten species of carcharhiniform sharks, most belonging to small catsharks and hound sharks. The fossil rays from these sites are truly spectacular. They include members of the extinct family of sawfish-like sclerorhynchids, plus typical rays like guitarfishes, skates, and sunfishes.

These Lebanese sites give us the most realistic snapshot of the diversity of sharks and rays living in the Late Cretaceous seas of the Northern Hemisphere. They show for the first time that modern kinds of sharks—those classified with living family groups—begin to dominate the seascapes at this time, with only a few archaic forms persisting.

Other sites with similar spectacular preservation of sharks of the same age as Lebanon are in Morocco and Mexico. The latter country has recently produced one of the most bizarre sharks of the entire Cretaceous period.

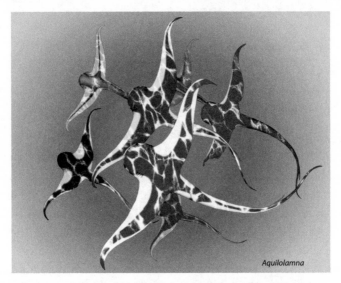

The incredible "manta-shark" *Aquilolamna* from Mexico was a Cretaceous filter feeder that probably lived like a manta ray   ARTWORK BY JULIUS CSOTONYI

*Aquilolamna* was a 4-foot-long shark resembling a manta ray stuck onto a long shark body. Unlike rays, it had a powerful tail that thrust the shark through the water. The teeth are not visible on the fossil, so they must have been minute. *Aquilolamna* and its kin went extinct at the end of the Cretaceous, and its loss was almost immediately filled by early members of the modern manta ray family.

## Deadly showdowns in Late Cretaceous seas

The seas were invaded by yet another predatory group of reptiles, the mosasaurs, about 82 MYA. They played a major role in reshaping the food chains at the end of the Cretaceous. They evolved from large lizards closely related to the monitors (varanids); indeed, they looked a lot like a Komodo dragon with paddles rather than arms and legs. At first they were small forms, around 6 to 10 feet long, and for this brief time, sharks like *Cretoxyrhina* still ruled the roost. However, it was only a few million years later (around 90 MYA) when large powerful mosasaurs like *Tylosaurus* emerged as top dogs of the seas. Growing up to 39 feet long, these were far larger than any sharks of the day, and their stomach remains also testify that they feasted on sharks as well as other large fishes, ammonites, and marine reptiles.

Mosasaurs dominated as the marine apex predators for the last half of the Late Cretaceous. Some of them, like *Globidens,* developed rounded teeth for cracking open ammonite shells, as evidenced by shells showing tooth bite patterns that precisely match the teeth of mosasaur palates. The very end of the Cretaceous seas harbored the most gigantic of all marine reptiles, with *Mosasaurus* possibly reaching 60 feet long. It had powerful, stout conical teeth for killing whatever it could find—large sharks, huge bony fishes, and a platter of various marine reptiles.

The large predatory sharks like *Cretoxyrhina* and *Cretodus* couldn't compete with the gargantuan mosasaurs, so they went extinct around 78 MYA. The largest shark in these final few million years of the Cretaceous might have been the hexanchid *Xampylodon,* whose large teeth were found on Seymour Island, Antarctica. It might have been around 16 to 20 feet long. Its teeth are often found with the skeletons of long-necked plesiosaurs, suggesting that it was scavenging on their carcasses. The highly predatory nature of large living hexanchid sharks, which sometimes prey on seals and small whales, suggests another alternative— that they were hunting and killing the plesiosaurs. We may never know which view is correct unless we find a pathological clue in a future fossil find. *Cretalamna* would have been around 10 feet long at the very end of the Cretaceous. It was a very important species at this time that would survive to carry on its line to the largest of all sharks—but not before the Cretaceous world was plunged into chaos.

At 66.4 MYA, an asteroid about seven miles across struck the Earth, and havoc ensued, resulting in the extinction of the large dinosaurs on land, the pterosaurs in the skies, and the huge marine reptiles such as mosasaurs and plesiosaurs in the oceans. The ammonites and many other marine creatures would also die out. How would sharks fare through all this chaos, and would the survivors be capable of making it through the next 66 million years to today? Let's find out in the next chapter, as we also encounter the rise of new predators in the marine world.

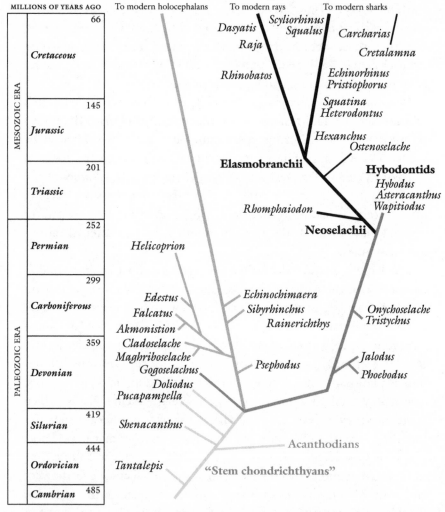

A simplified tree showing some of the main highlights of shark evolution up to the end of the Mesozoic era   DIAGRAM BY AUTHOR

# PART 4
# THE AGE OF THE MEGASHARKS

CHAPTER 12

# SHARKS AFTER THE IMPACT

## The Rise of Modern Sharks

———————

## Attack of the Killer Asteroid

Around 66.4 MYA, at the end of the Cretaceous period, a massive as-teroid crashed and exploded into what is today the Yucatán Peninsula of Mexico, leaving a vast crater whose rim, about ninety-three miles across, is still detectable through seismic profiles. The explosion evapo-rated unimaginably large volumes of seawater and vaporized billions of tons of crustal rock, hurtling debris thousands of miles into the atmo-sphere and creating vast dust clouds that encircled the globe. Common seafloor minerals like gypsum were also vaporized, which led to envi-ronmental chaos as rapid acidification of the oceans, and a temperature drop of around 14°F owing to blocked sunlight by atmospheric aero-sols and dust, brought about an immediate collapse of Earth's ecosys-tems. It might have taken another ten thousand years or so to wipe out most life on the planet. During this time, 75 percent of all species went extinct, including large dinosaurs, flying pterosaurs, and any land-dwelling animal that weighed more than about fifty pounds.

In the oceans, giant marine reptiles, the abundant squid-like am-monites and belemnites, and many kinds of plankton that formed the foundation of the ocean's food chains, along with the giant inoceramid clams and horn-shaped rudist clams that built massive reefs, all disap-peared, along with 45 percent of all corals.

It was also a time of great devastation for sharks at the species level.

A detailed study of how sharks and rays fared across this extinction event used a database of 3,200 shark and ray fossils representing 675 species. The authors, Guillaume Guinot and Fabien Condamine of the University of Montpellier, France, calculated that those sharks and rays eating hard-shelled prey, like the hybodonts and rays, were hit the hardest by the extinction event. Regular predatory sharks declined by around 62 percent, while the shell-crushing forms of both sharks and rays lost around 72 percent of their species. The victims included hybodonts, the scavenging crow sharks, some archaic sand tiger sharks, the sawfish-like sclerorhynchid rays, and about four other families of ancient rays.

Where you live, how mobile you are, and what you eat are the three biggest factors for surviving a global mass extinction event. The species that survived mostly included those with wider geographic ranges that were free-swimming forms rather than species dwelling on or near the seafloor eating hard-shelled prey. Sharks that had previously lived in moderately deep water, in the mid- to outer-shelf seas, were much less affected than the coastal shallow-water dwellers.

The surviving sharks would give rise to the most abundant sharks and rays that dominate the seas today. One surviving lineage of Cretaceous lamniforms would develop into the most spectacular guild of megapredators Earth had witnessed since the dinosaurs.

Sharks now faced increasing competition from two new challengers throughout the Cenozoic era—the 66-million-year period from the asteroid impact to today. First, whales, seals, and sea cows (manatees) would soon appear in the seas and take up major positions in the oceanic food chains, some as gigantic apex predators. Second, an explosive rise of more teleostean ray-finned fishes would reach new peaks of diversity and take on extreme body shapes (think sunfishes, eels, and deep-sea anglerfishes). This means they could challenge sharks in nearly all ecological niches in all habitats, from the bright, briny surface waters to the deepest, darkest abyssal plains of the ocean floor.

These fishes would evolve into abundant medium-sized predators like marlin and tuna, and smaller forms would evolve into massive schools of filter feeders like anchovies and sardines. Others take over the newly emerging coral reefs using a myriad of colorful and unusual body shapes. Putting this into perspective, today there are around thirty thousand species of teleosteans—the most abundant backboned creatures on our planet—while sharks, rays, and chimaerids are represented by around twelve hundred species.

The sharks that survived the end-Cretaceous extinction event can be divided into two categories: those that had lived for many millions of years, including early and archaic forms that had already peaked in the Mesozoic seas and were on the way out, and the newly emerging groups that first appeared not long before the extinction, including many species classified in modern shark families. The dominant sharks of the Cretaceous in both size and diversity were the lamniforms, which included more than a hundred different species in both extinct families, like the crow sharks, and living groups like the sand tiger and goblin sharks. Today, the living lamniforms contain just fifteen species of sharks in eight families, much less diverse today than they were in the Cretaceous. The major driving force in lamniform sharks of the Cenozoic would be the race for bigger body size, ultimately leading to a continuum of changing species culminating in the largest shark that ever lived: megalodon.

## Old and new sharks of the new era

After all the end-Cretaceous chaos, the Paleocene period (66.4 to 56 MYA) marks the beginning of the Cenozoic (meaning "new life") era. In their 465-million-year history, we are now in the home stretch for sharks, the last 14 percent of the total time they have been around.

The most diverse shark group alive today are the carcharhiniforms, which include many kinds of catsharks, houndsharks, requiem sharks, hammerheads, and the tiger shark. Today there are 291 species of carcharhiniform sharks grouped into ten families. One of these, the deep-sea catsharks, contains a whopping 111 species, making it the most diverse family of sharks of all time. While carcharhiniform sharks are typified by a wide range of body shapes, from deep-sea bottom dwellers to the massive great hammerheads and tigers, they all share a special feature of the jaws—the teeth are largest on the sides of the jaws rather than at the front. The carcharhiniforms first appeared in the Early Cretaceous, yet they were not very abundant until right at the end of that period, when there were around forty species. They survived the end-Cretaceous extinction to emerge as the new rising stars of the shark world.

So how did this massive change from the Cretaceous dominance of lamniforms to the Cenozoic triumph of carcharhiniform sharks occur in the aftermath of the devastating extinctions? To answer this question, I spoke with the leading expert on the subject, a sharp young shark paleontologist named Mohamad Bazzi from the University of Uppsala in Sweden.

Mohamad was born in Beirut but grew up in Sweden. He is today one of the leading experts on the extinction of sharks at the end of the Cretaceous. His novel approach combined a rigorous mathematical study of shark tooth shape variations (**morphometrics**) with statistical modeling to reveal the trends in shark diversity across the mass extinction event. One of his key findings was that the overall tooth shapes of sharks did not vary much before or after the extinction event, except for the large species with triangular cutting teeth like the large-bodied lamniforms including the crow sharks like *Squalicorax*. The deep-sea sand tiger sharks (odontaspids) underwent a burst of diversification immediately after the extinction event.

The variations in shark tooth shape through time also tracked the changes in habitats and resource availability. Tooth shapes correspond closely with diet in most extant shark species, and the post-Cretaceous declines in certain lineages like lamniforms can be attributed to their becoming increasingly more specialized in their choice of prey. While the modern lamniform sharks are far less diverse than their carcharhiniform rivals, they show more disparity—the range of variations in shape—in their existing species.

Next, Mohamad analyzed shark tooth shape variations matched against various environmental factors to see if there was any correlation between variations in shark teeth and changing oceanic and climatic conditions. He compiled a new database containing nearly 4,000 fossil and extant shark teeth. The results found that the curve of sea level changes and ocean surface temperatures closely matched the curves for lamniform and carcharhiniform tooth-shape disparity over time—disparity being the range of variation seen in the tooth shapes. The end-Cretaceous extinction greatly affected the large-bodied lamniforms, and with their absence, giant hexanchid sharks filled their vacant niche at the start of the Paleocene period. We know that living hexanchids sometimes invade the predatory niche of the lamniform white sharks, so this is not such a surprising result. The extinctions were also heightened in lines of sharks that were specialized predators, possibly corresponding to the simultaneous loss of large marine reptiles going out at the extinction.

**The piked dogfish was the world's most abundant shark at one time. The genus *Squalus* first appears in the Late Cretaceous around 100 MYA**   FISHES OF AUSTRALIA (FISHESOFAUSTRALIA.NET.AU)

Aside from lamniforms and carcharhiniforms, many other sharks started to appear at the start of this new era. The Paleocene is a highly significant time because the vast majority of living shark genera began appearing in our oceans, according to the fossil record of their teeth. Here are some examples of well-known living forms and their brief ancestral history.

The squaliform sharks include well-known species like the abundant dogfishes and the giant Greenland shark. They are a highly diverse group comprising 135 species, including the dogfishes, the deep-sea lantern sharks, the sleeper sharks, the gulper sharks, and the kitefin and cookiecutter sharks. The piked dogfish, *Squalus acanthias,* was once probably the world's most abundant shark species and the target for large commercial shark fisheries. *Squalus* first appeared in the Cretaceous, about 100 MYA, and several new species appeared at the beginning of the Paleocene. While its fossil diversity remained low throughout the Cenozoic, today we have some thirty-four living species of this iconic shark. This is much the same pattern for the other squaliform groups, as so little is known of the fossil record of the living shark genera.

The deep-sea kitefin shark, *Dalatias,* first appears in the Paleocene too. It has a poor fossil record and so is known only from teeth. Closely related cookiecutter sharks (*Isistius*) first appear in the late Paleocene, about 59 MYA, based on lower jaw teeth. They have on rare occasions taken a bite out of humans swimming in open waters. At the time they first appeared, whales had not yet invaded the oceans, so it's likely that cookiecutters first evolved to prey on large bony fishes like tuna and marlin, or on other large sharks. The oldest billfish, *Hemingwaya,* also appeared in the Paleocene, appropriately named after Ernest Hemingway, who liked his big game fishing.

The living sawshark, *Pliotrema,* makes its first appearance in the mid-Paleocene, no doubt taking advantage of the demise of the giant fossil sawfishes (a group of rays), like *Onchosaurus* (in the previous chapter). Sawsharks would continue diversifying in both in open marine and near-shore habitats, frequently invading estuaries.

The modern family of white sharks and kin, the Lamnidae, makes its first appearance at the start of the Paleocene, represented by the genus *Isurolamna,* with teeth up to 1.3 inches high, found in Europe and North Africa. It must have been a large shark similar in shape and size to today's makos, perhaps reaching up to 12 feet long. The other line of lamniforms that emerged at this time was the *Otodus* group that evolved seamlessly from one of the several species of *Cretalamna* with similar-looking teeth. It is represented by abundant teeth of the giant *Otodus obliquus* found nearly all around the globe throughout the Paleocene. More about this shark and its reign in the next chapter.

The electric *Torpedo* rays appeared in the Paleocene with some species reaching enormous sizes, far exceeding living forms, based on tooth sizes. Extinct forms like *Eotorpedo* from Europe and North Africa might well have been at least 10 to 13 feet wide and capable of discharging very strong electric charges to kill large prey. Luckily for us, these giant electric rays are no longer around to shock us unexpectedly.

## A thermal crisis hits the oceans

Just as new kinds of sharks and rays were getting established in this new post-asteroid world, another global crisis was about to take place and once more brutally shake up marine food chains. The Paleocene-Eocene Thermal Maximum event is marked by ocean temperatures suddenly skyrocketing by about 11°F over the course of perhaps five thousand years, causing rising seas from thermal expansion of the water, and increased acidification. This may not sound particularly

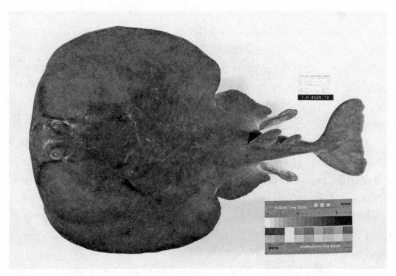

**Electric rays like this *Torpedo* appeared at this time too**
BY ROBIN MCPHEE, COURTESY OF CSIRO, AUSTRALIA

rapid, but in geological terms, this was a warp-speed warming—which then lasted for around two hundred thousand years. The cause is debated, but we suspect massive volcanic eruptions from the North Atlantic Igneous Province, which would have caused enormous amounts of carbon dioxide to be released into the atmosphere. Whatever the cause, the effect was what some scientists refer to as a **biotic crisis**— a severe reduction in abundance of certain marine species without the extinction rates that define the big global mass extinction events. It struck the oceans just 11 million years after the asteroid strike that killed the dinosaurs.

This second crisis had a dramatic effect on nature's smallest organisms—an effect that trickled up to larger species when as many as half of marine benthic foraminiferans, a common amoeba-like microorganism with a calcareous shell that lived on the seafloor, went extinct. These creatures formed the base of the food chain for many species. Yet all was not lost: Other microorganisms and some plankton thrived in the higher temperatures, which meant that not all marine food chains had been destroyed. Sharks had to adapt as best they could or face the consequences of extinction. They rose to the challenge.

My friend David Ward, a shark paleontologist based in London, told me that his extensive collection of shark fossils from Egypt from the

Paleocene-Eocene Thermal Maximum event contains filter-feeding species, suggesting that some sharks enjoyed this warmer phase of the oceans. This implies there was plenty of plankton in the oceans, so food chains were being recharged from the bottom up. Other evidence from shark denticles and other microremains throughout the event shows an increase in shark biomass at the start of the warming, returning to pre-warming levels at the end of the two-hundred-thousand-year-long event. Tracking shark teeth shape changes across this event, Mohamad Bazzi found that the event triggered a shift toward more specialized diets in sharks, probably associated with the decline of certain reef-dwelling bony fishes at the time. Overall, there were no dire effects on sharks across the event—it was more of a party time for them.

Just as things were recovering from the events of the Paleocene, another phase of global warming was set to begin.

## Sharks really get warmed up

The world was much warmer in the following geological time period, the Eocene (56–33.4 MYA), when global temperatures increased once more by a whopping 16–25°F (9–14°C) between 52 and 50 MYA. This second warming event in the Early Cenozoic is called the Early Eocene Thermal Maximum. A good example of just how much warmer the world was at this time is the presence of alligators living on Ellesmere Island in far northern Canada, which even then sat at latitudes between 75° and 78°—well within the Arctic Circle. Eocene forests in the state of Oregon resembled those found in the Central and South American jungles today.

An ancestral filter-feeding whale shark called *Palaeorhincodon* and the modern basking shark both appear between the Early and Middle Eocene (48–40 MYA) respectively, possibly as a response to oceans that are now bursting with productivity—there was much planktonic food around. It's as though someone topped up the oceans with big booster shots of nutrients, and the large filter-feeding sharks quickly evolved to take advantage of them.

Deep-sea lantern sharks appear at this time, venturing into the dark mesopelagic zone (656 to 3,281 feet deep) where it was relatively safe from surface water predators. It's a great place to live for a small shark,

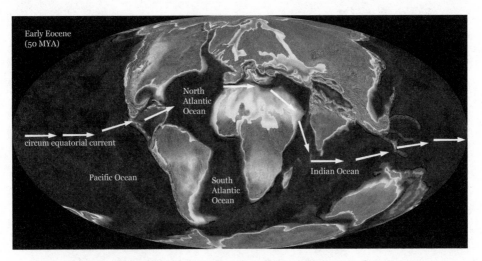

**The world in the Early Eocene, when temperatures were much warmer than today, around 50 MYA** RON BLAKEY, DEEP TIME MAPS (DEEPTIMEMAPS.COM)

considering that most of the world's squid also live there. They flourished down there, even developing their own wild fluorescent lighting systems in their skin for having crazy disco parties (more for hunting, evading predators, and courtship). Today there are some forty-five known species of this genus.

The first living wobbegong sharks, *Orectolobus*, appeared in the mid-Eocene alongside the first of the zebra sharks, *Stegostoma*. Wobbegongs get their name from an Australian indigenous word meaning "shaggy beard" on account of the peculiar fringes and tassels coming off the front of their mouth region. They are found only in the western Pacific and Indian oceans surrounding Australia up through to Japan in the north, represented by a dozen species. They are small sharks; most are about 3 feet long, with the largest species, the spotted wobbegong, occasionally spanning up to 10 feet. They first appear about 45 MYA in the fossil deposits of the U.S. state of Mississippi, so the group clearly once had a much wider global range.

Wobbegongs occupy the niche of bottom-dwelling, opportunistic feeders who use their sensitive barbels and electroreceptors to detect prey items like small crabs and shrimps. Zebra sharks, also called leopard sharks because of their spotted appearance, have long bodies with a tail almost half their entire length. This tail shape is perfect for slow-moving forms that flit along the seafloor. They are medium-sized sharks up to II feet long that are sluggish feeders by day, becoming very active hunting around reefs at night. Their distinctive teeth have three spear-shaped prongs, perfect for pulling out small crabs, shrimps, and snails from crevices. Today they live only in the Indian and western Pacific oceans. The oldest zebra shark fossil is a single tooth found in the Eocene deposits of Seymour Island, Antarctica.

Aside from *Otodus,* other lamniforms appearing at this time include the very first thresher sharks, although we have no idea whether the oldest forms had very long tails, as do living species, because we have only their isolated teeth. The living genus *Alopias* appears in the Lower Eocene of England and Morocco, based on teeth. Today its teeth are up to 0.6 inch high, with maximum body lengths around 20 feet for the common thresher. Threshers are so named for their trick of leaping out of the water and slapping the water with their long tail to stun small fishes so it can then eat them. Sometimes threshers work collaboratively in groups to attack large schools of fishes. The modern goblin shark, *Mitsukurina,* also first appears in the Early Eocene. Other lamniforms from the Eocene onward will be dealt with in more detail at the end of this chapter when we investigate the rise of the giant megatoothed sharks.

The ungainly goblin shark haunts the deep seas catching fishes with its sharp, spindly teeth. This family of sharks first appeared around 100 MYA   BLUE PLANET ARCHIVES, LLC

The carcharhiniforms kept diversifying throughout this time, with modern whaler and requiem sharks in the genus *Carcharhinus* appearing at much the same time as the first of the modern tiger sharks, *Galeocerdo*. I focus on these groups later as their story is particularly important to the big picture of where sharks were heading.

While most of these fossil records are from isolated shark teeth, there is one quite exceptional site that entrances us with complete, perfectly preserved snapshots of the fossil sharks and rays of early Eocene age, during the peak climatic warm period. It is at Bolca, in northern Italy, a magical place for all fossil fanatics.

## The awesome sharks of Bolca

I first visited the Natural History Museum in Verona, Italy, as a student in 1982, on my first overseas trip. I had heard about an old museum full of spectacular fish fossils, so I wrote a letter to the curator, Dr. Lorenzo Sorbini, who responded welcoming my visit. On the day I visited he kindly showed me around the museum's displays of fossil fishes, and he later sent me a copy of his beautiful color book on the Bolca fossils. I wasn't able to visit the Bolca fossil site on that trip. So, when I began writing this book, I felt the urgent need to go back to Italy and see the fossil site for myself and take notes on the fossil sharks and rays in the museums.

In November 2022, my wife, Heather, and I went back to the museum in Verona and made a special visit to see the Bolca fossil sites and its town fossil museum, assisted by a local paleontologist, Cesare Papazzoni (who on the same morning had kindly driven us to see the giant Cretaceous sharks at Sant'Anna D'Alfaedo). We headed east, upward into the beautiful Lessini Mountains toward Bolca. It was a long, winding drive, and we arrived at Bolca just in time for a late lunch on a gorgeous Sunday afternoon. Fortified with a superb meal of homemade pasta and ragout washed down with the obligatory glass of local vino, we headed out to the fossil site a few miles away in the national park. From the parking lot it's about a mile walk down a pleasant, forested track to the bottom of the hill, locating us in a narrow basin, known as La Pesciara di Bolca, or "the fishbowl of Bolca."

Here we admired a small hill riddled with tunnels where the fossils had been extracted over the past five centuries. You read that correctly—

The famous Pesciara di Bolca (Fishbowl of Bolca) site north of Verona,
Italy, where fossils have been mined from the hill on the right since the
1500s   AUTHOR PHOTO

five hundred years of fossil collecting. A local doctor, Andrea Mattioli,
published work on the fish fossils in 1555. Then, in 1571, the town of
Verona set up the first museum housing and displaying the natural history of the region, featuring these amazing Bolca fossils. Over the next
two centuries, private collecting of the fossils increased. By the end of
1792, Giovanni Gazzola had more than twelve hundred specimens of
fossil fishes on display in his private museum. Then, in 1797, Napo-
leon's armies invaded Verona and confiscated some six hundred spectacular fish fossils from Gazzola's museum. The French emperor gave
these fossils as spoils of war to the Natural History Museum in Paris,
where they were later studied by Louis Agassiz; they remain there
today. I saw a miniature scale model of Gazzola's remarkable museum
in the Natural History Museum in Verona, complete with tiny cabinets
filled with minute painted models of the fossils—it was like some crazy
scientific dollhouse.

The Museo dei Fossili is located in the little town of Bolca not far
from the fossil sites. It is managed by the quarry owners, who have
kept the quarry started by the Cerato family since the 1700s. It is
jointly run with town council support, so that the very best fossils

found in the region are carefully prepared, studied, and placed on public display. The fossils are not allowed by law to be sold, to protect the sites from vandalism, with the sole exception of a limited number of slabs with fishes on them that are allowed to be turned into souvenir clocks. I feel this is a justified exception as so many of the fossils found there are of the same common small fishes that are now well represented in many museum collections around the world. And what better clock to have than one with a 48-million-year-old fish on it to remind you of our short existence as a species on Earth, when measured on the scale of deep time?

**TOP: Ancestral species of *Galeorhinus*, a small, perfectly preserved triakid shark. BOTTOM: *Eogaleus*, an extinct carcharhinid shark. Both specimens from the Bolca site** PHOTOS BY AUTHOR, COURTESY OF THE NATURAL HISTORY MUSEUM OF VERONA

The Bolca site is exceptional, with color patterns and pigments visible on some of the prize fossils. While the sharks and rays are near perfect, with their complete body outlines, skin, and fins clearly defined, for me it was one small fossil that brought a tear to my eye, perhaps representing an ultimate perfection of form in a long-dead living creature. If fossils were art, this one would be a *Mona Lisa*. It is a tiny, complete little octopus with eight curved legs, and in its head you can see a teardrop-shaped black ink sac. It highlights the delicate nature of fossil preservation at the site, which has now yielded some 250 different species of fishes alone, along with a host of distinct marine creatures like perfect crabs, lobsters, shrimps, corals, clams, snails, worms, cephalopods, and many other small animals. We can determine clearly from all these exquisite remains exactly what the sharks and rays were hunting. The site also contains a vast number of terrestrial plants, insects, spiders, and some larger animals, like crocodiles, turtles, and snakes, that all lived on land or in rivers but whose remains got washed into the sea to be buried.

Around 48 MYA this region was an archipelago of small islands in a tropical sea on the northeastern margin of the Atlantic Ocean. A current flowed around the world's equator at this time, keeping the seas in these higher temperate latitudes quite warm. The Bolca fossil site represents one of the earliest collections of coral reef fishes. It contains some eight families of ray-finned fishes that are found only on coral reefs today.

Two species of carcharhiniform sharks are found at Bolca, both known from complete specimens, plus another rare form, a lamnid, known from a few isolated small teeth. The largest complete shark is *Eogaleus bolcensis,* which reached about 5 feet in length. Its teeth are well preserved and typical for small requiem sharks (in the carcharhinid family) with coarsely serrated lower teeth that almost resemble those of the tiger shark. The smaller complete form is *Galeorhinus cuvier,* a small school shark in the triakid family. The dark pigmented tips on its fins and tail fin are clearly seen in some of the best specimens on display at both the museums in Bolca and Verona. It grew to almost 3 feet long, although a recent study pointed out that all the known specimens are juveniles, so adults were likely quite a bit larger.

Studies of the environmental settings that both sharks came from show overlapping yet different ecological niches. *Eogaleus* lived a free-swimming lifestyle near the coast, while *Galeorhinus* favored a lagoonal-coastal habitat. Sharks in these families today occupy a wide range of environments from coastal seas, lagoons, and coral reefs, so the fossil species at Bolca exhibit a conservative habitat preference through time.

The other Bolca shark is a sand tiger shark called *Brachycarcharias,* known only from some slender curved teeth that show that it grew to around 5 feet long. As in living sand tiger sharks it clearly hunted fishes, of which there were many different species to choose from in the warm seas around ancient Bolca.

The stingrays of Bolca are equally spectacular, with perfect outlines of every detail from their stomach contents to mothers bearing unborn embryos and others with patches of orange skin pigments. Several kinds are found there as complete individuals. The commonest forms are known today as living forms such as the whiptail stingrays *Trygon*, and *Dasyatis*. These have powerful, venomous spines in the tail. They hunted on the seafloor chasing fishes and small invertebrates. The extinct eagle ray *Promyliobatis* found here is not too dissimilar to the common living eagle rays. Electric rays, or numbfishes, like *Titanonarcine*, are also found at Bolca. Today many of these forms feed on specific prey like sand worms (polychaetes), but one of the Bolca species had a unique diet. It ate masses of *Alveolina*, a microscopic foraminiferan that were buried in the seafloor. These are single-celled amoeba-like creatures that secrete a calcareous shell around themselves. The fossil ray had a dense package of them in its gut, signaling a unique feeding strategy for rays that is not known today. The full range of Bolca ray fossils show that 48 MYA, the rays of the northern Tethys Ocean included ancestral species of the most common rays around today.

Bolca sharks and rays are exciting to look at because some are just

**Essentially modern rays like this early eagle ray were abundant by the Eocene. Natural History Museum of Verona** PHOTO BY AUTHOR, COURTESY OF THE NATURAL HISTORY MUSEUM OF VERONA

like living species complete with their color patterns and pigments of the round eyes pressed onto rocks. When I viewed them in 1982, it was the first time I had ever seen a perfect fossil of a complete shark. To me it resembled an embalmed blacktip reef shark with its telltale dark patches of pigment on the tips of all its fins (although we now know it is a different extinct species). That memory, now reinforced from the new spectacular specimens I examined at Bolca, has stuck in my brain, forever reminding me that long-dead creatures can be sublimely beautiful when trapped in their eternal death.

Yet they still cry out to us to heed the warnings for our future. These ancient coral reefs of Italy were once resplendent with life, and through the natural cycles of nature they are now gone. They remind us that our living reefs are currently under grave threat from rising sea temperatures caused not by the rhythm of nature's cycles, but directly by us humans.

## Serious predatory sharks appear in our seas

During the Eocene, the oceans became the site of a vicious arms race when new animals invaded the sea and became prey for large predatory sharks, who started appearing in the oceans at this time. Let's discover how some of these large sharks emerged. Then, in the next chapter, we will continue to track this arms race between the first whales in the seas and the lamnid sharks called *Otodus*.

The tiger shark and bull shark are two of the three large predatory sharks that have caused most human fatalities in close encounters, along with the white shark. Today, tiger sharks range from southern Australian latitudes to the northern seas of Iceland, and they forage in a wide range of habitats including tropical reefs, open oceans, and estuaries. Growing up to 22 feet long, they take their name from the distinct striped pattern on their backs. They mature between four and thirteen years old, their longevity varying according to the region they grow up in. Because these sharks are well known for their willingness to eat just about anything they can fit into their mouth, they have been dubbed "garbage bins with fins." They have been reported eating fishes, jellyfishes, sea snakes, marine iguanas, seabirds, turtles, marine mammals, people, carrion, rubbish, and many other species of sharks

Tiger sharks appeared in our oceans about 45 MYA, and some extinct forms grew to about 25 feet long    PHOTO BY NIGEL MARSH

and rays, including their own pups. However, as much maligned as they are as "man-eaters," the following true story from Australia shows they are not always guilty of manslaughter as accused.

In 1939, a 17-foot tiger shark was caught on a bait line about 1,100 yards off Coogee Beach, Sydney, and was placed in an aquarium. Some eight days later, much to the horror of onlookers, it vomited up a bird, a rat, and a muscular human arm. The police came in to investigate and at first thought that the shark must have eaten someone, perhaps a suicide victim. The arm bore a distinctive tattoo of two boxers in red trunks facing off. The coroner examined the arm and saw that it had been cut cleanly off at the shoulder by a sharp instrument, certainly not the work of the ragged, hooked teeth of a hungry tiger shark. After they posted a photo of it in the newspapers, the victim's brother came in and told the police it belonged to a James Smith. The fingerprints were then matched to those of James Smith, who was already on file for an illegal betting charge. He had been murdered before being cut up and his parts were tossed from a boat into the sea. After a series of arrests and charges, no one was convicted of the murder. It was argued in court that one arm alone is not enough to prove a murder, as it is possible that the one-armed survivor might be out there somewhere. The poor tiger shark was the real victim in all this—he never killed anyone, but he remained confined and ill after being caught. Just three days after coughing up the arm, he died.

So how and when did tiger sharks first get their start in the oceans? Have they always been non-fussy oceanic omnivores—eaters of all things? The fossil teeth of tiger sharks start appearing in the Middle Eocene, around 45 MYA, showing that the group arose at a time of increased ocean productivity brought on by the Eocene Thermal Optimum. Julia Türtscher of the University of Vienna studied the many extinct species and culled the list down to just six valid species. They can use their can-opener teeth to first crunch down hard on, then hack open the shells of turtles, one of their favorite meals. Up close you can see that some of the larger serrations on their teeth actually have smaller serrations on them, so they are supremely adapted for sawing through flesh. Unlike many other carcharhiniform sharks, whose upper and lower jaw teeth differ significantly, tiger sharks have similar teeth on both jaws.

The living tiger shark, *Galeocerdo cuvier,* started its career in the garbage industry in the Pliocene period, about 4 to 5 MYA, from teeth found in Europe, the United States, and South Africa. Tiger sharks' teeth of Miocene age from Bone Valley, Florida, have been posted on the internet with sizes of up to 1.4 inches long, way larger than the biggest modern tiger shark teeth, so perhaps these represent ancient tiger sharks up to 25 feet long. Tiger sharks are distinctive enough to be placed in their own family, with recent studies of their DNA showing them to be most closely related to the hammerhead sharks—all of which are most closely related to the requiem or ground sharks (*Carcharhinus*). These sharks appeared in our oceans at exactly the same time as tiger sharks.

The tiger sharks, hammerheads, requiem sharks, and many kinds of carcharhinid reef sharks all took off at the start of the Eocene period, immediately carving out their own niches in the world's oceans. Today the requiem sharks inhabit a wider range of habitats than any other group of shark, some having the capability to invade rivers to travel thousands of miles inland in search of prey. They are typified by species like the common bronze whaler or copper shark featuring a sleek, streamlined body with large fins and a mouthful of small wide teeth bearing fine serrations. They feed primarily on fish—though some species like the bull shark are aggressive hunters.

The carcharhinid family is one of the most notorious groups for known human fatalities. This is why they are also called the "requiem sharks," from the French word meaning rest (or death), alluding to their hunting skills that kill prey efficiently. They are a highly diverse group of at least fifty species (including thirty-five *Carcharhinus* species), and the most nefarious one is the bull shark. While there are few cases of their directly killing humans, most encounters occur when they bite people on the lower limbs, often in shallow water or when surfers trail their legs in the water. Their razor-sharp serrated teeth inflict very deep wounds; people can quickly die from shock and rapid blood loss.

The oldest member of the modern genus is *Carcharhinus underwoodi*, named after my colleague Charlie Underwood. It was found from the Middle Eocene deposits on Madagascar, dated at 45 MYA. Its teeth are very similar to those of the modern bull shark. The oldest member of the same family is a strange, coarsely-serrated-toothed shark named *Abdounia* from the lower Paleocene of Morocco (65 MYA), showing that they first appeared immediately after the mass extinction event. They were similar in size to modern bull sharks.

The milk sharks and sharpnose sharks (*Rhizoprionodon*) are common species in the Pacific, Indian, and western Atlantic oceans. They are small predatory sharks less than 5 feet in length that hunt bony fishes, squid, and crustaceans in shallow coastal waters and shelf seas down to about 600 feet deep. They first appeared in the Early Eocene (c.50 MYA) in Northern Africa, Europe, and the United States, and are known only from fossil teeth.

Hammerhead sharks are a recently evolved group of carcharhiniforms that developed specialized head shapes from rounded scalloped saucer-shaped heads (bonnethead sharks) to heads bearing extended wings with the eyes at the ends in typical hammerhead species. They are known from two genera and nine living species, the largest being the great hammerhead, reaching up to 20 feet in length. The wide bonnet-shaped or hammer-shaped heads allow the sharks to develop streamlined profiles for cutting through the water quickly as well as allowing larger surface areas for their electroreceptors for hunting prey. The winghead shark (*Eusphyrna blochii*) is the most extreme example of

Hammerhead sharks appeared in our oceans about 30 MYA. They are closely related to tiger sharks   PHOTO BY NIGEL MARSH

all hammerheads—the wings of the head extend wider than the out-stretched pectoral fins, or about half the shark's total length.

Teeth of the living hammerhead, *Sphyrna,* first appear in the early Oligocene period about 30 MYA, while teeth of the modern smooth hammerhead species start in the Miocene of Portugal, about 15 MYA. Hammerhead sharks have two types of teeth, for clutching slippery prey like octopus or crabs, as seen in bonnethead sharks, or for cutting into large prey using teeth with finely serrated edges, as used by the great hammerhead, which mainly hunts rays and large bony fishes.

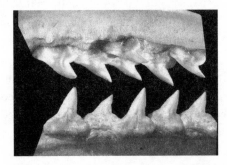

Great hammerhead teeth are perfect for gripping and saw-ing prey with their serrated edges   PHOTO BY ROSS ROB-ERTSON, SHOREFISHES OF THE TROPICAL EASTERN PACIFIC IN-FORMATION SYSTEM

While all these sharks were establishing their own particular way of eking out a living in the seas, another line of lamniform sharks, the **otodontids,** that started with *Otodus* was taking over as the absolute top predator of the world's oceans. As richer new food resources became available in the oceans, they grew larger and more voracious.

At around the same time as the Bolca deposits were being laid down in Italy, a long-snouted wolf-sized mammal in Pakistan named *Pakicetus* began swimming in the sea hunting for fishes. It was the first archaic "walking whale," heralding a long line of seagoing mammals that would ultimately lead to the largest living animals on Earth. The dramatic rise of whales in the oceans would be both a blessing and a curse for sharks.

In the next chapter we will follow the family line of megatoothed sharks called *Otodus* through their arms race with the ascent of whales, culminating in the largest predatory whales and most ferocious shark predators to ever grace our oceans.

CHAPTER 13

# ASCENT OF THE SUPERPREDATORS

*How Shark Predators Got Very Large Very Quickly*

―――――――――

## Krazy Kazakhstan and the mystery of the megatoothed sharks

For most of the last century, the origin of the gigantic megatoothed sharks culminating in *Otodus megalodon* was a mystery hotly debated by paleontologists. One group of scientists thought they must have evolved from serrated-toothed ancestors; others thought they might have gained serrations on their teeth at some critical stage in their evolution. Like all mysteries in paleontology, evidence of such transitions was required by finding fossils in layers of known precise age when the evolutionary transition was taking place. It's exactly like how Louis and Mary Leakey spent decades searching in Africa looking for the missing link—the earliest ancestors of humans—because they realized it was the right place with the right geological strata and representing the correct time period to find such fossils. They just needed to go and search. It took them many years, but they eventually succeeded.

To solve the mystery of when the megatoothed sharks began their rise, someone had to go and search the right area of the world, exposing rock from precisely the right time zone, to pinpoint the exact layer showing when an earlier species of *Otodus* first gained serrations. Serrated edges were the key to solving the mystery because they suddenly appear at a precise time in the evolutionary line. It was a hunt for a missing link—to discover a new fossil shark species showing serrations only partly developed in a large sample of its teeth. That would link the smooth-toothed *Otodus obliquus* and the larger serrated-toothed species.

One of my friends, David Ward, from England, set out on a series of wild expeditions to the remote deserts of Kazakhstan to crack this juicy mystery. First, a little background on David, who is far from your average paleontologist.

David started his working life as a veterinarian but stepped back from his practice some decades ago to hunt for fossil sharks around the globe. He has contributed a large number of scientific papers and been instrumental in naming more than ninety new fossil species. Aside from his impressive achievements in the field, perhaps the most inspiring thing about David's story is that it shows you can make a significant contribution to science without necessarily having formal qualifications in the field. He regards finding new fossil sharks as relatively easy compared with the intense work of sorting the data and publishing it. He is an excellent example of the twentieth-century gentleman naturalist.

David was invited by his colleague Victor Zhelezko to work in western Kazakhstan starting in the 1990s, at a time when the Soviet Union was disintegrating and Kazakhstan was newly independent. In the 1920s, fossil sharks' teeth had been discovered from the spoils of a well dug on the banks of the River Emba at Aksuat. David had seen similar exquisitely preserved teeth from the legendary site of Aktulagay on a visit to the Geological Institute in Ekaterinburg. The largest teeth, about 1.3 inches high, were from the common lamniform *Otodus obliquus.* Remarkably, some of the teeth showed the beginnings of serrations developing along their cutting edges, while other teeth had none. A third group of teeth from here were heavily serrated, like a bread knife. David was intrigued. The field geologists who collected these teeth had thought they had eroded out of the same layer, so represented only one species with a lot of variations—some smooth, some serrated. This idea provided no hard evidence for a sequence of shark species changing their teeth over time.

In David's mind, the only way to solve this problem was to go back there and collect the *Otodus* species layer by layer to see if there was a definite change as the sedimentary rocks became younger. This was extremely important, because at the time, some scientists thought that *Otodus obliquus* was an ancestor of the megalodon, which was then thought either to be closely related to the white shark, *Carcharodon,* or to belong in another genus, *Carcharocles.* David saw this as an unprecedented opportunity to head back to the field to find the source of this ancestral line of the megalodon. For paleontologists at the time, it was like an explorer seeking the source of the Nile.

Here's the story he recounted to me of his remarkable field expeditions that brought home stunning scientific results that finally solved this major mystery.

Kazakhstan, May 2000: Together with his geologist friend Chris King, David set off from the UK and eventually landed in Aktau, on the northeast coast of the Caspian Sea. Following some tense discussion of who was supposed to bring the map showing their destination, David remembered he had a sketch of it in his field notebook drawn by Victor Zhelezko. It gave the all-important distance and compass bearing of the famous Aktulagay site from a bridge over the Emba River, the only reliable landmark in this vast desert area within several hundred miles.

From there the drive took two days on harsh gravel roads, the last thirty miles off-road over rolling chalk steppes, but they eventually

David Ward takes a break after working in the hot, remote desert of Kazakhstan finding fossil sharks' teeth   PHOTO SUPPLIED BY DAVID WARD

found the Emba bridge where the sharks' teeth had originally been discovered in the 1920s. This area was arid semidesert in some places, with intermittent marshy areas along the base of the ridges. These dry areas, known locally as *sors*, allowed the team to drive quite quickly, so long as they kept a close eye out for muddy patches. After a couple hours, Chris spotted a range of dark hills on the horizon, and they knew at once they had found the legendary site of Aktulagay (a Shangri-La for shark paleontologists). They arrived close to dusk and made camp on the base of the cliff, at the junction between the Cretaceous layers and the younger Cenozoic white chalks.

David reminisced that it was a truly wonderful place. The soft chalk valley floors were festooned with tall white blossoms of wild garlic and a scattering of bright ruby-red dwarf tulips. Every few yards, a dark orange belemnite fossil (the internal part of a fossil squid-like creature) poked out of the cream-colored soil. The small artesian wells drilled in the area were usually alkaline or saline, not drinkable by the wildlife, so saiga antelopes and wild horses were not abundant here. The trade-off was that the shade-giving cliff sections were not trampled by these animals, so the fossil shark teeth that had weathered out of the chalk remained undamaged, ripe for the picking.

Field work started at first light and ended at dusk, fourteen hours later. The weather was pleasant, with temperatures in the seventies under a cloudless sky. Once Dave and Chris got their eyes tuned in to spotting the black teeth against the dark clay, they started finding lots of fossil shark teeth. As they crawled over the weathered clay surfaces, a pair of eagles circled above, occasionally casting shadows over them.

On the first day, they found a mix of fossil shark species that were once living just above the seafloor as well as deep-water sharks, but none that lived on the seafloor. This valuable information suggested that the seafloor was oxygen-deprived, which explained why the teeth were so well preserved, as scavengers had not disturbed the carcasses when they fell to the sea bottom. As David worked his way farther from the camp, moving higher up the cliffs, the age of the rocks became younger and the big teeth of *Otodus* became more common.

Then David found a wonderful area of outcrop where clearly nobody had previously collected. Fossil sharks' teeth littered the ground. But the higher he went in the cliff section, the more heavily serrated the teeth became. The biggest discovery was that over a 16-foot thickness of sediment, representing just the blink of a geological eye, the *Otodus* teeth had changed from the non-serrated *Otodus obliquus* to the partially serrated *Otodus aksuaticus* to the fully serrated and much larger *Otodus auriculatus*. Here was the evidence he was seeking, showing how rapidly the first giant serrated megatooth shark had evolved. Even better was the fact that the dark clay could be correlated, using marine microfossils, to the same marine layers in Denmark, which had been accurately radiometrically dated using minerals trapped in their volcanic ash layers. The transition of these three species was extremely well dated, showing that it took place between 52 and 48 MYA. The first fully serrated large megatooth of the species *Otodus auriculatus* appeared close to 48 MYA, at the very end of the Eocene period. That gave a stupendously accurate time scale to gauge these evolutionary changes.

**Shark teeth fossils collected from Kazakhstan by David Ward, showing the transition from the non-serrated teeth (left) to partially serrated and fully serrated types (right)** AUTHOR PHOTO

For David, this was his "trip of a lifetime." Yet it was not over, and the coming trials of the next site would test him to his limit. The next day they headed off to another site about four rattling days' drive away called Kolbay, where limestone was present and there were beautiful fossils to be found. Kolbay was the steepest of the canyons along a four-mile cliff line. The sample of loose sand he was digging was good, with hundreds of shark tooth fragments emerging as the silver sand drifted through the sieve. He was absorbed in digging when suddenly the sand in front of him erupted and a hideous creature, half spider, half lobster, emerged. Its body was about 6 inches long, yellow and dark brown with long chestnut brown lobster-like legs. Where some of us would have screamed and possibly fallen down the cliff, instead David grabbed a long-handled trowel and scooped the creature up to examine it. He found out later it was a giant solifugid, or camel spider, a nonvenomous form but reputed to have a very painful bite (and contrary to the lore about how camel spiders got their name, they do not crawl up the legs of camels and eviscerate their bellies). The site was named Solifugid Ridge after that event.

After many expeditions to Kazakhstan over several years, David and his colleagues had finally located the precise layers where the transition of the non-serrated *Otodus obliquus* teeth began developing serrations, culminating in the rapid appearance of the first large coarsely serrated species *Otodus auriculatus*. They found several other new species of sharks from the sites. They also mapped the geology of the site in detail and accurately dated the entire geological sequence. These finds mark the beginning of the megatooth line.

What drove these sharks to grow to even larger body sizes, ultimately peaking with a possibly 66-foot-long megalodon? New kinds of rich, nutritious creatures had been invading the seas steadily since the end of the dinosaur age, and no doubt some of these helped fuel the increase in shark body sizes.

## Penguins and whales invade the seas

Immediately after the non-avian dinosaurs died out, we see a rapid rise in birds and land mammals, which soon after began their own invasions of the oceans. While early birds had invaded the seas since the Cretaceous, another major invasion of birds into our oceans occurred straight after dinosaurs went extinct—the first penguins. They arrive in the early Paleocene with small forms like *Wimanu* from New Zealand around 63 MYA, then shortly afterward we have quite large forms like *Kumimanu*, towering over 5 feet high and weighing around 200 pounds, by the end of the Paleocene. The largest of all penguins was *Palaeeudyptes*, slightly taller but much bulkier (6 feet 8 inches, 250 pounds) and certainly large enough to be of interest to any big predatory shark of the day, like *Otodus*, if it ventured into such cold southern waters. Fossil penguin remains have been found on Seymour Island, Antarctica, and in New Zealand, so they evidently favored the cold southern oceans, which were rich in plankton and small fishes at the time. Penguins had the southern oceans to themselves without any direct competition from other birds or mammals for the first 15 million years of the Cenozoic era.

Gigantic penguins invaded the seas around 65 MYA, providing another high-energy food resource for sharks ARTWORK BY THE AUTHOR

   The next biggest invaders into the seas were the mammals, which had begun diversifying by occupying the ecological niches on land left vacant by dinosaurs, growing steadily larger in the absence of reptilian predators. Some of them preferred hunting in rivers and shallow seaways and soon evolved adaptations like broad swimming limbs that aided them in adjusting to a semiaquatic lifestyle, much like the hippopotamus today. It would be only about 6 million years later that this new marine invasion picked up speed and gigantic whales started to abound in the tropical and temperate oceans, not just near shorelines but also in fully aquatic areas.

   The early "walking whales" like *Pakicetus* and *Ambulocetus* still retained arms and legs with hands and feet. They evolved from archaic long-snouted hoofed animals called **ungulates**, which by 49 MYA had adapted to a life hunting prey in the shallow seas. Whales are most closely related to the living hippopotamus. When this idea first came out it was dubbed the "Whippo hypothesis," and it has since been confirmed by molecular similarities in the DNA of living whales and hippos. Back in the early Eocene, 50 MYA, a small hoofed mammal was the last common ancestor to a split after which one line would take to the seas and evolve into whales and the other would evolve into hippos.

These early walking whales would not be a threat to sharks for another 9 million years, when at about 41 MYA they attained large enough sizes to become serious open-ocean predators. The whale lineage all started out as predators, having sharp teeth for feeding on large fishes and possibly other sharks and smaller marine mammals—including other very early whales and manatees. This kicked the evolutionary arms race between sharks and whales into top gear and resulted in some whales specializing into gargantuan filter feeders.

*Ambulocetus,* an early walking whale about 10 feet long, appeared in our seas 47 MYA   ARTWORK BY THE AUTHOR

It's amazing to think that whales are so much younger than sharks in our oceans, yet today they share space with sharks and include several species that are much bigger than the largest living sharks. But that wasn't always the case, as we shall soon see.

Let's look now at what new other kinds of lamniform sharks besides *Otodus* and its kin were appearing in our oceans.

## Giant thresher sharks

Sometimes making a big discovery in paleontology is not the result of a carefully planned expedition to a remote site. Sometimes, if we are lucky, it just happens serendipitously when we least expect it. This is one such case also involving David Ward, who was part of a team that found the largest thresher sharks ever known. This was a momentous find, as it shows that another giant lamnid predator existed at about the same time that megalodon was first appearing in the oceans. The hunt for the giant fossil thresher was initiated at a rock and mineral show in Tucson, Arizona.

Some years ago, a commercial dealer at a rock show showed David a tooth of a large Miocene age thresher shark. Rumors of such a species, a "Bigfoot" of the fossil shark tooth world, were rife among the private collectors. However, the species had never been formally studied. Most museums have fossils that were collected or acquired by someone who is not necessarily an expert in that special area of research (like ancient sharks). A dinosaur expert might collect shark teeth from a site as a by-product of digging up a dinosaur skeleton in marine sediments. Such undiscovered fossil treasures can lie dormant in collections for decades, or even centuries, before someone with the right expert eye spots them and recognizes their true scientific significance. David went back to the drawing board and meticulously searched through many of the major state museum collections in the United States. He eventually found more specimens of the elusive giant thresher by simply knowing what he was looking for. The tooth was at least twice the size of the maximum recorded in living forms, so it must have come from an enormous shark about 25 to 28 feet in length and lived in the first half of the Miocene period (20–13 MYA). In some cases, these teeth were so large that they had been mistaken for the side teeth of a megalodon, or had been wrongly labeled as something else.

David teamed up with Bretton Kent from the United States, and in 2018 they named this new giant thresher *Alopias palatasi*. Due to its gigantic body size, it is unlikely to have had the long thresher tail that characterizes all three living species. It was simply too large to have behaved like a modern thresher. Its deadly serrated teeth tell us it had adapted the same feeding strategy as the white shark, hunting seals and whales. It likely evolved from another large thresher with non-serrated teeth, *Alopias grandis*, found in slightly older Miocene deposits in the United States. That species may have reached 18 feet long. It demonstrates that like the *Otodus* line, the thresher sharks also changed their dentition from smooth-edged teeth to serrated teeth within a single lineage. The giant thresher *Alopias palatasi* was far larger than any living white shark, and the second-largest shark predator in the oceans of its day, alongside megalodon. The huge abundance of whales, manatees, and seals in the seas at this time clearly sustained all these major predators living side by side.

These finds have pulled us up into the Miocene period (23–5.3

The thresher shark (left) has a long upper tail lobe to stun its prey (fishes). Some ancient forms had serrated teeth (right) for eating marine mammals, and might have grown as large as 25 to 28 feet in length  PHOTO BY NIGEL MARSH / TOOTH IMAGE COURTESY OF DAVID WARD

MYA). It was a highly significant time globally for marine faunas, when whales were peaking in diversity and sea levels and temperature fluctuated. Several different lineages of sharks reached their peak body size—threshers, tigers, *Otodus* species, and ancient white sharks, among others. What factors drove them all to get so darn big?

In the Early Miocene, Africa collided with Europe and Asia, closing the passage between the Mediterranean Sea and the Indian Ocean and altering ocean currents. A short phase of global cooling at the start of the period drove phosphate deposition in the eastern Atlantic Ocean, then the world began warming again from around 18 MYA during an event called the Middle Miocene Warm Interval. This increased nutrients in the seas, and the large amounts of plankton created an immense food supply to bolster populations of large filter-feeding whales, the baleen whales. These whales were preyed upon by the largest sharks that ever lived, like megalodon and the giant thresher and tiger sharks. They were also hunted by a terrible mammalian killer, a giant predatory sperm whale with large teeth in both its jaws, called *Livyatan* (an old Hebrew word meaning to twist and coil, used for the biblical Leviathan or sea monster). So even though they had an ample food supply, the baleen whales were under pressure to stay alive.

Let's now look at the entire evolution of the megatoothed sharks leading up to megalodon. What changes took place in the seas, and what other factors drove these sharks to their extraordinary gigantism?

# Rise of the megatoothed sharks

As a kid hooked on fossils, I used to collect fossilized sharks' teeth from sites all around my home state of Victoria, Australia. I will never forget the day in 1969 when I found a large well-preserved shark tooth at an abandoned limestone quarry near the city of Geelong in Australia. It was my twelfth birthday, so my father had granted me my wish for him to take me fossil hunting for the day to some sites that were a couple hours' drive from home. I collected loads of interesting fossils that day, including well-preserved extinct heart urchins and big fat clamshells, but my heart skipped a beat when I spied the edge of a big serrated tooth poking out of the quarry wall. It was buried in soft crumbly yellow limestone that formed about 23 MYA (Early Miocene) on the bottom of the shallow seafloor during a time when the oceans were slightly warmer than they are today.

I dug the tooth out carefully with my trusty penknife. Later, at home, I cleaned it up. It was about 2 inches high with razor-sharp serrated edges. It belonged to *Otodus angustidens*, the giant predatory shark immediately ancestral to the big one, megalodon. It was an awesome shark tooth, easily the best in my young collection, and a sublime birthday present. Later, when I got my first position in paleontology as a curator at the Western Australian Museum, I was happy to donate my private fossil collection to this museum so it would have a permanent, safe home. I later chose this prized tooth to go on public display in a gallery devoted to the evolution of life. It continued to give me great joy every time I took visitors and their children into the gallery and showed them the special tooth I'd found on my twelfth birthday.

*Otodus angustidens* tooth I collected on my birthday trip at age twelve AUTHOR PHOTO

The tooth I found that day came from the largest shark in the oceans of its time, one of a series of *Otodus* shark species that over tens of millions of years slowly changed into larger species. The big question here is what drove this series of changes, and how much we do know about each species as an apex predator, because some species in this dynasty are known from far more than just their teeth. The evolution of a lineage of species tracked through such clear yet subtle changes as seen in their teeth are called **chronospecies** (from the Greek *chronos,* meaning time). Let's go back to the beginning of the megalodon family line to see how it all started and what changes in the ocean's environments and faunas kept it driving along to bigger and more successful descendant species, culminating in megalodon.

The line started with the Cretaceous lamniform shark *Cretalamna,* which we met briefly in the last chapter. It appeared about 110 MYA in the Early Cretaceous as a small form about 5 or 6 feet long and diversified into several species by the end of the period, including some larger forms reaching maybe 15 feet long. It successfully sailed right through the extinction event 66 MYA and thrived until about 46 MYA. Most scientists accept that *Cretalamna* and *Otodus* diverged sometime close to the start of the Paleocene, maybe 65 MYA. One branch of this split-off from Early Cenozoic species of *Cretalamna* leads to the living porbeagle and salmon sharks (*Lamna*), while the other branch becomes *Otodus,* which first appears at the very start of the Paleocene some 65 MYA.

Although there may be slight regional differences, it's important to this story to know that at any one time, globally, we only ever have one species of *Otodus* around in our oceans until we reach the very end of the line and megalodon appears in the Miocene period.

While most scientific books and papers refer to *Otodus obliquus* as the first species of its line, my colleague David Ward has suggested that an older, smaller precursor, *Otodus medaivia,* was possibly the ancestor to *Otodus obliquus.* This smaller ancient form known only from teeth lived from 66 to 62 MYA in the Atlantic seaway that stretched into what is now the central United States. If David's theory is correct, this species likely gave rise to *Otodus obliquus* around 62 MYA.

*Otodus obliquus* teeth measure up to 4 inches high, with vertebral discs around 5 inches wide. Both of these remains suggest its maximum body size was around 26 to 30 feet, making *Otodus obliquus* the largest predator in the oceans for about a 16-million-year reign (62–48 MYA). Its teeth are incredibly abundant around the world, having been found in England, Europe, Africa, and Australia, a testimony to its ultimate success. There were no whales or large reptiles in these seas when this shark was alive, so it was likely a specialist living on large species of bony fishes like billfish or early tuna, which first appear in the seas at this time. Other large predatory lamniform sharks coexisted with *Otodus obliquus* including *Palaeocarcharodon,* a white-shark-sized beast with serrated triangular teeth. I suspect it hunted large bony fishes or big rays, using its serrated teeth to help saw off chunks of flesh with each bite. Two giant sand tiger sharks also lived at this time, *Jaekelotodus* and *Striatolamia,* both around 13 to 16 feet in length; they probably fed on slender fish as primary prey.

*Otodus obliquus* and *Xiphiorhynchus*   ARTWORK BY JULIUS CSOTONYI

The next species in the megalodon line is the one David found in Kazakhstan, which has now also been found in the United States, *Otodus aksuaticus.* Its teeth are still quite similar to those of *Otodus obliquus* except they have irregular serrations along the cutting edge but lack them at the tip. These teeth have ragged side cusps that aid holding the struggling fishes while slicing into them. Juvenile white sharks have similar side cusps on their teeth, losing them at maturity when they switch from primarily eating fishes to marine mammals. *Otodus aksuaticus* lived straight after *Otodus obliquus,* appearing around 48 MYA. At this stage, the teeth of these *Otodus* species are all the same size, so there is no significant change in body size.

The next step is when things get serious quickly. *Otodus aksuaticus,* now armed with serrated teeth, suddenly grew a lot larger, morphing into a new giant species that appeared on the scene in the Early Eocene about 49–48 MYA. *Otodus auriculatus* had large, robust teeth up to 4.5 inches high with well-developed serrations, equipped for tearing large chunks of flesh from its prey, probably large bony fishes and other sharks, with small whales present, but not yet well established, when it first appeared. It grew to at least 31 feet long, and its teeth are much more massive than for *Otodus aksuaticus,* suggesting that it was a much beefier, robust shark, perhaps twice the weight of its ancestor species. It reigned as the ocean's top predator until around 43 MYA.

*Otodus sokolovi* appeared around 43 MYA, replacing *Otodus aksuaticus.* It differs from that species in having the serrations running all the way along the edges to the crown tip. Clearly the predatory strategy of piercing and cutting mammal flesh had caught on. *Otodus sokolovi* was slightly larger in mass than its ancestor, maybe reaching around 31–33 feet long. This species had plenty to feed on as there was a rapid increase in predatory large whales like *Basilosaurus* and the **squalodonts**, or shark-toothed dolphins.

*Basilosaurus* grew to around 60 feet long and was clearly an apex predator that fed on fishes and sharks, as proven by stomach contents in one fossil specimen. Many different groups of whales, dugongs and manatees, and a guild of smaller dolphin-like whales up to 12 feet long, called squalodonts, easy prey for megatoothed sharks, were now becoming abundant in the seas of the world.

*Otodus sokolovi* might also have hunted gigantic whales like the 65-foot-long *Perucetus,* arguably the largest whale that ever existed. *Perucetus* appeared in the seas off Peru between 39 and 37 MYA. Artists' reconstructions show it resembling a monster maggot with a bloated, chunky body and presumably small head (although the skull is not known). Scientists calculated its body mass between 93 and 375 tons, with the average at 200 tons (today's blue whale reaches 219 tons). It was so heavy because it had a solid skeleton weighing 2 to 3 times more than a blue whale skeleton. Such dense bone functions as a weight belt for dugongs and manatees that feed on seagrasses in shallow waters. While we can't tell what *Perucetus* ate, we can guess from its dense bones that it wasn't a fast-moving predator and instead might have grazed on seagrasses, or scooped up mollusks from the seafloor. The oldest filter-feeding baleen whale was *Mystacodon*, which also appeared around 37 MYA. The earliest baleen whales were nothing like today's blue whales or humpbacks. They had teeth that were widely spaced with lots of small nerve and blood vessels emerging on the jaws, possibly for the first filter-feeding structures: fine baleen bristles, made of essentially the same substance as your fingernails.

In the Oligocene period (33.9–23 MYA), the seas were now full of many families of toothed whales (**odontocetes**, like sperm whales and dolphins) and early baleen whales (**mysticetes**), including the first members of the gigantic guild of living baleen whales like the right whales, humpbacks, minke whales, and blue whales. With so many whales and other tasty mammals abounding in the seas, the next step for large predatory serrated-toothed sharks like *Otodus sokolovi* was to grow much larger—and so they did. A grand old feast followed after that breakthrough!

The enormous *Otodus angustidens*—the species of which I had found a tooth when I was a kid—appeared at the start of the Oligocene period. It had teeth up to almost 4 inches in diagonal length, with the little side cusps now more reduced, trending to a larger, wider, and more robust triangular tooth type. This shark had a maximum body length around 36 to 39 feet. This was a true behemoth, the biggest predator in the oceans of the world, reigning between 33 and 22 MYA, just creeping into the first years of the Miocene period. We know a lot about this shark because its teeth and fossilized vertebral discs have been found at many sites, each associated with a rich diversity of potential prey items such as whales, dolphins, penguins, and seals.

One of the best-preserved fossils of *Otodus angustidens* was found in New Zealand in 2001, comprising some 165 teeth with 35 vertebral discs from one large individual close to the top of the species size range. At its earliest appearance, it likely fed on smaller-sized whales, and penguins that were much larger than species around today, like the giant extinct species of *Aptenodytes* that stood around 4 feet high. Another set of some forty-five well-preserved *Otodus angustidens* teeth was also found at Jan Juc in Victoria, Australia, which the local paleontologist Erich Fitzgerald said shows clear signs of its being scavenged upon by a pair of sixgill sharks (*Hexanchus*), because their teeth were found among the larger *Otodus angustidens* teeth.

Fossil remains found at Jan Juc in Australia suggest that the carcass of a giant *Otodus angustidens* was scavenged upon by sixgill sharks ARTWORK BY PETER TRUSLER

These saw-edged megateeth of *Otodus angustidens* reveal that the lineage specialized in hunting large mammals, just as white sharks do today. The differences in nitrogen isotopes, which accumulate in the dental tissues when sharks form their teeth, can be used to determine what they were eating, as certain heavier isotopes accumulate in the tissues as you go higher up the food chain. The results published in 2022 showed that in all the largest species of *Otodus,* from *Otodus au-riculatus* through to *Otodus angustidens* and *Otodus megalodon,* there is

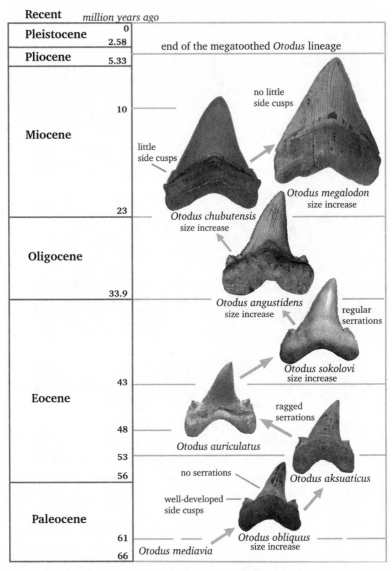

**Evolutionary sequence of *Otodus* teeth showing the evolution of larger and more serrated tooth types over time** DIAGRAM BY AUTHOR, WITH CREDIT TO DAVID WARD FOR INFORMATION AND IMAGES

a shift to higher Nitrogen-15 ratios, meaning these species fed on a richer diet of marine mammal prey, perhaps combined with occasionally feeding also on other apex predatory sharks (maybe including their own species).

As whales kept evolving and diversifying, this meant a lot more prey came into the oceans, so *Otodus angustidens* just kept growing larger. We have a larger, chunkier intermediate species called *Otodus chubutensis* appearing around 27 MYA that is extremely close to *Otodus angustidens* except that the teeth are slightly wider, and the little cusps that sit on each side of the big triangular main cusp are still present but merge more with the main triangular central cusp of the tooth. *Otodus chubutensis* lived at the same time as early megalodon populations, and disappears around 10 MYA. This has led some scientists to think it might well be a variation of the megalodon species that morphed into megalodon around 10 MYA.

*Otodus chubutensis* probably maxed out at 44 feet in length but more commonly was between 30 and 36 feet long. It preyed upon fish, sea turtles, whales, and sea cows. Between 27 and 11 MYA we find an overlap of the *Otodus chubutensis* teeth with another kind of even larger tooth lacking any side cusps, so the normal range of variations within a species could explain these small variations around the globe. Toward the end of its range the teeth completely lose these little side cusps and become even larger, more perfectly triangular, and quite thick. This distinct new species sits at the end of the line and the top of the food chain for *Otodus*. It represents the biggest shark of them all, *Otodus megalodon,* the ultimate apex predator in the entire history of life on Earth. It's highly likely that megalodon evolved from the older, smaller species in a rapid burst of evolutionary change as new food resources became more plentiful.

In the next chapter I will tell you all about this fabulous iconic shark, from its place in our early history to the most recent unexpected research findings.

CHAPTER 14

# MEGALODON

## *The Greatest Superpredator Ever*

---

### Seas off Peru, 6 million years ago

The group of long-nosed dolphins swam through the warm seawater, breaking the surface with occasional enthusiastic leaps into the stark sunshine, then diving crisply back into the clear water to join their compadres. Not far away, a pod of small baleen whales headed southward to feed in the nutrient-rich Antarctic waters. Three males and four females were clustered together in a rather tight formation as they swam along at their regular pace, always wary of imminent danger.

About 6 MYA, the seas off Peru were much warmer than today by at least 8°F. The mid-Miocene had been a warm time throughout the world, with sea temperatures soaring to 88°F in places. That was a time of great expansion of whales and other marine life as productivity soared in the oceans. In the final million years of the period, two of the three marine gateways connecting the warmer tropical seas together had closed, blocked by tectonic uplifts. This caused a dramatic rerouting of oceanic currents resulting in abrupt cooling of the oceans. Food resources were on the wane, so predators had to hunt harder for food. Gigantic predators had an even harder time, as they needed much more food each day than smaller creatures.

The little whales belong to an extinct species called *Piscobalaena*. They were slow-moving animals that looked a lot like a scaled-down version of today's southern right whales, with the oldest individuals reaching around 16 feet. The largest whales of today, such as the 100-foot-long blue whale, had not yet appeared, although early ancestral forms related to them had just appeared in southern oceans. There had been a surge in diversity of baleen whales, with many species appearing in the first half of the Miocene, culminating in a peak of species diversity that would start declining in the late Miocene and never recover.

The dolphins belonged to an archaic type named *Atocetus,* ancestors of the living La Plata dolphins that would later inhabit the same bountiful seas. They watched the pod of small whales out of the corner of their vision, as both groups moved on warily, their keen sense of echolocation sounding out the waters around them for both food and signs of danger. Suddenly the returning sound waves signaled danger. A large living mass was moving in the waters ahead, coming at a fast pace straight toward them. It was a menacing thing to detect when nothing at all was visible in the water ahead. The nervous dolphins began to speed up, moving quickly below the surface as the whales huddled closer together, changing course to head at right angles as quickly as they could away from the path of the large incoming creature.

What the dolphins saw next was a blood-curdling horror straight out of a nightmare. The massive 60-foot shark crashed through the gloomy waters to charge upward and breach the water, seizing one of the baleen whales in its powerful jaws. *Otodus megalodon* had made its entrance. This mature female shark could have weighed nearly 60 tons, so the much smaller whale had no way to escape her jaws—six feet wide and lined with unforgiving rows of thick, serrated triangular teeth. When coupled with the incoming momentum of the shark's massive body weight, the jaws rammed the prey and, as the beast crashed down into the water, with one powerful bite sliced the defenseless whale into two neat halves. Thick red billows of blood darkened the water. With a sharp turn of its head, the gargantuan shark swooped to finish the job, gulping the tail and body of the whale, taking the last chunk of bleeding flesh in one large swallow. Other large sharks, like the archaic white shark, *Carcharodon hastalis,* around 23 feet long, were now sensing the metallic tint of blood in the water, tentatively homing in from far and near to join in the feeding frenzy.

The *Otodus megalodon* swung around and began chasing one of the smaller female whales. The pod, confused and highly distressed, tried to dive deeper to escape the lurking danger. With a powerful thrash of its huge, forked tail fin, the killer shark sped swiftly down, snatching a second whale, biting a third of the animal off, before swallowing the chunk whole. It turned around and savaged the bleeding carcass, shaking it in its enormous head, scaring away the smaller sharks that had come in to share in the bloodfest. Within a few minutes, she had swallowed the second whale, not quite enough to slake her monstrous appetite, and headed back to clean up what was left of the first victim.

Smaller sharks had arrived on the scene en masse. They began tearing the bloody remains of the first whale into shreds when *Otodus megalodon* once more came thrusting in, shaking its gruesome head to scatter the smaller predators as a horse deflects flies. She chomped at what was left of the whale's shredded body, swallowing the last bloody pieces in one gulp. The smaller sharks, some up to 25 feet long, included enormous extinct species related to the mako and one of the first of the modern white sharks. They were mere minnows in the shadow of this beast. The monster was now satiated, bloated by the massive meal of two small whales. Slowly she moved away from the scene, leaving the other whales to flee hastily on their southward journey.

Meet *Otodus megalodon,* the queen of the killer sharks, the most powerful superpredator Earth has ever seen either on land or in water. Superpredators are what I referred to earlier as creatures capable of killing prey equal to or larger than themselves, and this shark clearly had that ability. Because there is a genus of clam called *Megalodon* (the name simply means "large tooth"), we will use "megalodon" (lowercased and unitalicized) as our common name for this giant shark.

You might think that we paleontologists make up scenes like this, but nature is more imaginative than anything our human brains can conjure. While it's true that I choreographed this frightening scene in my head, nature provided the dancers and told us what moves they could perform.

## Megalodon: Size matters

Megalodon was not just any predator, she was *the* predator, the most ferocious creature who ever lived, with the deadliest jaws of all time, spawned from the dark side of nature. The jaws of a maximum-sized female (females were larger than males) could deliver the strongest bite force of any creature that has ever lived, a whopping 40,000 pounds per square inch in each fatal chomp of its jaws. That is enough strength to crush a pickup truck and have some force still left in the megalodon's tank. She was three times more powerful than *T. rex*, the largest land predator of all time.

Megalodon appeared around 23 MYA and disappeared around 3.6 MYA. *O. angustidens* morphed into this new, larger form (including the variation called *Otodus chubutensis*) when ocean temperatures cooled slightly, providing an increase in oceanic nutrients which fueled larger populations of its favorite prey, large filter-feeding whales.

Here at last we have reached the pinnacle of 460 million years of shark evolution, the largest shark ever and Earth's biggest predator. People of all generations, young and old, have feared it since the first reconstructed gigantic toothy jaws of it began to grace our museums nearly a century ago. Even today it remains a powerful research magnet for some paleontologists who can't pull themselves away from the lure of its enticing fossil remains. We continue to study its fossils through new technological approaches to discover unexpected new things about how it lived, what it ate, and clues to why it went extinct. It also has a strong grip on the minds of "cyptozoologists"—people who still believe it might be alive out there in our oceans. I'll discuss all these topics soon, but first we need to consider the most commonly asked question about megalodon: How big was it?

If we simply use the relationship that exists between tooth size and body size in the great white shark (as a model) and extrapolate the same relationship to megalodon, we could derive an approximate maximum size for this monster fish at around 52 to 66 feet. Its mass would have been 25 times the mass of the largest white shark, maybe reaching 61 tons, much heavier than the biggest *Tyrannosaurus* (10 tons) or any other land-based predator you choose. In recent years, a new way of estimating the size of sharks uses the maximum jaw width as shown by the teeth. Using this new method gives new maximum size for megalodon of around 66 feet long. Another paper published straight after this one led by Jack Cooper from Swansea University calculated the size of different body parts for megalodon individuals of different lengths. It concluded that a megalodon reaching 52 feet would have a head about 15 feet in length and a tail fin that was 12.5 feet in height. The large triangular dorsal fin would have been about the same height as an average adult person.

**New restoration of megalodon based on recent research showing that it had a much longer body than previously thought**   ARTWORK BY JULIUS CSOTONYI

You can see a full-sized megalodon model hanging in the foyer of the Smithsonian National Museum of Natural History in Washington, D.C. It was built in 2021 under careful scientific direction and spans nearly 52 feet. It is impressive to see its massive body and huge jaws bristling with triangular serrated teeth. Other stunning models of megalodon, or its reconstructed jaws, are displayed in several other museums around the world.

For the past century, every classic reconstruction of megalodon, including many recent ones in the movies, show it as a scaled-up white shark, since it was commonly thought to be an ancestor of the living species. The most recent data about megalodon's body shape was published in early 2024 by a consortium of twenty-six renowned fossil shark experts led by Phillip Sternes of the University of Washington. They showed that its body shape was more elongated than that of the white shark, and the tail would have had a longer upper lobe. Its body might have looked more like that of a basking shark, with a longer tail for slower ocean cruising. This study was based on spotting an error in a previous interpretation of the vertebral skeleton of the articulated megalodon from Belgium (more on that specimen soon). It demonstrated that all previous reconstructions of megalodon that used the white shark as a model for its body shape are incorrect. This has serious implications for future Hollywood productions, as every B movie ever made about this shark used scaled-up white sharks to make their CGI megalodons (back to the drawing board, guys!).

**The Smithsonian National Museum of Natural History's new megalodon model is 52 feet long, representing a reasonably large but not maximum-sized individual**  PHOTO COURTESY OF HANS DIETER-SUES, SMITHSONIAN NATIONAL MUSEUM OF NATURAL HISTORY, WASHINGTON, D.C.

This iconic extinct shark is known principally from its huge fossil teeth, and sometimes rare articulated remains of large, calcified vertebral discs, like the superb fossil from Belgium now in the Royal Belgian Institute of Natural Sciences, which comprises 141 associated but slightly disarticulated vertebral discs and no teeth. It measures around 35 feet long yet is incomplete, and from the growth rings in the discs we find it was forty-six years old at death. It must belong to megalodon, as no other shark of this age grows to such large size. Recent study of megalodon's scale shapes based on a partial skeleton found in Japan revealed that it was generally a slow-cruising shark capable of an occasional burst of speed to catch its prey. Kenshu Shimada estimated from the scales that it swam through the ocean at around 1.74 mph, which is slower than most of the living lamnid sharks, including the white shark, which cruises at between 1.8 and 2.2 mph.

Catalina Pimiento, whom I call Queen of the Megalodons (she approves), is the world's leading expert on this species. Her work helps us understand why researching the long-dead remains of a gigantic shark is useful today, leading us to better understand the factors that are now changing our world. Her research has shed new light not only on the biology and lifestyle of this giant shark, but also on how it coped with major shifts in global climate and oceanic circulation in the Miocene and Pliocene periods (between 23 and 2.6 MYA). And she has made the most detailed study of the factors that drove it to extinction.

Catalina grew up in Bogotá, Colombia, keeping her head low and going to school during a time of violent civil unrest that resulted in hundreds of thousands of victims. Her undergraduate work was on living whale sharks, and this became her hook into marine biology, leading her eventually to the University of Florida. While she intended to work on living sharks, her supervisor suggested she might look at a project on fossil sharks.

**Catalina Pimiento, aka Queen of the Megalodons, is a renowned expert on megalodon and its paleobiology** PHOTO COURTESY OF CATALINA PIMIENTO

Catalina agreed and began working on a site in Panama where middle-late Miocene shark teeth were found, so she collected and described this new collection of fossil sharks. She is the only person I know who has written two postgraduate dissertations on megalodon.

Her postgraduate works thus dragged her willingly into the world of fossil sharks, gravitating around the big body mass of megalodon and the mystery of its extinction.

Her first question to solve was what the average size for a megalodon was and whether this size changed over geological time. Catalina looked at thousands of fossil megalodon teeth from around the globe to discover new insights into the daily life and ultimate demise of megalodon. She proved megalodon had an average body size around 33 to 35 feet for the 14-million-year period the samples represented.

We also know from her work that mother megalodons gave birth to young sharks about 7 feet long in specific nursery areas in the eastern Pacific, based on sites in Panama where there are large accumulations of small megalodon teeth. Recent work has now identified at least eight other megalodon nurseries spanning between 16 and 3 MYA, in the Pacific and Atlantic oceans and the Caribbean Sea. Young sharks at about a month old were around 13 feet, and they remained juveniles even when at 36 feet long (we can tell as the teeth retain side cusps in juveniles that are lost in adults). Like white sharks they likely took more than a few decades to sexually mature.

The megalodon was huge, so how did it fill its belly in its endless search for food and energy?

## Megalodon: the ultimate predator

We know of two nearly complete sets of megalodon jaws—one from Saitama, Japan, and one from the Yorktown Formation in Maryland. The latter was used as the basis for a complete reconstruction made by the Smithsonian National Museum of Natural History in 2020. That reconstructed life-sized set of megalodon jaws is 9.5 feet high by 11 feet wide and contains nearly full rows of fossil teeth in every jaw position (182 teeth in all, although a complete megalodon jaw is estimated to have had around 276 teeth).

We have many thousands of good examples of megalodon teeth from many sites around the world, including some that made their way to the surface as part of the dredging of the deep ocean seabeds. We bone detectives can deduce a lot about megalodon's lifestyle from studying the teeth and deciphering the geological context of where they were deposited. For example, they tell us where it lived and at what time, and what it ate, from toothmarks it left on bones or from the teeth it left embedded in them.

The biggest megalodon teeth are fat triangular teeth measuring just over 7 inches long, adorned with sharply serrated biting edges like a jagged steak knife. They have been found in every part of the world except Antarctica (where we do have Pliocene fossil deposits with fossil whales but no megalodon teeth). This tells us that even with its huge body mass there were still parts of the world where it would not venture because of the extremely cold seas.

A large, perfect megalodon front upper jaw tooth from South Carolina. The left-hand view shows the outside face, the right the tongue side. Bar scale is one inch  IMAGE COURTESY OF DAVID WARD

So, with such powerful and terrifying jaws, what did megalodons like to eat? In our scene at the start I imagined the shark hunting the small *Piscobalaena* whales, because we have direct evidence of megalodon tooth marks incised into the bones of this fossil whale. Other bones of seals show megalodon tooth cuts as well. Their big teeth leave distinctive cut marks that look like deep trenches of dotted lines caused by the tooth's serrated edges slipping over the bone surface. The fossil whale was attacked from the side, targeting the head region, as the cuts are found on the lower jaws, while the seal was bitten around the shoulder, suggesting the top half of the body might well have been severed by the powerful bite.

Megalodon tooth scars have been found on the tooth of a large extinct sperm whale that lived 5 MYA. The deadliest and most spectacular of all the predatory whales at this time was one whose name harks back to biblical and literary fables, *Livyatan melvillei* (honoring Herman Melville and *Moby-Dick*). It had deadly curved tusks lining both its jaws (modern sperm whales have only lower jaw teeth). It reached a maximum body length of around 57 feet. I have no doubt that both these giants preyed on each other at different stages of their lives—adult megalodons feeding on juvenile *Livyatan*s and vice versa.

We can also determine what megalodon was eating from the chemistry of its fossil bones and teeth. Zinc is a necessary element for formation of mineralization of the skeleton of animals, especially teeth. It is acquired by eating certain types of prey, and certain isotopes of zinc can be traced in the shark's teeth. By analyzing the ratio of zinc isotopes scientists can plot what level of the food chain ancient sharks were occupying. Using this method, a team led by Jeremy McCormack of the Max Planck Institute in Germany was able to determine that megatoothed sharks were eating high-level food resources like marine mammals. The ratio of certain isotopes of zinc in the tooth enameloid of fossil sharks determined whether the shark was a top predator, based on comparing the same zinc ratios in various living sharks. What Jeremy and his team also found was that during the Pliocene period (5.3 to 2.6 MYA), the emerging guild of white sharks occupied the same niche as megalodons—in other words, eating much the same prey.

While this finding might suggest that the increase in abundance of white sharks at this time could possibly explain the demise of megalodon from stiff competition for the same food resources, there is no hard evidence that this was really the main cause of megalodon's extinction. For example, today we find that orcas coexist with white sharks and still eat similar large prey items—small whales, dolphins, and seals—but one is not driving down the population of the other as far as we can see; rather, populations of both are kept in balance by the amount of prey resources available. Fossils confirm that orcas and white sharks have coexisted happily for at least the past 6 million years, a similar time that white sharks swam alongside megalodon.

The lamniform sharks include the only shark species that can generate higher body temperatures than surrounding seawater by capturing heat generated from their powerful body muscles. Therefore, they are essentially warm-blooded or endothermic. Today, these are the white shark, makos, and thresher sharks. Paleontologists have speculated that as megalodon is one of the lamniforms with a lifestyle similar to that of the white shark, an apex predator hunting marine mammals, it also would likely have been warm-blooded. Some suggested its body temperature might have been as high as 95°F. The advantage of being endothermic is that it would have made it easier for megalodon to cruise across oceans with fluctuating water temperatures or venture into colder, higher latitude seas rich in nutrients where whales like to feed.

This was all speculation until a breakthrough was announced in mid-2023 that provided hard evidence that megalodon was warm-blooded. Using the state-of-the-art method of **clumped isotope paleothermometry**, which measures the temperatures of chemical bond formation in the tissues of the teeth (see the source notes at the end of the book for more details), this technique showed that megalodon really did have a highly elevated body temperature of around 80.6°F. The researchers also calculated the temperature of the seawater from oxygen isotopes formed in clamshells found with megalodon teeth in the same fossil deposits. This showed the giant shark had a body temperature about 12° to 13°F higher than the seas it lived in. Today's whale sharks have a similar body temperature to that calculated for megalodon, but they do this through living in warmer seas—they do not generate their own internal body heat. The megalodon's warm-blooded metabolism could have been a key driver of its growing large so quickly. One study suggested that its body heat was needed for digestion of large pieces of ingested whale meat as well as for absorbing and processing nutrients.

Another, somewhat darker theory has recently emerged that could explain why sharks like megalodon and its ancestors could have grown so big and been warm-blooded. The lamniform sharks, the group that megalodon belongs in, are today characterized by their intrauterine cannibalism. This occurs when the young hatch from eggs inside the mother and begin to eat the unhatched eggs, or even other baby sharks, still inside the mother's womb. Because the large lamnids like white sharks, makos, and thresher sharks are warm-blooded, they have higher energy requirements than most other sharks. The intrauterine cannibalism raises their body temperature slightly higher than the seawater in which they live, enabling them to hunt in colder waters and to swim faster. According to Kenshu Shimada, this warm-blooded condition in lamnid sharks could have "pushed up the internal heat" levels required for a big megalodon, also speeding up its growth rates.

We have now seen how abundant and successful megalodon was—so what caused its ultimate demise?

# Megalodon: end of the good life

How does such a mighty shark at the apex of the ocean's food chain go extinct? Was it a global climate change trend, or did its prey suddenly die out? What evidence do we have for assessing its extinction? The sudden disappearance of megalodon sometime around 3 MYA remains one of the great mysteries of shark paleontology.

Let's start by looking at what was going on with respect to continents moving, currents shifting, and climates changing during the last few million years of megalodon's reign.

At the start of the Pliocene (5.3 MYA), global temperatures rose to about 5° or 6°F higher than they are today. Then things turned around at 3 MYA, when there was a dive in sea temperatures as the Arctic ice cap formed and cold currents flowed up from the Antarctic. The warm equatorial currents that flowed around the globe ended when the long, slow collision of North and South America finished around 2.7 MYA as the Isthmus of Panama rose from the seafloor. At this time we find giant land animals moving between South America and North America—ground sloths, elephants, saber-toothed cats, and many other creatures—in what is called the Great American Biotic Interchange. Seaways that might have been migratory pathways for big sharks like megalodon suddenly closed.

Using her database of many hundreds of megalodon teeth finds from around the world, Catalina Pimiento compared the abundance and geographical range of teeth with causal factors like sea level changes and consequent habitat loss. Her conclusion was that the reduction of coastal habitats as sea levels fell because of growing ice caps at the poles caused a collapse of marine ecosystems necessary for sustaining the entire food chain of the oceans. This environmental change caused the extinction of one-third of the marine megafauna—not only megalodon but also many species of whales, sharks, large sea turtles, and huge flying seabirds—at the end of the Pliocene, around 2.6 MYA. The collapse of these ecosystems implies that megalodon lost many of its primary food sources during this time as many species of whales went extinct, driving it to its demise.

David Ward thinks that megalodon could have bumped out a little earlier than this date, as he doesn't know of any megalodon teeth that have been reliably well dated younger than 3.7 million years old. Either way, we can say with utmost confidence that the last megalodon went extinct sometime between 3.7 and 2.6 MYA.

The reasons for megalodon's exact demise remain unknown, but they could relate to either ocean cooling starting about 2.6 MYA, causing the habitat loss discussed above, or biological factors like the events surrounding the evolution and migration of whales to colder Antarctic waters where the sharks could not go.

It seems likely that the growth and huge size of modern baleen whales, the largest animals on the planet, could well have been driven by predation pressures from megalodons. Their ability to endure and feed in near-freezing Antarctic waters might have been a factor in why megalodon went extinct. Whales cope well in Arctic or Antarctic waters, but no sharks today can survive in waters that are as low as 32° to 29°F. The annual migration of large filter-feeding whales to feed in rich Antarctic waters might have been the last straw in the megalodon ecosystem collapse. It likely went extinct with a whimper rather than a bang.

## Megalodons in archaeology and history

Peoples around the world have been fascinated with megalodon teeth for many thousands of years and have used them for sacred, medicinal, and ritual purposes. North American First Nations people living around Maryland in the United States attached megalodon teeth to handles for making a cutting tool sometime between 2,500 and 1,000 years ago. The Mayans offered megalodon teeth to their gods as they have been found at the temple in Palenque, Mexico, dated between AD 200 and 600. The geographical boundaries of the Mayan world were home to the fabled deities of the sea and supernatural energies. Fossil megalodon teeth were likely links to the deities, so held in very high esteem.

Europeans have known about megalodon teeth for thousands of years. Pliny the Elder referred to "glossopetrae" or "tongue stones" in his work *Historia Naturalis*, written around AD 73. He believed they fell from the sky at the eclipse of the moon and were helpful for appeasing "winds"—that is, curing flatulence. The name came from the similarity between the forked tongue of the snake and the two-pronged roots of some fossil lamniform teeth, like those seen in all *Otodus* species, which was also thought to endow them with special powers.

To explore this further I must relate the work of the English fossil shark expert Chris Duffin, whom we've already met through *Saivodus* and *Ostenoselache*. Chris, a multitalented guy, also studies the uses of fossil shark teeth in art, medicine, and archaeology. He has found some unbelievable uses of megalodon teeth in medicine. Up until around four hundred years ago, the megalodon and other prehistoric shark teeth found on the island of Malta were thought to be petrified tongue stones formed by imprints of the tongue of Saint Paul when he visited the island around AD 60.

Important Europeans in the late medieval times lived in constant fear of being poisoned by their enemies. They believed that the fossil tongue stones (shark teeth) held the power to counter the poison. Chris found a passage extolling this power in the sixteenth-century *Sloane Lapidary* manuscript: "Tongs of Adders . . . should be set in silver, both for kings [&] lords at their meate, so yt they mey be kept ye safer from poison." They crafted little elaborate silver trees called *languiers* (French) or *Natterzungen* (German, adder's tongues) to hang the teeth from so a guest could dip a tooth into their wine cup before drinking. The idea was that if there was poison present, the tooth would sweat or change color. There is some chemical truth to

A *Natterzungen* from around 1500, holding sharks' teeth for dipping in wine to detect the presence of poisons
PHOTO BY CHRIS DUFFIN, UK

this precautionary ritual, as traces of cyanide or arsenic in the drink would make the metal clasp react and turn it a darker color, warning the imbiber that their drink was poisoned.

One of the best of these little silver and gold trees shows an unexpected link between Christianity and the megalodon. It can be seen in the green room of the Staatliche Kunstsammlungen in Dresden, Germany, and comprises a silver base with six long pedicels emerging through a canopy of silver leaves, each terminating in a flower from which a tooth of a fossil mako *Isurus* is suspended with a gold band holding it on. In the crown of the tree we find Mary with the baby Jesus in her lap, leaning up against a large megalodon tooth. Another *languier* held in the treasury of the German Order in Vienna is almost a foot high with fourteen megalodon teeth hanging from its branches. It is clear from these and many more examples that exist in European museums that megalodon teeth were highly regarded both for their capacity to protect important people's lives and as symbols of Christianity.

In these early days, people didn't believe these were ancient sharks' teeth. It took a special kind of genius, the Danish cleric-scientist Nicolas Steno, to demonstrate that these artifacts were not in fact tongue impressions made by a visiting saint.

In 1667 a large white shark was caught by fishermen off the coast of Liverno in Italy, and the Grand Duke Ferdinando II de' Medici ordered the head to be sent to Steno for study. Steno examined the teeth and noted how similar they were to the Maltese "tongue stones," and in his work published that same year he compared fossil shark teeth to the white shark teeth. Steno declared that what were being called "tongue stones" were really the teeth of ancient giant sharks. However, the fact that the fossil teeth were found inside rocks bothered him deeply. This led him to study geology and examine how layers of rocks were formed by the slow accumulation of sediments. He was eventually able to deduce clear explanations for how the teeth of marine creatures could end up being encased in limestone rock.

Steno was a founding pioneer of the science of geology and a clever pioneer paleontologist. His research into the anatomy of living sharks was also first-class work that will long be celebrated.

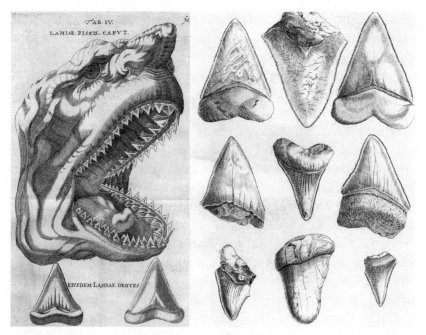

Images of white shark head and megalodon teeth by Nicolas Steno in an unpublished Vatican catalog from 1667 identifying the triangular objects as the teeth of ancient sharks, not Saint Paul's "tongue stones" as previously thought    PUBLIC DOMAIN

## Hunting for megalodons

It's clear that people have been collecting megalodon teeth for thousands of years. They have always been highly prized. So how hard is it to find a megalodon tooth?

As a young boy hooked on collecting fossil sharks' teeth, I used to dream of finding a big complete megalodon tooth. I lived for that surge of adrenaline you get turning over a rock on a lonely beach to find a perfect large shark's tooth waiting for you, untouched and unseen by humans for millions of years. On my desk sits a 3-inch-long brown shining tooth from a giant extinct shark called *Carcharodon hastalis*—one that I collected at nine years of age. It still gets my heart racing to pick it up and run my fingers over its shiny surface. As a teenager I traveled all over my home state of Victoria searching for fossil sharks' teeth. Paleontology was my drug of choice. Sometimes, even now in my sixties, I still have that same wonderful dream I first had as a child about finding the perfect fossil. I am walking along a beautiful beach searching layered limestone cliffs that bristle with all kinds of interesting fossil bones. Then, unexpectedly, I see it. Just an edge of sharp serrations sticking out of the rock, the sky-blue enamel of the exposed megalodon tooth shimmering in the sunlight. I pluck the tooth from the soft rock, flap my arms, and slowly fly off into the sunset (well, it was a dream, after all).

When I was eleven years old, my cousin Tim and I eagerly collected fossil shark teeth from the Beaumaris Beach site in Melbourne. We took our prized specimens to the National Museum in Victoria so we could meet with the paleontology curators to help us understand what we had found. They identified our fossils and gave us valuable information about other sites around the state where we could collect similar fossils. One of these sites was the district around the town of Hamilton, in western Victoria, where large nicely preserved megalodon teeth had been found. We itched to get there and search those sites, but it was more than two hundred miles from our home, so it was a big undertaking. We begged our parents to take us. The museum curator had kindly given us contact details for a local fossil collector, Lionel Elmore, and we wrote to him for advice about the sites and inquired if he would be kind enough to guide us. His letter arrived back a few weeks later, agreeing to take us to the sites and show us his private fossil collection, and with that confirmation in hand, Tim's parents agreed to drive us to Hamilton.

When we arrived at Lionel's house some weeks later, we were excited to finally be near the sites where megalodon teeth could be found. Lionel showed us his amazing collection of fossils. He had many kinds of sharks' teeth, large whale bones, rounded whale ear bones, many kinds of fish remains, and peculiar head-sized fossil cowry shells. This treasure trove was the reward of decades of hard work searching the creek banks and gullies of the farmlands around the county. He had several complete megalodon teeth. One was very special indeed, not because it was unusually large or well preserved, but because of where it was found—far from the known fossil sites.

Lionel told us that this megalodon tooth was a **manuport**, an archaeological artifact carried by the Bungandidj indigenous people of the region. It isn't surprising, really, because megalodon teeth are unusual and beautiful objects that have been a source of fascination for humans well before science was invented. In this case, it was transported about seven miles from the fossil site, probably because it was valued as a useful object. Why wouldn't it? Megalodon teeth are nature's steak knives, the perfect tool for stripping skin or meat off the bone of an animal.

On that trip we found many fine shark teeth of the Pliocene age, about 4 million years old. Tim bagged a good-sized megalodon tooth, spotted with his sharp eyes and quickly plucked from the bottom of the Muddy Creek streambed. I was pleased with the collecting trip but still yearned to find one of these prized teeth for myself one day. I eventually found one at Beaumaris Beach a few years later, but it was just half of a large tooth. One day I still hope to find a complete one.

Hunting for megalodon teeth is a hobby for some, big business for others. South Carolina is home to some of the best sites in the world to find big megalodon teeth, and the best specimens are found by diving off the Atlantic coast. The teeth erode out of the soft Pliocene muds of the Yorktown Formation as currents winnow away the sediments, leaving the heavy objects behind to settle on top of the muddy floor. The best teeth are found by pushing your hands into the soft mud and feeling around for large triangular objects. In addition to megalodon and other kinds of shark teeth, the underwater sites yield a wide range of fossil bones including parts of ancient whales through to mammoth tusks.

When I was living in Los Angeles I made several visits to the famous Sharktooth Hill site in the Sierra Nevada foothills near Bakersfield, California. The sedimentary deposits here are called the Round Mountain silt, a soft crumbly greenish layer that can easily be excavated for teeth. Large perfect megalodon teeth are found here by lucky fossil hunters on rare occasions. This is because these are quite old sites, dated at 16 to 15 MYA, a time when megalodon was just taking off. The digs are a commercial operation where you pay a fee to dig and what you find you can keep. I found other sharks' teeth and fossil whale bones at this site, but no megalodons.

Perhaps one of the greatest collectors of megalodon teeth in living history was the late legendary Vito Bertucci, "the Megalodon Man," an Italian American who was enticed to South Carolina by the megalodon teeth. As a manufacturing jeweler, he was attracted to the fossil teeth not only for their intrinsic beauty and prehistoric appeal, but also because they were a highly sought-after item for fossil collectors around the world. One of Vito Bertucci's greatest accomplishments was to make a reconstructed life-sized set of megalodon jaws measuring 9.5 feet high by 11 feet wide, with four full rows of actual fossil teeth in every jaw position (182 teeth in all). This effort represented around 16 years of collecting underwater in often dangerous conditions to find all these teeth in perfect condition. A smaller replica of Bertucci's reconstruction was eventually acquired by the Calvert Maritime Museum in Maryland in 2018. Vito unfortunately lost his life when he suffered a heart attack on a dive searching for fossils in Ossabaw Sound in Chatham County in late 2004. His body was found with a bag full of prize megalodon teeth. I recall meeting him many years ago at one of the exhibitions at an annual Society of Vertebrate Paleontology meeting. He was a very lively and amiable fellow, who loved to talk about meg teeth. All the time. Scientists I know had a deep respect for him because of his unabashed passion for the great extinct shark.

Megalodon teeth can be easily bought from mineral and fossil shops around the world or purchased online. Most of the commercial specimens come from well-known sites in North and South Carolina and Maryland in the United States, and recently megalodon teeth from southeast Asia are appearing on the market. They are rare indeed, unless you are willing to go diving in the right places off the coast of South Carolina or some Florida beaches. I still haven't found one. Good luck, but keep an eye out for curious bull sharks.

## Could megalodon still be out there?

One of the urban myths surrounding megalodon is that it might still be alive today. Could it be lurking out there in the deepest, darkest parts of our world without having been detected?

So how did the myth of megalodon still being alive first start? I have my own theory based on a specimen I've seen in the ichthyology collections of the Natural History Museum in London. I believe it all started with a large white shark caught off the coast near Port Fairy, Victoria, Australia, in 1836, whose massive jaws were brought back to London and placed in the United Service Museum, and later transferred to the collections of the Natural History Museum. In 1840, Richard Owen published his groundbreaking work on the evolution of teeth called *Odontography*, and on page 30 he discusses the jaws and teeth of the white shark, based on the large specimen from Port Fairy. He states, "In the United Service Museum there are preserved the jaws of a Carcharodon, of which the upper one measures four feet and the lower one three feet eight inches, following the curvature. The length of the largest tooth is two inches, the breadth of its base is one inch nine lines: the total length of the shark was thirty-seven feet."

Thirty-seven feet! That's almost twice the length of an extremely large white shark. It implied that other monster sharks could still be alive in the seas off southern Australia.

About twenty years later, the British Museum's fish curator, Albert Gunther, noted that the tag attached to the jaws stated they came from a 36.5-foot shark, which is close to what Owen had published. It was either a handwritten typo for 16.5 feet, which accords with the actual size of the jaws, or a deliberate joke played by the label writer. As we knew very little about the maximum size of white sharks then, he must have believed the label and the size of the shark. No doubt he would have spoken to Owen about it, as Owen was then his boss, the director of the museum. Gunther published this information about a 36.5-foot-long white shark in his 1859 *Catalogue of Fishes in the British Museum of Natural History*. Eleven years later Gunther published a

An erroneous estimate of body length based on these two-hundred-year-old white shark jaws from Australia may have started the idea that megalodon could still be alive today  PHOTO BY HEATHER ROBIN-SON, COURTESY OF THE NATURAL HISTORY MUSEUM, LONDON

second edition of the same catalog; this time the maximum size of the white shark was "rounded up" to 40 feet in length! These seemingly authoritative publications claiming outlandish sizes of white sharks must have played a major part in fueling the myth that megalodon is still alive out there in relatively recent times.

A 2013 Discovery Channel mockumentary entitled *Megalodon: The Monster Shark Lives* argued that megalodon could still be out there today, even though it bore a short disclaimer that the film was fictional. Nonetheless, it seems to have sparked a lot of subsequent interest in the topic. A recent survey of Shark Week viewers found that 70 percent of those who responded to their megalodon survey believed these monster sharks are still alive in the deepest unexplored parts of our oceans. The Mariana Trench, at the western edge of the Pacific Ocean near the island of Guam, is our deepest ocean trench, at around 36,000 feet, and some folk believe megalodons are hiding down there. Let's dig into this idea to see if it has merit.

The notion of undiscovered megalodons alive at the very bottom of our deepest oceans is based on the premise that they inhabit largely unexplored areas. This is simply incorrect. The first successful descent to the bottom of the Mariana Trench was achieved by the French explorer Jacques Piccard in the bathyscaphe *Trieste* in January 1960, and in 2012 the Hollywood film director James Cameron, of *Avatar* fame, also went to the bottom of the trench in his *Deepsea Challenger* submarine. In fact, there have been more than twenty-two manned descents into the trench and another seven descents by uncrewed submersibles, all with cameras exploring the deepest and most remote regions of that trench. No megalodons were sighted down there, or on the way down or back, nor did these efforts uncover any evidence of the recent existence of megalodon, such as large decomposing carcasses.

Our deep-sea drilling and seafloor dredging programs occasionally bring up petrified fossilized megalodon teeth. These teeth are heavily impregnated with minerals, thus providing no scientific evidence of their being from recently living sharks. It's most likely that they are millions of years old. The notion of megalodon's being alive today lacks any reliable, testable evidence.

# After megalodon

After megalodon went extinct sometime close to 3 MYA, our oceans changed dramatically. There was a 2.6-million-year period of successive ice ages, comprising seven phases when sea levels rose and fell by hundreds of feet as polar ice caps grew and glaciers expanded in both northern and southern regions of the continents, then melted causing sea levels to suddenly rise. During these cyclic ice ages, sea levels dropped by up to 400 feet, creating land bridges between some major continental regions (Russia and Alaska, for instance), and between significant landmasses like England and Europe. These new configurations naturally affected the pathways of oceanic currents, changing water temperatures and nutrient abundance in certain parts of our oceans. The net result was to stress the ecosystems, causing the largest, most specialized predators to die out first. The large *Carcharodon hastalis* and slightly smaller *Carcharodon planus* went extinct at this time, along with the killer sperm whale *Livyatan* and several kinds of large baleen whales, leaving the current species of white shark, *Carcharodon carcharias,* to take the reins and become the new apex predatory shark in our modern oceans.

The 20-million-year reign of megalodon, the world's largest predator, was finally over, and the oceans would never be the same again (luckily for us). Megalodon could not afford to maintain its sumptuous lifestyle when climates and ocean currents started to change and ice caps formed at both poles. Megalodon has taught us that even with size on your side, and a revved-up warm-blooded metabolism to cope with changes in water temperature, it's the factors beyond your control that will bring unexpected chaos into your domain and cause your ultimate demise. Once an apex predator disappears, others race to take its crown at the top. The subject of our next chapter is the winner of that race: the iconic white shark.

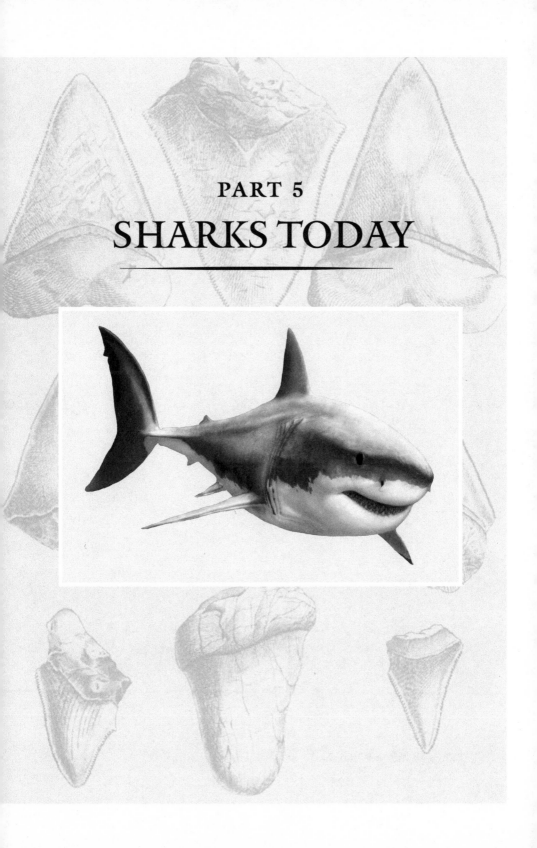

# PART 5
# SHARKS TODAY

## CHAPTER 15

# WHITE SHARK

## *A Natural and Cultural History of an Iconic Living Shark*

---

**JANUARY 11, 2023, NORTH NEPTUNE ISLANDS, AUSTRALIA.** We are anchored in a cove about a hundred feet offshore from the North Neptune Islands, forty-five miles south of the Eyre Peninsula in South Australia. I'm here at one of the best places on Earth to see white sharks in their natural habitat. The islands are now a conservation park named after the famous underwater filmmakers and shark conservationists Ron and Valerie Taylor, the first people to film white sharks in the wild.

It's a breezy hot day, a bit choppy in the waters, but perfect for what we are about to do. As we say in Australia, I'm wearing my brown underpants. I'm in a black wetsuit about to climb alone into the submerged metal cage, the first person to do so on our trip, prompted by the fact there is an impressively large white shark out there circling our boat. A weight vest is attached to me so I won't bounce around in the gently rocking cage, then a crew member hands me a regulator to breathe through. I step down backward blindly, cautiously, into the cage, into the water, into the white shark's domain. Immediately I feel a rush of irrational fear that the powerful shark is out there just waiting for me, intending to charge and destroy the cage and eat me, like the cage scenes in *Jaws*. It's nonsense, of course; the shark is simply there in its natural habitat, at one with the tranquil deep blue seascape.

The ocean's largest and most powerful predatory shark suddenly enters my view. It's a breathtaking moment. The shark is a smaller individual, around 12 feet long. It turns around and swims casually straight toward me. With the knowledge that they can accelerate suddenly to bursts up to 35 mph, I feel a rush of adrenaline as it slowly moves closer until its face is a few feet from mine, only the metal cage between us. It moves in a curving arc around me, its muscular cylindrical body beautifully streamlined, its dark gray-brown top layer meeting its milky white underbelly along a jagged boundary, like a snowy mountain range against a stormy sky. Up close I begin to feel pity for the shark rather than fear, when I see that it is marked with individual battle scars and patterns. This fellow hasn't had an easy life. During the next three days in the cage, after watching several white sharks, I learn that each shark is tattooed or pierced by scars that tell its own personal story of survival.

I become relaxed and begin enjoying my time just watching the shark, observing its behavior. There are baits of tuna gills and guts that the crew throw out to attract the sharks closer to the cage. Strict regulations govern the shark tourism industry here, so the sharks are not harmed or affected in any way by the presence of the boats. The rules do not allow the tour operators to chum the water with blood of animals, only fish parts, deemed to be part of the shark's natural diet, up to a limit of 220 pounds per day. The tour operators do not feed the sharks, nor do they interfere with their natural behavior. The idea is simply to lure the sharks closer to the boat by pulling the baits in slowly so that the divers in the cage can observe them up close, without allowing the sharks to take the bait.

Eventually the shark becomes interested in a freshly tossed chunk of bait, and as the crew pull it closer toward my cage, I witness an awesome yet frightening sight. It is the stuff of every beach swimmer's nightmares. The almost grinning toothy face appears out of nowhere, coming straight at me, head on, with its large mouth poised to open. It is focused as an apex predator on its prey (the moving bait, not me), and when it is ready to have a go at the bait it opens its mouth revealing the upper row of teeth. Seconds later, as the bait is whipped out of the water, the shark leaps out at it, but misses, crashing back in the water next to my cage, then quickly rolls around and swims away into the deep blue seascape. I'm pumped full of adrenaline. It was a truly awesome couple of seconds to see this powerful charge, alone in the cage, the shark in action doing what it has evolved to do so well over several millions of years.

I tallied up some seven hours in the cage observing nine different white sharks, from baby sharks only 5 feet or so long, to mature males who have been coming back to the same spot year after year. My new friend "Imax" is a 14-foot male with very distinctive dark sooty gray surface color and nicked dorsal fin. He has a distinct attitude, not as flighty compared to some of the other sharks I observed, far cruisier in his mannerisms. Sadly, this lordly shark has a mild case of scoliosis, a twisted spinal column near his tail. To me he is kingly; he is the Richard III of sharks.

Imax, a 14-foot-long male white shark. Each shark shows its own personality when you have time to observe them at length in their natural habitat  AUTHOR PHOTO

Sharks are rarely aggressive; they become more active because they are just hungry, pumped up, and wanting a feed. Aggression is shown only when they are threatened territorially or in courtship. In such cases they might arch their backs and push their pectoral fins downward. Imax just cruised around in deference to the tasty baits around him, checking me out. He has been coming back here to the same spot every year for the past fifteen years, and he is estimated by shark biologists to be at least thirty years old. One time he slowly swam toward me, then slipped silently under the cage, only a couple feet below, and for me it felt like the opening sequence of the *Star Wars* movies in which the huge spaceship takes forever to cross the screen. I watched this large male then slowly disappear down into the hazy blue gloom, like a nightmare vanishing as the morning sun hits your face. As I breathe through the oxygen tube, I settle down and carefully watch the great fish at ease in its domain. The watery silence is dampened by the heightened sound of your own breathing, the crashing thump of your beating heart, and the gentle background trill of bubbles escaping the regulator.

The graceful maneuvers of this powerful fish win me over. These are not monsters, an illusion created in our media-frenzied minds by the grizzly and unnatural thought of sharks always wanting to attack and kill humans. No, they are intelligent and sentient creatures that embody the apotheosis of evolution, fishes that after 450 million years have used natural selection to continuously improve their body design and are now once again at the top of their game.

An exciting incident occurred in the cage on the third day. A female shark about 11 feet long came charging straight toward us to grab the bait. By putting on a powerful burst of acceleration at the last minute, she breached out of the water, then came crashing down with a mighty splash, her huge body pushing up against the side of our cage, jolting us suddenly sideways. There were two of us in the cage when this happened, and we both lost our balance, rocking around dizzily inside the cage, our gaze focused intently on the huge beast lying up against the bars only a foot or so away. She thrashed around, then quickly righted herself and moved off into the distant water. I will never forget those few seconds of being up close and personal with a white shark.

To see a large white shark with its serrated triangular white teeth and massive jaws only a couple feet away from your face is a dichotomy of terror and wonder mixed with a super-dose of adrenaline pumping through your brain. I had come eye to eye with one of the greatest predators on the planet, and we took each other's measure.

White sharks have eyes that appear black in most photos, but the irises are actually a very deep blue, as you can see up close. Those eyes look into the core of your very soul. I felt privileged to observe the sheer power of its large high-angle tail fin thrusting gracefully through the water, while admiring its sharply pointed snout incised with rounded grooves for the nostrils. The reason for the name "great white shark" or "white pointer" becomes apparent, although these days scientists prefer to call them just "white sharks." Like watching lions on the Serengeti, or tigers in the jungles of India, observing a massive apex predator at home in its domain is an awesome experience for anyone who appreciates nature. The white shark presents nature at its wildest and most terrifying, while at the same time relaying a calming sense of gentle power.

Before I tell you about the origins and evolution of white sharks based on their fossils, we first need to learn some basic facts about them as living creatures, about their amazing sensory systems and hunting abilities. Then we are ready to explore some of the remarkable recent research discoveries about their behavior and lifestyle. I will start with a historical anecdote about how they were given their scientific name.

## A white shark primer

The historical discovery and naming of the white shark is also an interesting tale that brought together two of the world's top biologists of their day.

Juvenile white sharks like this one prefer to feed on fish, later switching to warm-blooded mammals like seals and whales as they get larger   PHOTO BY ANDREW FOX

Carolus Linnaeus (1707–1778) was the Swedish father of modern taxonomy, who gave living things their binomial scientific names (genus and species—for instance, we are *Homo sapiens*). Linnaeus was the first scientist to formally classify the white shark, along with myriad other species of animals and plants, although he didn't have much expertise on sharks. He put the white shark in the same genus he created for all sharks—his genus *Squalus,* originally erected for the dogfish, a beast totally different from the white shark. Nowadays we place these two creatures in different orders and families of sharks, but in 1758, Linnaeus simply christened the white shark *Squalus carcharias.* The name couldn't be more straightforward: *carcharias* is taken directly from the Greek meaning "shark," while *Squalus* is from the Latin meaning "shark"—thus his taxonomic name *Squalus carcharias* comically translates as "Shark shark." I think it's apt. There's nothing more sharky than a white shark, in my opinion.

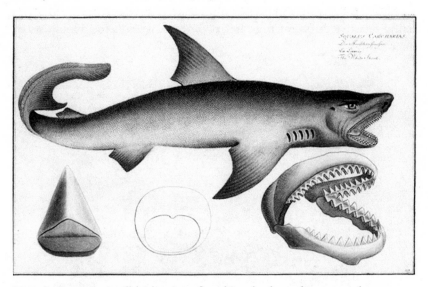

**Linnaeus's 1778 woodblock print of a white shark used to name the species** PUBLIC DOMAIN

It took an intrepid young Scottish physician named Andrew Smith to sort out Linnaeus's taxonomic muddle. Smith had been posted to the Cape Colonies in South Africa in the 1830s to supervise the medical care of soldiers. In his spare time, he studied the region's natural history with a view to documenting as many species as possible. In 1836 he met the young Charles Darwin during his landmark voyage of HMS *Beagle* on his visit to the Cape, and they soon became firm biology bros. Meeting Darwin really fired Smith up to complete his five-volume magnum opus, *Illustrations of the Zoology of South Africa,* the first volume of which came out in 1838. This volume featured the first detailed scientific description of a white shark, which Smith determined was very different from Linnaeus's dogfish genus *Squalus.* He erected a new genus name, calling it *Carcharodon* based on its teeth, meaning "sharp or jagged tooth," which was then coupled to Linnaeus's species name. Hence today the name holds fast as *Carcharodon carcharias,* the white shark.

What do white sharks like to eat with those teeth? It is well established that white sharks will feed primarily on squids and bony fishes when young but switch to higher-fat diets involving seals and whales when they mature. Recently the feeding preferences of juvenile white sharks living off New South Wales, Australia, were analyzed from forty sharks. Their stomach contents, combined with historical records of prey, showed that they feed primarily on bony fishes like the Australian salmon, eels, groupers, trevally, and whiting, as well as squids, giant cuttlefishes, rays, and other sharks (including hammerheads and requiem sharks). The young male white sharks here also eat more stingrays than the females, with ray consumption on the increase in recent years.

Another study looked at juvenile white sharks off Mexico and analyzed their tissue chemistry to see what they were eating. It found that foraging for prey, like tuna and other ray-finned fishes, near the surface waters was the major feeding strategy for the very young sharks. As the juveniles got older, they switched to foraging in deeper waters and on the seafloor. Overall, these sharks found most of their food (45 percent) from foraging close to the shore on the continental shelves. They ate free-swimming bony fishes like tuna, and small bottom-dwelling sharks and rays. Those that hunted in open seas derived 35 percent of their food from free-swimming fishes like tuna, and some seals. Adult white sharks primarily prey on seals, dolphins, and whales, but they also occasionally take large bony fishes, rays, seabirds, and many other kinds of prey, according to location or season. Despite humans' fear of them they definitely didn't prey on people.

White sharks have an amazing tolerance to heavy metal toxins that would kill other creatures. Their bodies accumulate toxic methyl-mercury in large amounts. Seventy-five percent of this comes from prey caught in the deep mesopelagic zone of the ocean (660 to 3,280 feet), while about 25 percent of the methylmercury in their system comes from eating seals in shallower waters. When mercury from pollution in the atmosphere is deposited in the ocean, a fraction of it gets converted to toxic methylmercury by microbial action. It then gets incorporated into the entire marine food chain. Predators like sharks accumulate more of it than others by eating large amounts of fish and squid in the mesopelagic zone. The concentrations of methylmercury in these sharks and other predators is predicted to increase in future years as a result of rising global temperatures, which increase the production of the deadly toxin.

White sharks have the most powerful bite of any shark, perhaps more powerful than that of any land animal alive today. A team led by Sasha Whitmarsh of Flinders University in South Australia actually measured the bite forces in the field using Futek donut load sensors (yes, that's right) from six white sharks ranging in size from 10 to 12 feet. These bites ranged from 822 to 4,657 Newtons (184–1,046 PSI), but are all from midsized sharks. If this is scaled up to a 21-foot shark, it would generate 9,320 N (2,095 PSI) at the front of the jaws, and a whopping 21,367 N (4,803 PSI) at the back of the jaws. For comparison, we humans bite at a maximum force of 162 PSI, and the bite of the saltwater crocodile (the hardest biter among land animals) has been directly measured at 16,460 N (3,700 PSI). The force of the crocodile bite is higher than that of actual recorded small-to-medium white shark bites, but lower than the calculated maximum bite force for a hypothetical 21-foot-long white shark.

Most white sharks use an ambush-and-strike attack method when preying on seals. A powerful bite will either kill or mortally wound its prey, which can then be eaten at leisure when it's dead. The spectacular breaching of a white shark out of the water is the result of its targeted attack while accelerating straight up from underneath a prey item. It can even use these jaws to best effect when taking down prey much larger than itself, like whales (discussed in more detail soon).

Early metabolic studies done in 1982 suggested that a white shark could subsist for 1.5 months off around 66 pounds of whale blubber, clearly pointing to why they prefer fatty mammals to fish as they get older. A more recent study used a different approach to base metabolic consumption on the daily energy usage calculated from their swimming speeds and concluded that the metabolic needs of a white shark were several times higher than found in the first study, and that they need to feed far more often than previously thought, probably every three days if they are eating high-energy foods like seals. In the Neptune Islands off the coast of South Australia, the sharks feast on young Australian fur seal pups over the summer when they are in season, then move to other offshore areas where large bony fish aggregations occur. It is interesting that when hunting for seal pups, the sharks use a special motion search pattern, indicative of pattern movement when prey is abundant, then switch to another, more random movement pattern using lower energy to prey on bony fishes. This means that hunting the large bony fish schools in open waters uses less energy and could be more efficient for the shark than staying around the shallower island seas to try its luck hunting adult seals.

So now that we know what they like to eat, let's see what hunting strategies they have evolved to best effect. And what do we know about their mating behavior and growth rates?

## White sharks on the hunt

In February 2017 a large white shark was observed attacking and killing a juvenile humpback whale in Mossel Bay, South Africa. It was witnessed by the crew of the Oceans Research Institute vessel, one of whom used an aerial drone to capture live footage of the event. The humpback whale was found entangled in ropes, in a poor condition, its emaciated body covered with barnacles and lice. Soon an 11-foot-long white shark slowly approached it, and after observing it for a couple minutes, inflicted a strong bite on its left side behind the flipper. The shark immediately let go and moved away without taking any plug of flesh to eat, just causing a severe wound. The shark then waited while the whale slowly bled for another 42 minutes before it came in and bit the whale a second time, under the tail, also immediately letting go without removing any flesh from the whale.

A second, larger shark estimated at 13 feet long then approached and attacked the whale on the tail base, and this action prompted the first shark to move away from the whale. The large shark then attacked the whale with a fourth bite, targeting the already wounded base of the tail. This was a more aggressive bite in which the shark violently shook its head and tried to push the whale under the water. By now the whale had lost a lot of blood and was very weak—it lived just another thirty minutes. One shark left the scene at this time, while the other waited around until the carcass sank to the bottom. The sharks were not sighted again at this point; presumably both were busy feeding on the carcass below.

While white sharks are well known as scavengers who will readily feast on dead baleen whales, this account was the first record of white sharks attacking and killing a living humpback whale. It showed they cooperated using a "bite and split" tactic rather than expending more energy to kill the whale quickly with many attacks. It gives us a good idea how these sharks probably operate by using measured, strategic hunting tactics rather than just charging in with brute strength. It also indicates that ancient white sharks of similar size could have brought down large baleen whales in the wild. When they first appeared in the seas about 6 or 7 MYA, baleen whales were much more abundant in the oceans and represented more species than today. Even with megalodon prowling around, there would have been plenty of whale and smaller manatee prey for the first white sharks to hunt.

White sharks are known for their predatory instincts that encompass both brawn and brains. In fact, scientists have recently shown that these sharks have evolved specialized hunting behaviors. Dr. Charlie Huveneers, a shark biologist based at Flinders University in South Australia, works just a few doors down the corridor from my office. Charlie is more at home wearing a wetsuit, immersed in the blustery, foaming waters off the coast of South Australia, than sitting at his desk. He is a leading researcher on white sharks who first became interested in the subject when he was growing up in Belgium. Charlie became interested in sharks at age eleven, when he did a school presentation about them at the suggestion of his mother, who bought him a book on sharks. The editor of that book was Dr. John Stevens, a legendary shark researcher based in Australia. Charlie would later work with him, tagging sharks, and today he is head of the Southern Shark Ecology research group at Flinders University.

Charlie has spent hundreds of hours on boats and in cages underwater observing, tagging, and tracking the large population of white sharks living off the Neptune Islands. His team's research is shedding much new light on the attack behaviors of the white sharks. Just recently they made more than four hundred observations on how they approach and attack their prey. Surprisingly, they found from their statistical analyses that white sharks always attack with the sun behind them, giving them the advantage of coming out of the light where visibility is poorest for the prey. Most shark attack victims simply don't see the shark coming at them. The white sharks are using their innate knowledge of astronomy to develop the most successful attack strategy.

Charlie Huveneers researches white sharks to help save them and to learn more about how we can live with them. Here he is about to take a biopsy sample from a white shark   PHOTO BY ANDREW FOX

Research carried out by Michael Martin and colleagues from the University of Miami in 2009 showed that white sharks use stalking methods similar to those of human serial killers. The research included input from Kim Rossmo, a criminal justice expert who is a specialist in the geographical profiling of criminal acts. Rossmo uses the locations of linked crimes to predict the next location of a criminal's "anchor point," or the region where the next attack might be initiated. Their study of some 340 attacks on fur seals showed that the white sharks hunted seals in areas known to be more prevalent—they were using an "anchor point" to launch attacks that would be more successful than using random places to attack. By the white sharks' effectively developing a strategy to plan an attack, they were behaving the way human serial killers do in their methodical approach to successful kills of their victims, keeping maps in their head of their underwater hunting grounds.

Another study looked at white shark vision, testing the "mistaken identity theory" that sharks attack swimmers and surfers because they see them as seals. Most attacks on surfers are from juvenile white sharks. The shape and motion analyses proved that images of humans swimming, or on surfboards being paddled or towed, could not be distinguished in the retinas of the shark from outlines of their natural prey, the seals. However, we know the sharks have a powerful sense of smell, so in many cases this may be the main sensory input that deters a shark from randomly biting a human in the water.

**White sharks often breach when chasing seals, as this successful hunter is doing off Neptune Islands, South Australia**  PHOTO BY ANDREW FOX

White sharks have amazing superpower senses of smell and electroreception. These fishes have poor eyesight, but highly attuned noses that can detect substances in water diluted by one part in 10 billion— they can smell a plume of water tainted with blood and tissue carried on a current from at least four miles away. Their sensitive electrosensory organs, called ampullae of Lorenzini, appear as a myriad of tiny pore openings covering the snout and side of the head. These are the openings for flask-shaped cells, each containing a network of delicate sensitive hair-like structures filled with a conductive jelly. These organs sense the electric fields of prey when other senses fail in the fog of bloodied water, to home in on struggling or fleeing prey. These ampullae are an important tool on the shark's sensory Swiss Army knife, as they do a range of other important jobs for the shark, like determining water temperature and finding magnetic fields.

## Mating and growth in white sharks

Estimates of how long a white shark can live are based on growth rings in the vertebral discs of dead sharks. Radioactive elements falling out from nuclear bomb tests in the Pacific in the 1950s got incorporated into the food chain and act as a distinct marker horizon at the time when sharks grew their vertebral discs. The radioactive elements were detected in the growth rings of the vertebral discs of two sharks used for the study, as these rings mark the year of the nuclear bombs. Counting the annual growth rings gives the age of the shark. The results showed that one female shark was forty years old, and the oldest male measured was seventy-three years old, so they most likely live a lot longer than this in ideal conditions. It makes me smile to think that the sharks filmed in the *Jaws* live cage scenes in 1975 could still be out there today.

While we don't have evidence how they mate, we have identified some potential nursery areas around the globe where numerous juvenile white sharks appear each year such as the Port Stephens area in New South Wales, Australia, and off Montauk in the Cape Cod region of the United States. They are thought to be able to navigate their way back to specific nurseries across the globe for giving birth each year using their electrosensory organs to tune in to Earth's magnetic field.

Scientists study populations of white sharks by finding out where they like to feed, such as near large seal colonies, where they can often be seen breaching out of the water with a seal in their mouth after a rapid dash to the surface. These special areas are where sharks migrate to every year or two, sometimes coming from far across the globe. White sharks have incredible transglobal migrations, which have only recently been revealed by trackers attached to their fins relaying their precise locations and depth ranges. We now know that at least one shark, nicknamed Nicole (after Nicole Kidman), has been tracked migrating from the Cape of South Africa across to the northwestern side of Australia, and that this return migration is the fastest of any marine animal, completing a twelve-thousand-mile journey in just nine months. Sometimes they don't journey alone. Scientists recently tracked two male white sharks between ten and fifteen years old, that they named Simon and Jekyll, some four thousand miles swimming together from the coast of Georgia in the southeastern United States to the Gulf of St. Lawrence in Canada over a period of eight months. We don't yet know whether they are siblings or just good friends, but I'm confident that further studies will be forthcoming on these compadres.

We also now know how to determine the age of living sharks without harming them by examining the size and degree of cartilage ossification in the male clasping organs. This method is giving us the first handle on the age of the populations of living white sharks. For example, a population of white sharks living off Kashima-nada, Japan, showed a faster rate of growth and therefore earlier maturity than for other populations, so it's likely to be variable for the species across the globe.

A recent study showed that male North Atlantic white sharks become sexually mature at age twenty-six, whereas females mature at age thirty-three, much later than humans. Estimating shark ages is challenging for older individuals, and some of my colleagues believe that certain populations of white sharks grow faster than this research suggested, maturing from ten years onward for males, or from fourteen years for females. The initial estimate of maturity was far longer than previously published estimates, highlighting the debate around aging sharks, but also signaling the urgent need for conservation of the species if they do in fact grow so slowly.

It's amazing how little we actually know about the mating habits of the white shark. For example, no one has ever seen them mating or giving birth. One potential mating area for white sharks is the area dubbed "the white shark café," a large area of ocean between Hawaii and Mexico. It's where many white sharks have been tagged heading each year, where they disappear for a few months and then reappear (with, one might imagine, a big smile on their face). The tagged male sharks were shown to bob up and down to great depths (1,200 feet) as many as 120 times each day, but not the females. Perhaps it is because they feed, mate, or give birth here, at great depths. The area has been recommended for a UNESCO world heritage site, but more data must be gathered to support the initial findings.

While humans have known about and feared white sharks for thousands of years, the white shark's deep-time origin story goes way back. So where, when, and how did white sharks originate?

## The origin of white sharks

Let's go back in time once again. As the oceans continued to cool during the Pliocene period—5.3 to 2.6 MYA—sharks weathered yet another environmental change: the end-Pliocene marine extinction event, which was marked by a dramatic diminishment of plankton. In the wake of this change, other large sharks and whales died off in droves, and great whites suddenly emerged as the last shark apex predators left standing (so to speak). How did they survive when other large sharks of the time did not?

The answer could lie in the great white's genetic makeup. In 2019, their complete genome was analyzed, unexpectedly revealing that it includes ancient genes that offer increased resistance to diseases, a fact that may have played a role in their long-term survival as a species. We don't often think about illness in wild animals, but when it comes to sharks, a recent study that surveyed the health of 1,546 living sharks found that 33 percent of them had some sort of inflammatory or infectious disease, about 47 percent had some sort of viral skin disease, and nearly 25 percent had nematode worm infections. This evidence leads researchers to suspect that such diseases and parasites were also common in prehistoric shark populations. So when the ecosystem collapsed at the end of the Pliocene period and the ocean's food chains were once again under stress, diseases were probably rampant. Among the stressed shark populations, those sharks with stronger resistance to these conditions may have survived when others went extinct— providing great whites with an opportunity to take advantage of the empty throne at the top of the food chain.

Now back to the fossils. While the fossil record of the first white sharks from the late Miocene (6–5 MYA) is mostly from teeth, there is one exceptional fossil head of a giant archaic white shark that unlocked an evolutionary puzzle for us. It was found in Peru by an extraordinary American fossil collector named Gordon Hubbell from Miami, Florida.

**Gordon Hubbell in his amazing shark museum**  PHOTO BY JEFF GAGE/FLORIDA MUSEUM OF NATURAL HISTORY

Gordon has made tremendous contributions to shark paleontology through his lifelong hobby of collecting fossil shark teeth as well as through his generous donations of the important fossils he has found to the Florida Museum of Natural History. Together with veterinarian and shark paleontologist David Ward in England, he demonstrates how significant fossil collectors are for advancing our knowledge of shark evolution.

In the mid-1960s, Gordon began traveling to the northern part of the Atacama Desert in Peru, a remote locality world-famous for fossil sharks and whales of the Late Miocene to Pliocene periods. During one of these trips in 1988 he obtained an outstanding fossil shark specimen that would cement our picture of how white sharks evolved. It was a new species of ancestral white shark, complete with all the teeth, jaw cartilages, and parts of the skull and backbone preserved. I asked him, now age eighty-eight, about his remarkable find, and he told me the full story:

"On one of our stops in the small town of Sacaco, Peru, I met Carlos Martin, a Peruvian farmer who had an olive orchard. He was also an amateur fossil collector, and he showed me a fossil skull of *Carcharodon* [the white shark] that he had just collected. He was busy trying to remove all the teeth from the specimen, and I encouraged him to stop, and offered to buy it as is from him. We shipped it back to the United States in a 55-gallon drum, and I finished preparing it when I got home. It is a fantastic specimen with 222 teeth and a lot of the cartilaginous skull preserved in place. The upper teeth were all in the proper position, and the lower teeth were a bit scattered. I eventually donated this specimen to the Florida Museum of Natural History, and it has since been studied and named after me. I made a total of 8 trips to Peru, and I have not experienced such fantastic fossil collecting anywhere else. There were places in the desert where you could not take a step without stepping on a fossil shark tooth."

So just how significant was Gordon's prize specimen? One might imagine a fearsome beast like the white shark has been in our oceans for a very long time, yet the fossil record tells us the exact opposite—it is a relatively recent species, in fact it is one of the youngest of all living sharks that have a fossil record. It emerged as a new species only about 5 to 6 MYA, with the oldest record of teeth similar to those of the living white shark being the ones from Gordon's prize find, *Carcharodon hubbelli,* which first appears, based on isolated teeth, around 6 MYA. Gordon's find was an ancient new species of the white shark, very close in its size and teeth shape to the modern white shark. It came from the famous Pisco Formation in Peru—the site I used for the reconstruction of the *O. megalodon-Piscobalaena* slaughter scene at the start of the last chapter. Samples of shells taken from the exact site that Gordon's fossil came from yielded firm radiometric dates of close to 6 million years old (Late Miocene age).

This ancient white shark had a minimum body length close to 16.4 feet long, and its vertebral disc growth rings show it was close to twenty years old at time of death, possibly not even sexually mature compared to living white sharks, suggesting this species could have grown much larger, at least to around 20 feet. Its teeth are typically shaped like

*Carcharodon hubbelli* jaws from Peru. At 6 million years old, *Carcharodon hubbelli* is the direct ancestor of today's white sharks   PHOTO BY JEFF GAGE / FLORIDA MUSEUM OF NATURAL HIS-TORY

those of the extant white shark except they have much lighter, finer serrations, and some of the biggest teeth at the front of the mouth dif-fer slightly in shape from those of living white sharks. It is regarded as an immediate ancestor of the living white shark, as simply by making a few minor tweaks to a few of its tooth shapes and positions, we arrive at the modern white shark with the same dentition. Its remains have been found around the Pacific rim, from Peru to California to New Zealand, in rocks dated between 8 and 5 MYA.

So what other sharks at this time were also close relatives of the white sharks? Living mako sharks in the genus *Isurus* are close relatives of white sharks. The giant extinct mako shark—which as a kid I fer-vently collected and knew as *"Isurus" hastalis*—shows that every tooth position is remarkably similar in outline and general thickness to that of the living white shark and *Carcharodon hubbelli*, except that they lack one feature: serrated edges. We have seen how rapidly in the Eo-cene, about 52 MYA, the non-serrated teeth in *Otodus obliquus* morphed into the serrated teeth of species *Otodus aksuaticus*. The same process took place with *Isurus* changing to *Carcharodon*. We also saw how non-serrated thresher shark teeth developed serrations in the gigantic Miocene form *Alopias palatasi*. This is the evidence that sharks with smooth-edged teeth can give rise to species with serrated teeth. There have even been several examples of *"Isurus" hastalis* teeth showing a few serrations along their edges, and these intermediate forms are the proof that it now rightly belongs in the genus *Carcharodon*, cementing its evolutionary position as the immediate ancestor of the living and fossil white sharks in the same genus. We know it fed on whales, as a tooth of this species was found deeply embedded in the fossilized lower jaw of a large baleen whale, also found in Peru. The scientists who studied this fossil think the shark was scavenging on the dead whale when it lost its tooth.

Large fossil teeth of *Carcharodon hastalis* that I found as a teenager at the Beaumaris fossil site in Australia (dated at around 6 MYA)
AUTHOR PHOTO

Another species of archaic white shark, *Carcharomodus*, lived between 14 and 4 MYA in the Northern Atlantic and is believed to have descended from the ancient mako *Carcharodon hastalis*. The largest fossil teeth suggest it had a body size similar to that of the living white shark. *Carcharomodus* also ate whales and dolphins, showing that sharks we once thought of as "ancestral makos" (renowned fish hunters) were switching their diet to marine mammals, the favorite food of the white sharks. This dietary change reflected in its serrated teeth confirms it is an ancient white shark, not a mako as first thought. There are other lines of evidence supporting the close relationship between white sharks and makos; it's all in their DNA.

These oldest known teeth undoubtedly attributed to the serrated form of *Carcharodon* are specimens from the late Miocene deposits of eastern Argentina, dated around six or seven million years old. The small number of specimens known are incredibly similar to the teeth of living white sharks, but we need more examples to scientifically prove this point. Nonetheless these finds give us a reasonably good date for the origin of the modern white shark.

Much of the white shark's preference for eating warm-blooded mammals is because they need to sustain a high metabolism. Let's learn a bit about their physiology and how they generate and maintain a high body temperature.

## White sharks are so hot right now

White sharks have a special body metabolism that enables them to switch between living in warm surface waters and diving to feed in cold deep waters. We know from their tagging with depth-sensory electronic devices that white sharks sometimes feed in the twilight zone (mesopelagic zone) at depths of 650 to 3,280 feet, and that the deepest white shark dive ever recorded was to an incredible 4,193 feet. It's very cold down that deep, so how do the sharks manage to keep warm and active? This ability is the result of another special physiological feature they have evolved that gives them the edge as apex predators.

It has long been known that white sharks have a higher body temperature than the surrounding seawater, supporting the misconception that they are "warm-blooded" like us. This concept is not correct; they cannot maintain a constant warm body temperature as birds and mammals can. Instead, they can just slightly raise their body temperature to a level above the surrounding seawater. They have a special kind of metabolism that heats up the blood, using a network of tiny capillaries that exchange heat generated in the muscles. These organs are called the **rete mirabile** ("miraculous nets") and are found in several parts of the white shark body, from deep inside the muscles near the stomach and also near the spiracular gill slit on the side of its head.

The effect of these organs in sharks is to raise the body temperature to around 8°F above surrounding seawater for the body musculature, or around 25°F for the internal stomach temperature. The liver might play a key part in raising internal temperature of the white shark's stomach, because the liver is filled with fats, which efficiently retain heat. Thermal conduction from vessels inside the liver also aid in keeping the gut warm. Such capabilities enable faster digestion, generating more muscular power as well as warmer outer body temperatures. A thick layer of fat below the skin enables the white shark to insulate its inner core more efficiently. Another smaller rete network located near the spiracle behind the eye raises the temperature of blood flowing into the eyes and brain, facilitating faster eye movements and quicker brain processing times, both much needed in an apex predator. White sharks are now regarded by biologists as **regional endothermic** sharks, meaning they have the ability to maintain higher body temperatures in certain regions where the waters are too cold for other predatory sharks. Their warm-blooded metabolism explains why they need to keep active all the time and need greater amounts of high-fat food. Their diet of primarily seal or whale blubber supplies the energy for their active lifestyle.

While it's "good to be king" as top dog in the oceans, there are still parts of the world where white sharks are the second tier in the food chain. There are bigger, more terrifying predators they fear. And many small ones that can take a toll on their overall health.

## What white sharks fear

Being the hunted, not always the hunters, is not a new concept for white sharks. We have seen how back in the late Miocene and Pliocene they lived in the shadows of much larger predators, like megalodon, *Carcharodon hastalis,* and orcas. The new information emerging about white sharks and their predators makes their future survival even more challenging than we previously thought.

In 1997, the marine biologist Alisa Schulman-Janiger from California documented the first case of orcas killing white sharks off the Farallon Islands just west of San Francisco Bay. The two orcas were observed to close in fast on the white shark. One grabbed it and turned it over to an upside-down position. This sent the shark into atonic immobility because it can't breathe through its gills when upside down, causing it to die of asphyxiation. The orcas like to eat the livers of the sharks, which are large organs, sometimes weighing more than a thousand pounds, and are very rich in oil and high in fat. To do this, they bite the shark near its pectoral fins, then squeeze the liver out through the wounds, like squeezing toothpaste out of a tube. This behavior has also been documented in South Africa, with drone footage showing that the sharks began to flee the area as soon as the orcas approached.

In 2015, *National Geographic* released a full-length documentary that dramatically captured this specialized behavior. In January that same year, a group of tourists witnessed a similar amazing scene while on a shark cage diving expedition at the Neptune Islands off South Australia. They watched a pod of some six orcas with two small calves in tow lure a white shark to the surface, then the orcas launched themselves upon it from above and below to kill it. The marine biologist Gina Dickinson, who witnessed the action as it happened, suggested to reporters covering the story that the orcas did this to teach the young calves their special predatory behavior. This is witnessing evolution in action. If orcas continue to teach these successful hunting techniques to their young, it will be only a matter of time before more and more orcas hunt great whites, and they could pose a serious new threat to the sharks' long-term survival.

In addition to the dangers from orcas, white sharks host a large number of smaller parasitic creatures that either feed directly off the shark or share its food scraps. A blood-sucking crustacean parasite about the size of a match head has been prevalent among white sharks living off Cape Cod in North America. This parasite, *Abdominius irrisionis,* is the cause of bleeding and small tumors on the underside of the sharks. The parasites are more active in the winter, when the sharks leave the Cape to range out over the open Atlantic Ocean.

We have seen already how white sharks use their knowledge of astronomy for hunting, but did you know they also have a keen interest in geology? It has been suggested that the geology of the Cape Cod area attracts white sharks there due to the coarse silica sands forming certain sand bars—these serve as an excellent surface for the white sharks to scrape the parasites off their bellies, and for the males to clean them off their claspers.

All creatures, large and small, have parasites, which by definition feed off their hosts without wanting to kill them. They can still reduce the energy level and overall fitness of the host, and in severe cases cause illnesses that weaken or kill them. White sharks are known to host at least thirty species of copepod parasites—small crustaceans with segmented bodies looking a bit like a flattened stubby shrimp. Parasitic copepods are found on nearly all large marine creatures, including baleen whales, which host species almost a foot long. It could be worse for the white shark, though, as long-living Greenland sharks often have a parasitic white copepod attached to each of their eyeballs. These sharks don't seem to mind, as they don't need their sight to hunt in the dark, deep Arctic seas. Many copepods inhabit the gills of the white shark to feed off scraps of food that pass in the water flowing over the gills. Others do the same by attaching to the dorsal fins or parts of the body.

In addition, white sharks have been found with several kinds of internal parasites including tapeworms. One species, called *Clistobothrium carcharodoni,* was found inside a large white shark caught off Los Angeles. The 15-foot shark had swallowed an entire young northern elephant seal. A closely related parasite occurs on southern elephant seals, so it was proposed that the white shark picked up the parasite from eating the seal. More recent genetic work has linked this same parasite as being one commonly found in its juvenile phase on striped dolphins and Risso's dolphins. The study went as far as suggesting that to pick up the number of parasites found on the white sharks being studied, they would have had to ingest anywhere from 8 to 93 Risso's dolphins, so the parasites tell us something about the white shark's predatory behavior.

Overall, parasites don't appear to bother the sharks much, as it's a mutually beneficial arrangement. Copepod parasites can keep the gill area clean of decaying organic matter, preventing infections. Parasites have little negative impact on the overall health of the sharks, and they are a part of evolution's way of recycling resources from one level of the food chain to another.

Undoubtedly the biggest threat to white sharks is us. White sharks suffered after the 1975 film *Jaws,* when thousands of these magnificent beasts were killed in what can only be described as a frenzy of fear. The director of the film, Steven Spielberg, stated recently in an interview, "I truly, and to this day, regret the decimation of the shark population because of the book and the film. I really, truly regret that." George Burgess, a former director of the Florida Program for Shark Research, claimed that about half the white shark population in the Northern Atlantic, several thousand sharks, were killed in the aftermath of the film. Other reports claim up to 79 percent of the white shark population declined in this area between 1986 and 2000, showing that the film's dire effects had a long-lasting tail. Today these sharks are protected in several countries and form the basis for an important tourist industry in this region and in California, Australia, South Africa, and most recently Nova Scotia.

Let's now take a brief look at how white sharks have played with our minds, and at times deeply embedded themselves into our hearts.

## White sharks in our hearts and minds

White sharks have played an important role in our imagination, inspiring major works of art, literature, and cinema and featuring in religious lore and many First Nations peoples' stories. Throughout human history, the depiction of the white shark has typically been a negative one, but sometimes they appear in the context of uplifting tales. The poet Hardwicke Rawnsley was one of the great British conservationists of the Victorian era. His poem "McDermott's Deed" (1896) was based on a true story that took place off Uzi Island, south of Zanzibar in Africa. While several men went swimming, the boatswain, McDermott, saw a large shark coming for them, so he jumped on top of the shark, scaring it away to allow the men to scramble to safety. The poem details features of the shark like its "nostril keen" and "eager eye," which suggests it was definitely a white shark. His summary of its key features in the poem are of course exaggerated when he described the mouth of the shark thus:

> His teeth are cruel as a saw
> His throat as hell is wide,
> Yea, cavernous as Death his maw,
> As Death, unsatisfied.

Perhaps the best-known depiction of the white shark in art is John Singleton Copley's painting *Watson and the Shark,* which was debuted in 1778 and now resides in the National Gallery of Art in Washington, D.C., with other original copies by Copley hanging in Boston's Museum of Fine Arts and the Detroit Institute of Arts. It records a white shark encounter with a teenage boy, Brook Watson, that took place when he was swimming in the sea off Havana in July 1749. The shark bit young Watson three times, eventually tearing off his leg. Watson survived and went on to lead a very successful life as a merchant and politician, becoming Lord Mayor of London and eventually

*Watson and the Shark,* John Singleton Copley, 1778. Watson, aged fourteen, survived this 1749 encounter but lost his leg. Later he became Lord Mayor of London   OPEN SOURCE IMAGE

knighted as Sir Brook Watson. He was easily identified around London by his wooden leg. The painting was also significant for advancing Copley's career because the huge publicity surrounding the painting was deemed to be a factor in his election to the Royal Academy. Copley never went to Havana. His painting was based entirely on the details of an eyewitness account of the encounter published by James Brayman in 1852.

The American artist Winslow Homer's painting *The Gulf Stream* (1899) is another famous depiction of white sharks threatening an imperiled victim, this time a Black man in a disabled sailboat. Homer was clearly sensitive to the plight of Black Americans. He has positioned the victim in a posture of despair with a ship angled in the background in such a way that, unlike Copley's work, takes away any hope of the victim's being rescued. Some critics see it as a metaphor for the future dangers facing African Americans in the post–Civil War period. Homer himself reflects that the man would be eventually saved. The historian

Alain Locke has argued that the painting is the first in the American art landscape to feature a Black man as the focus of a composition.

While there are numerous artworks we could discuss here depicting white sharks in a terrifying way, some artists have turned that image around to focus on the beauty and cultural significance of this iconic shark. Andrea Everhart is an artist living in California who spent some years in Hawaii; in 2016 she created a bronze sculpture titled *Ocean Beauty* that depicts a white shark with accurate anatomical details of its fins, nostrils, snout, and teeth. It is adorned with elaborate patterning across the shark's upper body, recalling Polynesian tattoo art. It links the power of the shark's body to the spiritual respect that some Polynesian cultures hold for the shark.

**The white shark as a subject for modern art:** *Ocean Beauty,* **bronze sculpture by Andrea Everhart. Its Polynesian tattoo themes highlight indigenous Pacific peoples' deep respect for the shark**
IMAGE COURTESY OF THE ARTIST (EVERHARTFINEART.COM)

The indigenous Australian Birpai people who live along the coast from Worimi to Point Plomer in New South Wales hold the white shark sacred as a totem. They feel a responsibility to look after the shark and thus don't fear it. They respectfully venture into the sea and are not troubled by the sharks, according to Nathan Moran, a Birpai and Thungutti man.

While there are many other First Nations peoples, especially around the Pacific, that hold white sharks in deep respect for their spiritual values, I think that any creature this majestic is likely to be worshipped by some and respected by all. Will they have a future in a human-dominated world?

## The largest white sharks and their future

White sharks were popular targets for big game fishermen throughout the latter half of the twentieth century, particularly after the series of four *Jaws* movies came out between 1975 and 1987. It was sickening to think that these magnificent sharks were just minding their own business and then yanked from the sea and killed for no reason other than a fisherperson (mostly men) could say they had caught a bigger shark than someone else. You generally can't eat white sharks because their flesh is often toxic with heavy metals like mercury, and their only asset is really their jaws and teeth for souvenirs or jewelry. Tragically, photos of fishermen with giant dead sharks of all kinds abound on the internet. The *Jaws* films inflamed the idea that really gigantic white sharks could be out there. This raises the question: Just how big *are* the largest white sharks?

Most great whites measure between 5 and 18 feet, with an average size of 15 feet, although the largest uncontested recorded length for a living great white shark is 19.7 feet, from a shark caught in 1987 off Ledge Point in Western Australia—its jaws ended up in Gordon Hubbell's collection in Florida.

Perhaps the largest white shark recorded in recent history is one caught off Cuba nearly eighty years ago. Its story is fascinating not only because of the impact it made at the time on the great writer Ernest Hemingway, but also because after it was caught, it was measured by a professional ichthyologist. Still, its size has always been doubted by current researchers. Why?

The shark, known as the Cojímar Monster, was caught off Cojímar, Cuba, in 1945, with a reported measured length of 21 feet and an estimated weight of 7,100 pounds (three and a half tons), as they did not have scales big enough to accurately weigh it (its liver, however, tipped the scales at 1,005 pounds). It caused such a stir that Ernest Hemingway, who had just moved to Cuba after his horrendous time at the killing fields of Hürtgenwald in World War II, was motivated to write a letter about it to his lover Mary Welsh in London. In it he makes vivid comparisons between the pugnacity of the shark and the enemy on the battlefield. The incident would later be revisited and reimagined in his classic short novel *The Old Man and the Sea,* using his first-hand knowledge of the Cuban fishermen and of the sharks that abounded in those waters.

While some scientists dispute the size of this shark, arguing that from photographs it looks around 16 feet or so in length, we must remember that original measurements were taken by a professional ichthyologist, Dr. Luis Howell-Rivero, from the University of Havana. He also sent the details to leading shark experts Henry Bigelow and William Schroeder at the Harvard Museum of Comparative Zoology. They clearly valued Dr. Howell-Rivero's work and published this record in their encyclopedic nine-volume compendium titled *The Fishes of the Western North Atlantic* in 1948. Furthermore, in 1974, Dr. Dario Guitart, also an ichthyologist, a student of Howell-Rivero's and the founder of the University of Havana's Instituto de Oceanología, interviewed the participants about the capture of the shark and was able to defend the original measurements.

**The Monster of Cojímar, Cuba, at 21 feet long reportedly the largest white shark ever caught (1945)** OPEN SOURCE IMAGE

The only remains of the Cojímar white shark today is a single tooth nearly 3 inches long, capped in tortoiseshell and suspended from a silver chain, owned by Dario Guitart's wife. It was examined by expert ichthyologists when they visited the Guitarts in 1998, and if it is scaled up according to what we know of the largest teeth in the largest accurately measured white sharks, then it is highly likely it was indeed from an extremely large white shark. In my mind the 21-foot measurement by Luis Howell-Rivero does have some credible proof attached to it after all.

Recent media shots of the gigantic white shark nicknamed Deep Blue have surfaced with claims that it is as much as 21 feet long. This large and sometimes gravid female travels a route back to Guadalupe Island via Hawaii every two years. She is famous for being caught on video with the American shark diver Ocean Ramsey. Ocean and her husband, Juan Oliphant, run a diving business in Hawaii, and both are active in shark conservation and educating the public about sharks' plight. Nonetheless, the claimed size of Deep Blue has been disputed, as it's never been accurately measured using a laser system, so her size is still only a guesstimate. David Bernvi, a grad student at the University of Stockholm, suggested that Deep Blue is around 17 feet in total length, using his own methods to calculate body size from the photos.

So, were white sharks on average larger back in the mid-twentieth century? What are the record sizes of white sharks caught by game fishing people?

Bob and Dolly Dyer were big-time celebrities on Australian media for many years. Bob hosted the country's favorite game show, *Pick a Box*, which started as a radio show in 1948 and became a big TV hit in 1957, airing until 1971. Bob and Dolly were keen big game fisherfolk who loved nothing more than chumming up the waters to attract large sharks so they could catch them using a rod. Both held world records for catching enormous sharks, mostly white sharks. Another Australian, Alf Dean, competed against them and eventually went on to break the world record in 1953 at Streaky Bay, South Australia. Alf gained an all-tackle world record by catching a 2,372-pound white shark on 54-thread line. He later broke his own record in 1955 with one weighing in at 2,536 pounds.

The largest white shark ever caught by hand was regarded as a bit dubious by some as the weight was never taken cleanly. This was when the American charter boat captain Frank Mundus from Montauk, New York, reputedly caught a 4,500-pound white shark using a harpoon (this well-rounded weight figure makes me immediately suspicious). He had a reputation for chumming the water by killing whales, and harpooning sharks, practices both outlawed today. In 1986 he and a friend caught a 3,247-pound white shark off Montauk, which is still the record for the largest fish of any kind ever caught by rod and reel. Mundus, who passed away in 2008, was rumored to be the inspiration for Quint's character in the film *Jaws*.

Clearly the sport of catching and killing white sharks was unsustainable at a time when we knew so very little about their biology, like their growth rates and time to sexual maturity. How many white sharks are left in our oceans is an important question to answer right now. In the waters off Australia and New Zealand, the total population is estimated to be between 2,500 and 6,750, including only 280 to 650 adult sharks. These figures are based on numbers existing in specific areas (known from tagging and observation) and from the limits of population variation gained from genetic sampling, then scaling it all upward.

Things started to look up for white sharks in 1992, when they were first protected by law in South Africa, and this was soon followed by Namibia, the Maldives, and certain parts of the United States. In Australia the sharks were first given some protection in 1996, and in 1999 their vulnerable status was extended to "vulnerable to extinction" status. The Australian white shark recovery plan was developed in 2002 and ratified in 2013 to ensure that this magnificent shark is protected and every effort is made to keep the species thriving.

Today some bold photographers now venture into the water with white sharks without a protective cage, claiming they know their behavioral patterns well enough to recognize when they are not in a feeding mode. Exquisite photographs have gone viral around the internet in recent years of Ocean Ramsey swimming alongside and sometimes touching the pregnant Deep Blue off the coast of Hawaii. It is known that some of these events took place after the shark had been feeding on a nearby whale carcass, so she had a full belly when the human-shark interactions took place. If sharks are pregnant, there are also hormones that kick in that stop them from feeding (this keeps mother sharks from eating their newborn pups), so this might signal another potentially safe time for daring humans to swim near them. Ocean is not the only person who does this. The Brazilian photographer Daniel Botelho has also captured stunning images while swimming with white sharks off the South African coast.

However, I would heed the warnings of the shark authorities, like Dr. Michael Domeier, director and founder of the Marine Conservation Sciences Institute, who recommends that divers *not* try swimming with white sharks, ever. Period. I'm with Michael.

Nonetheless, the passive behaviors demonstrated toward these free divers who confront white sharks do show us another side to the white shark: that they are not always the killers of our nightmares. White shark populations are slowly gaining ground, with some populations increasing while shifting farther northward, as in the Atlantic Ocean off Canada.

## Future white shark research priorities

Let's return to our white shark expert, Charlie Huveneers, to ask the big questions about our main concerns for white sharks in the future. His lab researches everything from human interactions with shark tourism and the effectiveness of shark repellents to general shark, ray, and chimaerid ecology and life cycles, and in particular, tagging and monitoring the white shark populations and their migrations off the coast of South Australia.

Charlie led a paper published in 2018 co-authored with forty-five of the world's top white shark researchers in which future research directions for the white shark were canvassed. The term "elusive" was used in the title as a reference to them being scarce in the 1970s–1980s. Now they are easier to study because we know so much more about them and their habits than ever before. Still, we have precious little data on their lifestyles, their mating habits, and their population trends. We simply don't know where sharks are moving to, and what the main factors are that drive them to move to different areas.

We know white sharks exist today as separate genetic populations based at different sites around the globe. This makes them more vulnerable in specific areas where they are less protected from fishing or face other survival pressures. Apex predators are like that—they exist in lower numbers than midlevel predators and they take much longer to sexually mature, also making them vulnerable if their populations begin to decline for one reason or another. The trends that are global or uniform for the species are harder to pin down. The paper posed ten main questions about white shark research, but I'm going to highlight the most important three here and recommend that you consult the paper for more information.

The top priority question concerning future white shark research was "What is the size and status of white shark populations?" It's because until we know the numbers of sharks out there, and whether their status is as healthy populations or endangered populations, we cannot design new policies or legislation to protect the sharks from future perceived threats.

The second big question was "What are the mechanisms driving shark distributions, movements, and migrations?" What factors, such as chasing food supplies, reaching mating or birthing areas, climate change, tectonic forces, or ocean chemistry, are the most significant ones that influence their mass movements? If we understood what drivers make sharks migrate large distances, sometimes across oceans at specific times, we could propose new marine reserves to protect them and help them successfully reproduce. Or better understand the seasonal timing of when not to go into the water for certain human recreational activities.

The third big question was "How can we quantify and alleviate current threats to adequately manage white shark populations and ensure their conservation?" This means we need to better understand all the threats that impact the life of the shark, both human-generated and natural causes or imbalances, to ensure their conservation. This encompasses a wide range of research to first identify all the threats and then seek appropriate actions that might help ensure the long-term survival of the species.

While we have come a long way in putting in place protective measures for the sharks in the sovereign waters of certain countries, we must remember that the sharks move around the globe on a frequent basis, so they are spending a large proportion of their lives in areas without any legal protection at all. There is much political discussion and legal work to be done before the species is anywhere near being safe. We have now heard the long and challenging story of the evolution of sharks, taking many twists and turns as sharks survived extinction events and reinvented themselves along the way. The figure below shows the summary of this long journey in simplified form.

Now let's take these thoughts about white sharks out to the bigger realm of all sharks, rays, and chimaerids. In the final chapter we will wrap up with a look at sharks in the modern world, their diversity, their roles in our ocean's complex ecology, the threats to their survival, and the reasons why we urgently need to take steps for their protection.

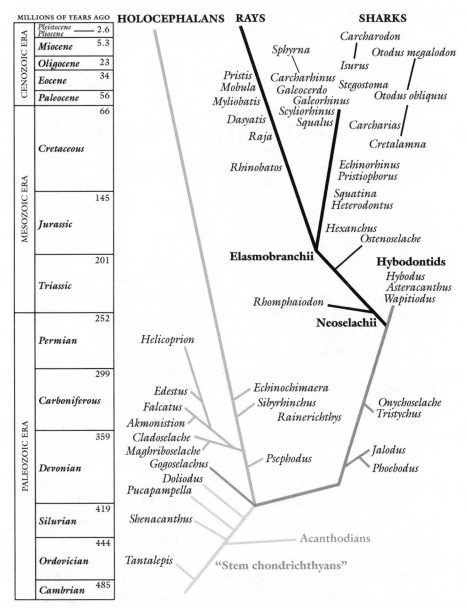

Simplified summary of shark evolution, indicating approximately when many iconic living forms first appeared   DIAGRAM BY AUTHOR

CHAPTER 16

# SHARKS AND HUMANS

*Can Sharks and Humans Live Together?*

———————————————

**465 MILLION YEARS AND COUNTING.** We have seen so far in our story how sharks over the past 465 million years have survived numerous threats from the natural world, including five global mass extinction events and several other major regional catastrophes in which oceans boiled or chilled, or major equatorial currents stopped flowing or changed course as continents slammed together. We have seen how recent ice ages shrank the seas and brought about marine extinctions. Yet despite all these factors, sharks have survived to the present day, though I wouldn't say they are thriving.

Today humans are driving the sixth global mass extinction event that is culling life dramatically on all continents and causing increasing chaos in our oceans. Sharks of all kinds, including the twelve hundred known species of living chondrichthyans (sharks, rays, and chimaerids), are facing their biggest threat yet. Let's take a sobering look into the history of sharks being used by humans for food or other products, as well as a reality check on how dangerous sharks really are to us humans. We can close our story by reflecting on the vulnerability of sharks right now, and the main challenges they face in the future from climate change, overfishing, and pollution. I will end our journey through the long secret history of sharks with a message of hope for the future as we consider possible solutions to these grave problems.

# Sharks as food and goods

Some shark populations are declining at an alarming rate, and several species are on the brink of extinction. They are primarily threatened by the fact that most of them are good to eat or provide necessary oils and cartilage materials for both local and global pharmaceuticals. They have become highly prized commodities. At least, some parts of them are. Let's consider how historical and future trends might direct our fondness for eating and using sharks.

Sharks have fed more than our hunger for stories; indeed, they've been an important food source that has kept many coastal communities alive. We know that the Cretans and Persians caught and traded shark meat more than five thousand years ago. Commercial fishing for sharks on a large scale did not really ramp up until after World War I. At that time, the belly meat of pickled dogfish was smoked and sold in Germany. In years when cod supplies were diminished, shark meat was frequently used in European fish and chips.

In 1925, in the United States, the newly formed Ocean Leather Corporation began tanning shark skins for bags and shoes, and the demand for sharks increased. Another use for sharks was discovered in the 1940s, when their liver oil was found to be a rich source of vitamin A. As the artificial vitamin trade grew, more sharks had to be caught to meet demand. More recently, their cartilage has been used as a medical resource for arthritis treatments. It is estimated that over the twentieth century, most sharks were caught for one or two resources only; about 80 percent of their bodies was usually discarded.

Today an even worse fate exists for sharks: shark finning. Live sharks are captured, their fins are cut off, and the dismembered, though still living, shark is tossed back into the ocean to sink and die a slow and painful death; unable to move or respire, it effectively suffocates. Shark fin soup has been considered a delicacy in China since the fourteenth-century Ming Dynasty. Today commercial fleets that catch fishes and sharks to supply restaurants around the world are responsible for killing some seventy to one hundred million sharks each year. Yes, you read that correctly. Between 2012 and 2019, the total value of the shark and ray trade exceeded $4 billion U.S.; shark fins alone account for $1.5 billion, with shark and ray meat worth around $2.5 billion. The fins of

**Shark fins at a market. It's estimated that 70 to 100 million sharks are killed by humans each year, most of them for their fins alone** ISTOCK IMAGE

rare giant sharks such as the gentle basking shark can fetch even higher prices, selling for up to $50,000 for a large fin.

Fortunately, there is a growing awareness of declining shark populations and the moral implications of this practice. Education has also played a role in slowing down demand for shark fins and shark fin soup. When the UK banned the practice of shark finning in August 2021, it also announced that it will soon ban all trade in shark fins. Such legal action doesn't stop the practice, but it restricts the market. Perhaps this step will encourage other countries to follow suit.

A 2021 study led by Nicholas Dulvy of Simon Fraser University in Canada found that one-third of all known shark, ray, and chimaerid species are now threatened with extinction due to overfishing. By examining and identifying the shark species sold at a Hong Kong shark fin market between 2014 and 2018, they found that eighty-six different species were involved in the trade, and sixty-one of these were threatened species—species that are at risk of extinction if their populations are not actively protected. That's roughly double the proportion of species on the threatened list than they found in an earlier survey between 2014 and 2015. What can be done in these cases?

Countries would need to agree on basic principles about why endangered species need be protected. Many countries manage sustainable shark fisheries, with approximately 9 percent of the current global catch of sharks, based on about thirty-three common species, being biologically sustainable. In my hometown of Melbourne, Australia, any fish-and-chip shop will sell you the standard meal of fish and chips, in which the fish is often "elephantfish" (yes, now officially one word). This is the holocephalan *Callorhinchus milii,* also locally called "flake." Today such resources are being carefully managed with catches monitored regularly and the species is in no danger yet from commercial fishing.

Many sharks, up to around 50 million individuals, are caught each year as unintentional casualties (called "bycatch") on long lines set out for edible fishes, mainly by large corporate ocean fishing operations. The fisheries do not provide any details of what shark species are caught and discarded, only an approximate weight for "sharks" as a general group, as they also do for turtles. Around 50 percent of all global taking of sharks is as bycatch. A 2019 report on the U.S. national bycatch for the National Oceanic and Atmospheric Administration reported that 54 percent of the U.S. bycatch came from four major fishing companies, and about 837 million pounds of fish and sharks were discarded as bycatch. That's around 12.5 percent of the haul, totally wasted.

Some people might justify this desecration of these magnificent creatures because of their perceived assault on mankind, some sort of revenge or penance for the fear of what they might do to us. But how dangerous are sharks, really?

## Shark "attack" statistics

Florida Museum maintains an International Shark Attack File that records all human-shark interactions and their outcomes. As you can guess, a "negative encounter" might include what we used to call a shark "attack," but it could also be a case of a person bitten by a small shark when trying to unhook it from a fishing line. In 2022, they investigated 108 negative shark-human encounters. The research team had two distinct categories for the type of encounter: provoked, which are caused by the behavior of humans, and unprovoked, in which the encounter was initiated by the shark. The 108 encounters included nine fatalities, of which five were unprovoked by humans, so that's five human deaths worldwide from sharks initiating the encounter. There were fifty-seven unprovoked bites on humans, thirty-two provoked bites. The latter category includes incidents initiated by a human doing something to the shark, such as divers who tried to touch a shark, attempted to feed them, or were unhooking a shark from a tangled fishing net. Another nineteen cases involved bites to vessels, or sharks inflicting bites on dead humans by scavenging on them. The major encounters were in the United States (41), with nine in Australia and one or two each reported from Egypt, South Africa, Brazil, New Zealand, and Thailand. Most encounters were from swimmers or waders (43 percent), with surfers accounting for 35 percent and free divers or snorkelers making up 9 percent. The rest were too varied to categorize but included rafting and scuba diving incidents.

These numbers are interesting. Of all the people on the planet who live on or around our rivers and oceans, only five were killed by sharks, and of the 103 who were otherwise harmed by sharks, almost half were behaving in a way that provoked the shark. Of all the ways to die—from heart attack, car accident, or plane crash—this does not seem like a huge risk. Deer kill more than four hundred people in the United States each year, mainly as a result of negative "deer-car encounters." In the twenty-first century we have new technologies to help keep sharks away from us when we enter their domain.

In recent years, electronic shark deterrents (ESDs) have been refined and tested with wonderful results. They work by sending out electric signals that confuse the sharks' electrosensory system. While manufacturers of such devices claim high deterrent success, only recently have independent tests been carried out by scientific teams without any vested interest in the products. Charlie Huveneers led a 2018 study that assessed the effectiveness of five personal shark deterrents for surfers, including two electric, two magnetic, and one chemical device, on white sharks in the wild. They tested each device with sharks taking bait with devices mounted on surfboard-like props that were substituted as the shark began closing in on the bait. Trials were repeated and tested until 397 trials had been run on 44 different white sharks. The best result was that one electronic product reduced the number of baits taken down from 96 percent to 40 percent. Other products worn as bracelets or leashes around the ankle, or as waxes smeared onto surfboards, had virtually no effect on deterring the sharks. Another ESD was tested against bull sharks and found to be significantly effective at deterring sharks, reducing the number of baits taken by 42 percent.

Today many laboratories based at universities, museums, and government agencies carry out research on sharks. In the early days, sharks were routinely killed to get the research data. Today more often they are tagged, measured, weighed, with small tissue samples taken, and then the shark is released. Some research specimens are collected from the bycatch of large fisheries operations or taken from shark nets.

Sharks are no longer intentionally killed for research. We need to research sharks for many reasons—to assess the health of shark populations, to understand more about their biology, their lifestyles, how long they take to reach sexual maturity, and roles of sharks in our ecosystems. We need to be able to gauge the sustainability of shark fishing and tourism, and to simply know how many shark species are out there. We paleontologists try and fill in the deep-time picture of how the modern shark fauna got to where it is today, and how sharks have reacted in the past to massive disruptions in their environments. The fundamental rationale for shark research is clear—we can't sustain any activity or create new conservation measures for sharks unless we fully understand the impact of that activity on shark populations. It's also vital that we comprehensively understand the role sharks play in our very complex marine and freshwater ecosystems, so that if species go extinct, we can anticipate the consequences.

## Sharks and human health

Sharks are also valuable to us in our quest to improve human health. Sharks have one of the oldest yet strongest immune systems ever to have evolved. They could well provide the cures for cancer, COVID-19, and many other diseases. This is a very real and exciting possibility. Aaron LeBeau is professor of pathology at the University of Wisconsin–Madison who studies the immune systems of nurse sharks, a docile species that is easy to keep in aquaria. Working with a Scottish biotechnology company, Aaron believes the answers to several human illnesses and conditions can be found in the blood of sharks. His team regularly takes samples from the nurse sharks to analyze their antibodies. They have found that shark antibodies are better than our own immune systems at neutralizing certain viruses that invade our bodies. The shark antibodies are smaller than our own, even smaller than mice antibodies. This means they can fit into the specific regions of proteins that mammalian antibodies cannot, making them more resistant to attack. Also, the shark antibodies are very robust, almost indestructible. They are effective also against a range of coronaviruses, such as SARS and MERS, as a study published in 2021 has shown.

**Nurse sharks have antibodies effective against many coronaviruses. Human health can benefit a lot from studying sharks.**
PHOTO BY NIGEL MARSH

Shark skin has also been shown to have powerful antimicrobial properties, allowing researchers to create antibacterial surface coatings, such as one called Sharklet AF. Such inventions help keep dangerous infections from spreading in hospital wards. But perhaps the most exciting discoveries made so far come from the latest research on shark genetics; white shark genes could provide breakthroughs for new medical treatments for cancers and other diseases. Think of that next time you watch *Jaws*—white sharks could hold the keys to saving your life one day.

By virtue of their long evolution, sharks also inspire new designs using biomimicry. We have seen how shark skin has inspired the production of new swimsuits and films to cover aircraft to reduce drag and cut fuel use and greenhouse emissions, but other uses of shark skin biomimicry include the use of antibacterial films. Professor Youhong Tang at Flinders University recently published a paper exploring the application of shark skin flow dynamics to model a new biomimetic film that reduces drag in water. This could be developed for ships and submarines to improve efficiency, cutting down fuel use and reducing global emissions as they transit the oceans.

Another example of shark research could lead to cures for human sight problems. This exciting potential comes from a recent study by a team at the Kobe and Osaka Metropolitan University in Japan, led by Kazuaki Yamaguchi. Whale sharks usually filter feed in well-lit surface waters, but they can dive to depths of around 6,200 feet. Pressures at such depths are immense, and seawater temperatures drop dramatically to a low 39°F. Kazuaki's team have shown that whale sharks have evolved a special "blue-shifted" rhodopsin pigment in the eyes that is turned on at different depths. This is triggered by temperature changes at depth, allowing them to see better by capturing more of the blue light in the darker depths. The research team then looked at the genetic coding causing this difference in whale sharks and found two gene mutations. One of these is the same mutation associated with a human retinal disorder known as congenital night blindness, whereby afflicted people have difficulty seeing in low light conditions. The mutation in humans was found to be unstable at certain temperatures. This means that understanding the whale shark's ability to turn their visual range on and off at different temperatures might signal a cure, with just a little more research into how the process is regulated.

Speaking of temperatures, it's now a well-known fact that we humans are changing the climate of Earth by making it dangerously warmer. How will this future scenario affect sharks in the oceans?

## Sharks and climate change

The climate is changing as a result of human-induced higher levels of carbon dioxide, and other greenhouse gases like methane, building up in our atmosphere. This is no theory, it's a proven fact based on reliable records of our global temperature steadily rising over each decade. This has the direct effect of increasing surface temperatures of our oceans. We will now look at one example of how climate change is affecting sharks' survival, then pose the big question: What do we need to do (or stop doing) to help them survive into the future?

The cute little epaulette sharks commonly found in tropical seas are the canaries in the coal mine of oceanic global climate change (there's a photo of one on p. 247). A 2021 study led by Caroline Wheeler of the New England Aquarium on the thermal regimes for epaulette sharks found that water temperatures affect their developing embryos. The team reared twenty-seven epaulette sharks under average summer temperatures (81°F) and for temperatures predicted under current climate change models for the end of the twenty-first century—around 84 to 88°F. The sharks reared under future climate temperatures were impacted by slower growth and reduced metabolic performance. At 81°F, the normal conditions for these sharks, the embryos hatched at around 125 days, fully developed and ready to feed, with a maximum metabolic rate at 100. At 84°F, just that little bit warmer, the sharks hatched after 110 days, with a higher metabolic rate of 120. At 88°F, just a little warmer than normal, the pups hatched at 100 days with a lower metabolic rate at 80.

So what does all this mean?

Under normal conditions, these are hardy sharks that have great survival ability against ocean acidification, acute temperature fluctuations, and lower oxygen levels in the water. However, if the temperatures rise to 88°F, the young sharks will have a very difficult time surviving. The *pejus* (Latin for "turning worse") temperature is the temperature that marks the beginning of a decline of traits, or genetic features, in a species. For epaulette sharks, this study showed its *pejus* temperature as 84–88°F—the temperature our tropical oceans will reach by the year 2100.

The long-term effect on these sharks, and probably many other species with similar physiologies, will be to restrict their geographic range as they try to stay within comfortable temperature zones, thus restricting the services they provide in their normal habitats, such as consuming other fishes and prey items that need to be kept in balance. Exiting their normal habitat would create a devastating ecological domino effect.

Another major threat facing all sea life is the growing mass of plastics and other chemical pollution growing steadily in all our major oceans. How severe is this problem, and how does it affect sharks and rays?

# Sharks and ocean pollution

There are five major gyres in our oceans—large, slow, circulating currents, like enormous slow whirlpools. The North Pacific Gyre concentrates floating debris into two enormous garbage areas, one to the southeast of Japan, the other off the west coast of North America. One of these swirling marine dumps is about 600,000 square miles in size. A 2018 paper by Laurent Lebreton, of the Ocean Cleanup Foundation in the Netherlands, showed that modeling of the rates of accumulation of plastic in the Great Pacific Garbage Patch within the North Pacific Gyre is increasing exponentially faster than in surrounding waters. The study also showed that between 87,000 and 142,000 tons of plastic were in this one garbage area. Microplastics (particles less than five-thousandths of a millimeter) accounted for 8 percent of the total mass but made up 94 percent of the estimated 1.1 to 3.6 trillion pieces floating in the sea.

The scale of the problem is staggering and should rightly scare all of us. It has been estimated that to clean up even one percent of this mess could take sixty-seven very large boats more than a year of work; 670 large boats with full crews might get it done in ten years. Yet, as it is accumulating faster each year, it would take much longer in real time. Or we could try to do something about stopping the generation of this horrible microtrash from its sources.

A recent study estimated that some 10 billion tons of plastic has been made by us since the 1950s, of which around 7.7 billion tons is no longer in use. While 9 percent of the disused plastic is estimated to have been recycled and around 12 percent of it was incinerated, the other 6 billion tons still exist either in landfill or in our oceans.

Plastics in our oceans are 80 percent derived from terrestrial sources and 20 percent from boats and marine sources. Large pieces of plastic are a menace to many kinds of sea creatures including whales, seals, turtles, large fishes, and sharks. Not only can these creatures get entangled by them, leading to death by starvation, but if they eat the plastics, they can cause intestinal or stomach injuries, or simply fill up the creature without supplying any nutrients, also leading to starvation. As plastic gets broken down by mechanical abrasion, it breaks into smaller and smaller particles. Nanoparticles are between one and

1,000 nanometers. These can cause significant harm to creatures of all kinds, as such incredibly small particles are easily absorbed directly into the creature's bloodstream.

Many recent scientific papers have been coming out showing the effects of these microplastics on marine life of all sizes, from microscopic phytoplankton to large whales. Microplastics have been shown to decrease the reproductive success of oysters and clams that filter-feed and absorb the smallest particles into their bloodstream.

A recent survey showed that of forty-six sharks belonging to four species caught in the northeastern Atlantic, 67 percent of them had plastic particles inside them, but the effects of the plastics on the sharks have not yet been determined. We know that populations of large manta rays and whale sharks from three big tourist sites in Indonesia were ingesting plastic particles as they filter-fed through the water. One manta ray vomit sample contained 26 pieces of plastic with a mean length of 0.3 inch. Half of the pieces were between 0.2 and 1.18 inch. Another sample contained 66 plastic pieces, some more than 3.5 inches long. This is a worrying observation, because such large plastic pieces would be expected to cause serious damage to the stomach and intestines. The fact that the manta rays were able to detect the plastic inside their mouth and vomit it up was one of the positive findings from this

Many sharks die each year as a result of ocean pollution, from microplastics in the water to ghost nets discarded from fishing trawlers
PHOTO BY VISIONDIVE (VISIONDIVE.COM)

2019 study by Elitza Germanov of Murdoch University in Perth, Australia. Much more research is urgently required to determine the biological effects on the life cycles of sharks and rays that ingest such plastics.

Discarded fishing nets called "ghost nets" are another cause of death for all large swimming marine life, including sharks, which often get tangled in them and suffocate. It has been estimated that around 640,000 tons of fishing gear are annually lost to the oceans. One research paper illustrated the gruesome death of small sharks, like nurse hounds, entangled in fishing nets. Shark deaths from human rubbish have long been documented. In 1931, a mako shark was shown dead entangled inside a discarded car tire off Cuba. In western Australia, shark fishermen noted that many sharks were killed by entanglement in the small plastic straps used to close fishermen's bait boxes. There are a hundred more examples I could discuss, but the point is clear: Human trash is a major hazard to all marine life, including sharks and rays. Something needs to be done about it urgently.

Clearly we see that sharks are being hit from all sides by overfishing, pollution, and climate change. What kind of future do they really have?

## The future of sharks

Sharks are truly under grave threat. Remember the recent study by Nicholas Dulvy (mentioned above) that showed that out of 1,041 species of sharks, rays, and chimaerids (chondrichthyans), one-quarter are threatened with extinction due to overfishing? The large shallow-water species were the most vulnerable. The most worrying conclusion they reached is that chondrichthyans have a much higher risk of extinction than nearly any other vertebrate group—more so than bony fishes, reptiles, birds, or mammals—but not as dire as our poor amphibians. The real concern about their future is summarized in a single line from this report: "There is still no global mechanism to ensure financing, implementation, and enforcement of chondrichthyan fishery management plans." Unless countries talk to one another and agree to seriously regulate shark fishing, and to actively prosecute offenders, it will simply not result in a realistic global protection of any vulnerable species.

Let's take a look at two examples of rare sharks that are listed as

The daggernose shark of northern South America is almost extinct, as there is no enforcement of the regulations that are meant to protect it  ARTWORK BY JULIUS CSOTONYI

"critically endangered" on the IUCN Red List (the International Union for Conservation of Nature—see Notes on Sources at the end of this book for more details) to find out what the main factors are that will determine their survival or extinction.

The daggernose shark, or *quati,* is one of the most unusual carcharhiniform sharks alive today, so named for its elongated, pointy snout. Growing to around 8 feet long, it is found in estuaries, river mouths, mangrove bays, and shallow coastal waters along a thirteen-hundred-mile stretch of South America from Venezuela to the Amazon coast. Its long, flat snout and small eyes are thought to be adaptations for living in turgid waters. The fossil record of this shark tells us it first appeared around 23 MYA in Europe and was living as far south as the Congo of Africa and in the Atlantic off North Carolina. Its current status as critically endangered was first registered in 2006, and since then, despite being an illegal catch, the shark is still caught and sold without any penalties being enforced. Some fisheries even target it deliberately, while other fisheries take them as bycatch. A recent study on its vulnerabilities led by Rosangela Lessa, of the Universidade Federal Rural de Pernambuco in Brazil, found that in the past three generations, these sharks have not been recovering in numbers, and it is now declared to be on the "slide to extinction" unless fishing can be more rigorously regulated to stop the species from being caught. The requirement to keep the fishing statistics necessary for the sustainable management of fisheries was abolished in Brazil in 2008, so no data are generated today on what is bycatch, or on the precise numbers of sharks being caught. It's not hard to conclude that human-driven fishing is clearly the main driver of its eventual extinction.

Another critically endangered shark is the Ganges shark, a carcharhiniform growing to about 9 feet long. It is a majestic, sleek shark that inhabits shallow and estuarine areas of the Indian Ocean around India and Bangladesh, with a former range including the Myanmar, Thai, and Borneo coasts. It frequents inland rivers like the Ganges and Indus and was sometimes confused with bull sharks that also live in these rivers, so it was blamed for attacks on people (although this has never been confirmed).

First discovered and named in 1839, by 1867 it was thought to be extinct. None were found until 1996, but even that specimen has since been disproved. Another shark was reported in 2006, and a 2019 study based on three new records of the species, led by Alifa Haque of the University of Bangladesh, used DNA analysis to positively identify shark fins of this rare species found at a local shark processing center. Again, although the species is now protected under Bangladeshi laws, these laws are simply not enforced.

The rapid decline of the species is probably the result of a combination of fishing and the severe pollution of its habitats. The current small population lives in the Bay of Bangladesh. The genus *Glyphis* has three similar-looking living species, including the New Guinea River shark (also critically endangered) and the Speartooth shark from New Guinea and northern Australia ("vulnerable" status). The genus *Glyphis* first appeared around 20 MYA in India, so it has been living in the same area since it first evolved.

Sharks are a vital component of all the world's tropical coral reefs. Common species that spend most of their lives on reefs or regularly visit reefs include the gray reef shark, white tip reef shark, black tip reef shark, Caribbean reef shark, silvertip, Galapagos shark, nurse shark, tiger shark, great hammerhead shark, and lemon shark. These and other sharks perform important services that keep reefs healthy, such as eating larger predatory fish like groupers. These gentle giants maintain the balance of the food web by allowing smaller fishes that groupers eat, like the wrasses and parrotfishes, to thrive. Parrotfishes eat algae on the corals, helping to keep the corals healthy, as too much algae inhibits the photosynthesis required for the zooxanthellae algae that live within the corals to thrive. Without the sharks' maintaining this delicate balance of nature, the reefs would die. They also help keep down unwanted species, like invasive forms that shouldn't be living there.

A study led by Aaron Macneil of Dalhousie University in Canada, published in 2020 (and co-authored with 123 of the world's leading shark experts), showed that reef sharks have been significantly decreasing in numbers in recent years. The study used fifteen thousand standardized baited remote underwater video cameras to count the numbers of sharks in the vicinity. They were in operation on some 371 reefs around the world over a three-year period (2015–18). On approximately 20 percent of all these reefs the team observed no sharks. In some reefs in several nations, the most common species of reef sharks were almost entirely absent.

The best results, based on expectations from past studies in comparison to what was counted, came from reefs off the Bahamas, around Australia and the Solomon Islands, and in the waters surrounding Micronesia and French Polynesia. Such regions are well governed with policies in place to ensure marine sanctuaries are protected from fishing. They are also the island nations where local indigenous groups have strong, intergenerational spiritual and cultural ties to sharks and rays. The worst-ranked places for the conservation of reef sharks were Qatar, the Dominican Republic, continental Colombia, Sri Lanka, and Guam, due to overfishing and poor enforcement of regulations. Overall, the study found that 59 percent of nations (24 of 58) had scores below 50 percent for shark abundance relative to expected populations. The loss of sharks was also correlated to nations with poor socioeconomic backgrounds; fishing for sharks is less regulated when it is a much-needed food source to sustain people, more so than in wealthier nations.

In June 2022, UNESCO declared Australia's iconic Great Barrier Reef, one of the largest and most diverse living structures in the world, to be under threat. The reefs are being bleached more frequently by warmer sea temperatures as a result of human-induced climate change. This is killing the algae that live with corals, stripping them of life, color, and viability as marine habitats. We need to think of reefs as the jungles of the sea, places where thousands of species coexist and depend on one another in complex ways. They are also the base of many food chains that feed the life of the vast ocean regions around them. Without the reefs, there are no reef sharks. Without reef sharks, there will be no reefs. In 2023 the "endangered" decision on the Great Barrier Reef was reversed as a political point, but that doesn't change the fact that the reefs are in dire danger. Only one thing will save them now—reducing our global greenhouse gas emissions.

There has been a great deal of new research about what we can predict about the effects of climate change on shark populations, and to be blunt, all of it is extremely grim. After weeks of scanning the scientific literature, I have not seen one published study that has shown a benefit for sharks under any models of future climate change. Most research conducted by thousands of oceanographers and marine biologists predict declines in the ability of sharks to reproduce and play their vital part in the food chain, as a result of perceived energetic imbalances, increased ocean acidification affecting aspects of shark metabolism, and the looming decline of prey species. Life is getting harder for one of the oldest and most resilient groups on the planet. Imagine the impacts on other life, too, on the many newer species we are only just coming to discover and understand.

Sharks need to be saved for several important reasons. They provide vital services that keep our oceans healthy, and some of these benefit us directly. Take this one example: One of the largest areas of seagrasses in the world exists today in Shark Bay, Western Australia. Green turtles feed on seagrass and can eat up to 5 pounds of it each day. Tiger sharks feed on the turtles, constantly keeping them on the move and thus ensuring that the seagrass prairies are not overgrazed. How does that benefit us? A quote from Sir David Attenborough, who filmed here for his *Green Seas* TV series, answers this question nicely: *"That has benefits for us all. A patch of seagrass can absorb and store 35 times as much carbon dioxide as the same area of rainforest. So, the prairies and their sharks are surprising allies in the fight against a warming climate."*

Another way of preserving sharks and our oceans for the future is to put a price on them that is higher than their current commodity value as food or products. We must evolve our thinking and get creative about what these animals mean for our future. Again, capitalism is playing a small part, but this time it is much more positive. Sometimes a simple piece of beautiful jewelry can make a big difference.

## Fossil sharks helping to save today's sharks

This is a story about a small souvenir company run by an Italian American named Vinnie Valle. After an initial career in computers, Vinnie decided to relocate from New York State to Venice Beach on the Gulf Coast of Florida. Venice Beach is known as "the Shark Tooth Capital of the World." Beachcombers, often using a long-handled scoop rather like a shrimping net, walk the sandy beach to scoop up shiny black fossil shark teeth from the surf line. Scuba divers moored offshore also screen the Miocene-age gravel seabed for fossil teeth. The more common teeth are from requiem and lemon sharks, but large lamnids as well as prized megalodons can occasionally be found.

Venice Beach has its fair share of busy gift shops selling all manner of colorful beach-related souvenirs that nobody really needs but they buy anyway. A corner in any shop has its fossil sharks' teeth, either in small bottles, stuck on postcards, or wire-wrapped and strung up as necklaces. The shark's tooth necklace is a peculiarly American icon, considered the ideal small gift to buy for friends, grandchildren, nieces, and nephews.

Fossil shark tooth necklaces help save living sharks that would otherwise be hunted for their teeth   AUTHOR PHOTO

The original basic shark's tooth necklace used a "white tooth" hung on a leather bootlace. A "white tooth" is one from a modern shark, obtained as a by-product of sport fishing, or caught deliberately to fulfill the demand from the jewelry industry. The species targeted include large mako sharks, tiger sharks, white sharks (which are quite expensive and protected), and large bull and dusky sharks. A large jaw can furnish fifty to eighty teeth, which, when sold to the jewelry industry, make killing a shark for its teeth alone financially viable. Looking back a generation, it would have been unusual to find any fossil shark's tooth in a necklace, except perhaps in or around Venice Beach in Florida.

When Vinnie moved to Florida, making fossil shark's tooth necklaces was just a cottage industry. Walking the beach with his partner, like many before him, Vinnie picked up teeth along the sand line and saw these as a possible income source. The first pieces he made sold slowly but steadily until he could not keep up with local demand and had to buy teeth from professional collectors. As demand grew further, Vinnie added different design combinations of teeth and beads, and his company, V&L Crafts, was born. The company continued to gain market share and soon outstripped the available supply of black Florida teeth. The solution was found in Morocco, where fossil sharks' teeth, a by-product of the phosphate mining industry, were available in sufficient quantities, so Vinnie's innovative use of small teeth in necklaces fitted in well. Supply kept up with demand, and thanks to V&L Crafts, a new branch of the Moroccan fossil trade was developed.

Importantly, from a shark conservation point of view, Moroccan fossil teeth undercut the price of "white teeth" sufficiently to render the hunting of living sharks for their teeth completely uneconomical. Now it is unusual to find a piece of jewelry with a white tooth for sale in a seaside gift shop. Vinnie's company is not the only firm making necklaces with fossil teeth; his success has spawned many imitators. When all manufacturers are added up, a rough calculation shows that the U.S. production of these fossil tooth necklaces saves tens of thousands of sharks each year from being killed *just for their teeth*. The partial banning of shark finning and Red Date Book plus CITES protection will also have helped. Even if this statistic is wildly optimistic, I still think that this is a most impressive number. Well done, Vinnie.

The biggest industry today involving sharks and rays where they are not harmed and most highly valued is, of course, tourism.

# Shark tourism and the future

Sharks are now at the heart of a new booming tourism industry that may be an important part of their salvation. A ten-year study done by an Australian team of researchers showed that the white shark tourism industry off the Neptune Islands, where I saw my first white sharks, is regulated well enough to maintain the industry as sustainable. The number of tour boat operators is limited so that the annual residency of the white shark populations is not affected. In 2013, Canadian fisheries scientists found that the global market for sharks as food was valued at around $630 million U.S. That same year, fueled by people's desire to observe sharks in their natural environment, the shark tourism industry was worth around $314 million. This study predicted that in twenty years (around 2033), if it continued to grow at these rates, the shark tourism industry could be worth $800 million U.S. each year. This economic boon could clearly overtake the total global value of sharks as food resources. This will make it more feasible to create more marine reserves to protect the now valuable sharks and cease fishing in those areas. The killing of sharks for food would continue, as most of the edible and bycatch sharks are caught in large commercial open-ocean fishing operations outside the realm of tourism. Still, some sharks are now more valuable alive than dead.

The value of shark ecotourism is proving to be economic as well as literally lifesaving for some sharks and rays. Giant manta rays were once hunted in the Indonesian village of Lamakera. They were sold at market for around $300 to $420 U.S. per ray, mainly for using their gill rakers in Chinese medicines. Conservationists launched an education program that helped the villagers by showing them the rays are worth far more alive. They brought tourists to the village to pay good money to see the rays in their natural environment. Since then, the villagers no longer hunt the rays to kill, but instead now revere them as their economic saviors. Using this economic data, a 2013 study showed that each manta ray kept alive was worth around $8,000 U.S. in tourism dollars for the village. That's an increase in value of around 2,000 percent per manta ray—enough to convince any villagers to save the rays rather than kill them.

Basking sharks and whale sharks are majestic creatures, the largest living sharks in our oceans, the latter reaching up to 62 feet long. They are the gentle, sentient giants of the shark world. Both have enormous tourism potential, which has been realized for whale shark tourists for a number of decades. The experience of diving with whale sharks is one of the top tourist attractions at Ningaloo Reef in Western Australia, as well as at various tropical localities throughout the world. Today, global whale shark tourism is worth around $50 million each year to local economies. However, in recent years the experiences of swimming with very large sharks is no longer limited to just the warm tropical reefs.

Every year, basking sharks frequent the cool coastal waters off Scotland and Ireland. For centuries the sharks were hunted for their large livers rich in squalene oil, even well into the twentieth century. Today there is a growing tourism industry around sighting the sharks off Scotland. Occasional boat collisions cause some human concern for the sharks in coastal waters, but most of the time they are easily spotted. They can form large groups when plankton is abundant in the warm current coming up from the south. The rugged Wild Atlantic Way runs to Malin Head on the northwest coast of Ireland and was recently featured on Discovery Channel's Shark Week as a hot spot for watching basking sharks.

**Increased interest in shark ecotourism, like this whale shark off Ningaloo Reef in Australia, increases our public knowledge about sharks** PHOTO BY ANONYMOUS SOURCE

Basking sharks display special social behaviors along this stretch of ocean, with the spectacular sight of these enormous sharks breaching the water driving tourists wild with delight. Scotland designated its Sea of the Hebrides as a new Marine Protected Area in 2020, recognizing the significance of the basking sharks for feeding and social interactions. They are also protected in the waters surrounding Ireland, as of late 2022. Scotland shark watching tours are most successful when the basking sharks aggregate close to the surface to feed, putting on quite a show. These docile filter feeders are not harmful to humans. You can enter the water to observe them feeding and experience what it feels like to be surrounded by a school of gentle giants swimming around you. It is a relatively safe shark experience up there, as the waters are devoid of known predatory large sharks that could negatively interact with humans. Definitely one for my bucket list.

The swim-with-basking-sharks experience can be encountered in the Hebrides and main sea of Scotland. In Scotland, marine tourism earned some £579 million in 2018, or about 0.4 percent of the entire economy. In 2021, three different agencies were offering "swim with basking shark" tours for around £720 a trip for a four-day experience. It is a growing attraction—a 2009 survey showed that up to 69 percent of wildlife tourists who visited the Isle of Man came because they were influenced by the presence of the basking sharks.

## A ray of hope

I will end this book on a positive note and a clarion call. Just think of the many benefits we humans gain by having sharks in our oceans: from sustainable food species to a growing ecotourism industry built by shark-loving divers, to how sharks regulate populations of marine organisms that would otherwise get out of control, and how sharks feeding at various depths circulate nutrients from one part of our oceans to another, keep ecosystems healthy.

Sharks' remarkable evolution—this secret history—proved them to be resilient in tough environmental conditions in the past, but they've never faced a predatory species like us before. They could possibly survive one of the many challenges they currently face—global warming, pollution, overfishing—but not all at once, and that is our doing. We must do more to help sharks survive and flourish in our oceans.

I would start by banning the gruesome international practice of shark finning for good. I would even go so far as to make the sale of shark fin soup and unproven so-called medical pharmaceuticals illegal.

We need to lobby our politicians to increase protections on shark species that are threatened, and to keep strong regulations of our mainstream fishing industries, remembering that millions of sharks are killed each year as bycatch by the large ocean trawlers that use deep netting. It's a no-brainer that we desperately need more marine parks to allow sharks and other fishes to have space to regenerate and have time to mature. This approach greatly benefits all fishing industries and will allow more sustainable harvesting of seafood resources in the future.

It's also important to recognize that sharks' strength comes from their diversity, and this reflects the various habitats they occupy across the globe. While the top-end predators—the bulls, the whites, and the tiger sharks—attract much negative attention from the world's media, we need to remember there are some 1,225 species of sharks, rays, and chimaerids out there. Most of these species have absolutely no interaction with humans but are impacted by our actions nonetheless. Many of these species are now endangered as a result of the collateral damage of our expanding human population and its overwhelming need for increased resources. The hope for shark conservation is in all our hands, but more so in the hands of some brilliant and passionate scientists around the world. I am lucky enough to work just down the hall from one of them.

The shark biologist Lauren Meyer dreamed of being a sea turtle as a young girl growing up in the inland city of Spokane, Washington, high in the mountains. She was unaware that marine science could even be even a career option. Then, when she studied medicine in the United States, she discovered she was more interested in marine biology, so she moved to Townsville, Australia, to pursue that field. She then joined the Southern Shark Ecology Group at Flinders University to complete her doctorate in shark ecology in 2019. She now works in a joint position there co-funded by the Georgia Aquarium in the United States. Her main area of study is **trophic ecology**—that is, how we can determine the dietary preferences of sharks from the study of the isotopes and fatty acids in their tissues.

Lauren has also studied shark tourism and the effect it has on shark diet. Her research is framing sustainable new ways for people to engage and become fascinated with sharks in the wild without altering their natural behavior. In 2018, Lauren cofounded Otlet, a global open access platform that enables more than six hundred researchers to utilize

**Dr. Lauren Meyer is a shark biologist who has set up an online resource, the Save Our Seas Foundation (saveourseas.com), to help scientists access sharks for research**  PHOTO BY ADAM GEIGER/SEALIGHT PICTURES

a collection of more than 22,000 biological tissue samples from some 322 species. This sort of enterprise facilitates new research, as resources are already there to work on. Most of the expense in the early days of shark research was simply getting out there to locate and catch sharks to take tissue or blood samples.

I think this last story gives us hope. The future of sharks will be in good hands if more young, dynamic researchers like Lauren pursue their dreams to be marine biologists and inspire a new generation of children to be enthralled with the awe-inspiring world of sharks and other life below the sea. One day, perhaps, people will overcome their fears and instead appreciate how much we have in common with sharks, and how much we will lose when we fail to see their value—to our communities, our oceans, and our future.

# THE WISDOM OF SHARKS

We have now witnessed how the 465-million-year story of sharks provides a powerful insight into the very mechanisms of how evolution works. Species do their best to change and improve their lot when faced with global environmental crises, or they go extinct. When climates and oceans are stable for long periods of time, some species are still under pressure to adapt; the growth in populations of other marine organisms can cause competition for similar food resources.

At the start of this book, I asked a big question: What is the take-home message that we humans can learn from the survival of sharks? Sharks give us three important messages from their experience surviving the five great mass extinction events of the past 500 million years.

First, sharks had to reinvent themselves after each devastating mass extinction event, developing new ways to be more opportunistic rather than becoming too specialized. This was a key secret to their survival success. Like sharks, we humans will need to reinvent our way of living as a sustainable global community if we want to entertain a future of hope. Today we face unpredictable, drastic changes to our environment. To maintain not just our standards of living but hope for survival, we must emulate the sharks and adapt to whatever resource bases are sustainable well into the future. Climate change is wiping out species around us and damaging trophic webs in marine and terrestrial ecosystems. The effects are too hard to predict when the factors are so complex. While I'm confident we will survive, it's how we live, the net quality of our lives, we are mostly concerned about for the future. It will ultimately depend on how we use our knowledge to stop such changes from reaching the irreversible tipping point.

Second, sharks had to regulate their size according to the resources available. For the first 140 million years of the Mesozoic era, following the Great Dying, sharks were small, most about 3 feet long, with none longer than about 10 feet. Only when oceanic food chains blossomed with new resources did sharks grow much larger, just before the second half of the Cretaceous period. Our lesson here is in our optimal population size. We've gone from approximately one billion humans on the planet in 1800 to more than eight billion today. We are destroying much of Earth's natural areas through expanding our human populations, extending our range and our never-ending consumption of resources. If we can willingly curb population growth, and stay small as the sharks did, then perhaps we can begin a new era of living sustainably.

Third, different groups of sharks survived several mass extinction events by moving from the ground zero of chaos, in the shallow-water shore zones, out into the deep open oceans. They later colonized shallow-water habitats again, and diversified when conditions were back to normal. Our lesson here is to keep moving to areas away from chaos, if we are to live healthy lifestyles. Crowded, densely populated, expensive cities are not optimum environments for all humans, but most of us are bound to live in them because it is where we are employed or where our families are. It's true that most of us in cities tend to live more stressful lives, caused by the daily pressures of commuting, congestion, pollution, and competition for housing and resources. Perhaps the future direction is for more of us to leave the ground zero of chaos that are our cities and seek out quieter, more natural settings for a healthier, more relaxed and contented lifestyle. More jobs working from home would facilitate this shift.

The oldest creatures on Earth are often associated with the gift of wisdom because they have seen the most winters, survived the biggest storms, and adapted to many challenges over a long period of time. Elephants, for example, are venerated in many cultures. But the oldest living backboned creature on Earth is not the elephant, nor is it Jonathan, the 160- to 180-year-old Seychelles tortoise. Deep under the seas of the North Atlantic live the majestic 25-foot Greenland sharks. Some individuals of this species, as mentioned in the start of this book, could now be close to or more than five hundred years old. They don't even sexually mature until age 150—that's a long time to spend worrying about puberty.

Just as the white shark that starred in *Jaws*, live scenes may still be alive today, perhaps some Greenland sharks are still alive that were cruising the North Atlantic around the time Sir Francis Drake made his epic voyage to the Americas. Imagine what those creatures might have seen, heard, experienced! The voyage of the *Mayflower*. The dawn and destruction of the slave trade. Early and nuclear submarines. The sinking of the *Titanic*. The highs and lows of the cod trade . . . World Wars I and II . . . all modern human history.

Sharks have witnessed our own evolution, predating our time on Earth by hundreds of millions of years, and are now at the mercy of our modern world. The longevity of the Greenland shark symbolizes the innate ability of these fishes to survive, despite all the challenges that have been thrown at them through centuries of human impact, from increased fishing and air and ocean pollution to the changing of Earth's climate. All this climate change is beginning to deeply shift the balance in our oceans and affect its food webs. As we've seen in the last chapter, it's likely that sharks are going to face their most difficult challenge for survival in the next century. And so are we.

The five-century scientific study of sharks and their prehistory has given us vast amounts of information, with most of it coming in swiftly increasing volumes over the past fifty years as new technologies have ramped up our ability to track, analyze, and investigate their anatomy, physiology, and population dynamics. We can now date fossils accurately, determine their past environments, and accurately model changes in climate, ocean currents, and Earth's tectonic movements over the past few billion years of Earth's history. We can study the deep-time evolution of sharks, fishes, and myriads of other marine creatures, both macroscopically and microscopically, right down to the molecular secrets of their DNA. We have corroborated the big picture of shark evolution through the matching of biological molecular data with geological fossil data.

This vast library of data, hard won through the cumulative work of science, has built a formidable body of knowledge that truly holds the power for us to change the world. We can fix climate change, we can stop ocean pollution, we can begin to restore denigrated environments—if we just draw upon this existing knowledge. It is the key to managing human well-being over the upcoming millennia, empowering us to live sustainably with nature. When knowledge is applied to solve problems, it becomes wisdom.

Now is the time to activate this wisdom, apply it, and protect our world's last remaining wild ecosystems while there is still time. If we can save the oceans and save the sharks, we can save the world.

# ACKNOWLEDGMENTS

Many people have assisted me in researching and developing this book over the four years it took to write. It stems from over four decades spent researching the evolution of fishes and sharks as my main day job, and countless discussions over the years at paleontology conferences and during field trips.

For generous hospitality while on the road, and time spent chatting about sharks over meals and bubbly wines, reviewing draft chapters, and for sharing your stories, I deeply thank David Ward and his wife, Alison. For reviewing draft chapters or parts of the text, and discussions about the content, I sincerely thank Michael Coates, Charlie Huveneers, Jürgen Kriwet, John Maisey, Mikael Siversson, Dick Lund, Eileen Grogan, Plamen Andreev, Catalina Pimiento, John-Paul Hodnett, Chris Duffin, Leif Tapanila, Kate Trinajstic, Zerina Johanson, Richard Cloutier, Héctor Botella, Philippe Janvier, Michal Ginter, Iris Feichtinger, Sylvain Adnet, Mohamad Bazzi, Jacopo Amalfitano, Charlie Underwood, Kenshu Shimada, Valéria Vaškaninová, Wayne Itano, Andrew Fox, and Lauren Meyer. Thanks also to Roger and Katherine Jones for their wonderful hospitality and for Roger showing me the fossil shark specimens.

For access to examine fossil and modern shark collections I thank Emma Bernard, Zerina Johanson, and Ollie Crimmen at the Natural History Museum (UK); Sam McLeod at the Los Angeles County Museum (USA); Hans-Dieter Sues at the Smithsonian National Museum of Natural History (USA); John Maisey (American Museum of Natural History); Lance Grande (Field Museum, Chicago); Matt McCurry (Australian Museum, Sydney); Tim Zeigler and Erich Fitzgerald (Museum Victoria, Melbourne); Helen Ryan and Mikael Siversson (Western Australian Museum); and Mark Tozer (Rodney Fox Shark Museum and Learning Centre, Adelaide), and the kind staff and volunteers of the Sant'Anna d'Alfaedo Museum in Italy. For help with collections enquiries I thank Caitlin Colleary (Cleveland Museum of Natural History, USA) and Roberto Zorzin (Museo di Storia Naturale di Verona, Italy). For guiding us to the fossil sites and museums near Verona, Italy, and discussion of these sites and fossils, I sincerely thank Cesare Papazzoni.

For helpful discussions about fossil or extant sharks, and other topics in this book, or providing information as needed, I thank Carol Burrows, Sue Turner, Gavin Young, Kate Trinajstic, Martin Brazeau, Per Ahlberg, Hans-Peter Schultze, John Maisey, Charlie Underwood, Moya Smith, Jack Cooper, Oliver Hampe, Alan Pradel, Rob Gess, Zhu Min, You-An Zhu, Lu Jing, Ivan Sansom, Humberto Ferron, Vachik Harapetian, Sasha Ivanov, Barbara Wueringer, Gary Johnson, Shaun Collin, Tim Smithson, Corey Bradshaw, and the late Alex Ritchie.

Special thanks to Julius Csotonyi for his masterful artworks and his patience, as new restorations of certain species meant changing draft images at the last minute. For use of images, or help sourcing them (even if they were not used in the final cut), I sincerely thank Fritz Pfeil, Jürgen Kriwet, David Ward, Andrew Fox, Nigel Marsh, Brian Choo, Shawn Hamm, Brian Engh, Michael Coates, Lauren Sallan, Leif Tapanila, Zhu Min, Michal Ginter, Philippe Janvier, Alan Pradel, Mikael Siversson, Diana Jones, Jesse Pruitt, Geoffrey Gage, Bruce McFadden, Gordon Hubbell, Christian Klug, Sylvain Adnet, Tetsuto Miyashita, Catalina Pimiento, Glyn Satterley, John-Paul Hodnett, Stefano Pisani, Roberto Zorzin, Leonardo Latella, Graham Nisbet, Neil Clark, Oliver Hampe, Chris Duffin, Megan Warwick, Alastair Graham, Ross Robertson, George V. Lauder, Andrea Everheart, Oleg Lebedev, John Megahan, Nobu Tamura, Peter Trusler, Peter Schouten, Jesus Marugan Lobon, Charlie Huveneers, Adam Geiger, Colette Beaudry, Ray Troll, R. Gary Raham, Rainer Schoch, and Miyess Mitri. The paleogeographic world maps were made by Deep Time Maps, with thanks to Ron and Dee Blakey.

For the breathtaking opportunity to see white sharks up close and personal on a trip to the Neptune Islands that changed my life forever, I sincerely thank Mark Tozer and Andrew Fox of Rodney Fox Shark Expeditions.

To the funding agencies that supported my research and field work on fossil sharks and fishes over the years, I sincerely thank The Australian Research Council, The National Science Foundation, The National Geographic Society, and Australian Geographic; and, for logistics support on my trips working in Antarctica, thanks to the Australian Antarctic Division and the United States Antarctic program. A Bettison and James Award from the Adelaide Film Festival also supported me in the early stages of this proejct.

For their support throughout the entire writing process and their understanding when deadlines got in the way of other family events, my love and thanks to Sarah Long, Peter Long, and Madeleine Cusworth. For constant support and interest in the project I thank Pat and Randall Robinson, Rohan, Magda, Michael and Jo Robinson, Rob and Di Ray, Margaret Brown, and Megan and Ben Patterson. For supporting flexibility in the workplace in my professorial role at Flinders University, enabling this book to be written, I sincerely thank Kathleen Soole and Alistair Rendell. For ongoing support and encouragement from the Flinders paleontology team throughout the project, I thank Alice Clement, Gavin Prideaux, Aaron Camens, Mike Lee, and Trevor Worthy.

The editorial team at Ballantine Books was terrific. I sincerely thank my publisher and champion editor, Mary Reynics, as well as Wendy Wong, and my thorough copy editor, Emily DeHuff, and the whole team at Penguin Random House, New York. Hearty thanks to my agent, Jane Von Mehren, at Aevitas Creative Management, for initiating the project and seeing me as the right author for the task, and for your generous help, support, and hospitality at every stage. Sincere thanks to Steve Brusatte for recommending me for this project, and for his kind words about the book.

Finally, my deepest thanks and love always to Heather Robinson, who took the photographs on our journeys to Italy, Spain, the UK, and the United States, reviewed all the draft chapters, did archival searches and image inquiries, helped compile the figure captions, and kept me sustained and loved while I pushed onward through every inch of this totally awesome sharky booky journey.

# Notes on Sources

Science is not about absolute facts, but about the process of arriving at the most plausible explanation for what we scientists observe in nature. I have tried to back up all my statements with the most recent peer-reviewed scientific papers, which are cited for specific parts of the book in each chapter. When I present my own opinions, these are made clear in the text. I'm aware from conversations with my peers that some will not think it's detailed enough—that I've skipped some of the(ir) most important discoveries and so on. It's inherent in nearly all scientific minds to try to tell such stories by including *as much detail as possible*. To them I say this: My original draft manuscript was far longer than the final book, so not everything I'd hoped to include has made the final cut. However, I've put some of my favorite cut sections in the following pages as extra text and discussions and included some biographical notes about fossil shark researchers, so please refer to the index for specific species, places, people's names, or other topics and you might well find the extra information you are looking for in this section.

Telling the long story of sharks evolution entails complex scientific terminology that I've tried my best to break down or avoid. My scientific colleagues will not necessarily agree with some of my decisions to oversimplify the terminology, so please refer to the glossary at the end of the book for more detailed explanation and discussion of terms used throughout. The following notes provide links to sources of more information on the specific topics covered in each chapter. Information on various fossil shark researchers that had to be trimmed from the main text is here included as additional biographical notes for each chapter. Paleogeographic maps used were sourced from Ron Blakey (deeptimemaps.com).

I have cited or referenced the following books:

Berta, A., and S. Turner. *Rebels, Scholars, Explorers: Women in Vertebrate Paleontology*. Baltimore: Johns Hopkins University Press, 2020.

Cappetta, H. *Handbook of Paleoichthyology*, vol. 3B, *Chondrichthyes II: Mesozoic and Cenozoic Elasmobranchii*. München: Verlag Dr. Friedrich Pfeil, 1987.

———. *Handbook of Paleoichthyology*, vol. 3E, *Chondrichthyes: Mesozoic and Cenozoic Elasmobranchii; Teeth*. München: Verlag Dr. Friedrich Pfeil, 2012.

Carrier, J. C., J. A. Musick, and M. R. Heithaus, eds. *Biology of Sharks and Their Relatives*. Boca Raton, Fla.: CRC Press / Taylor & Francis Group, 2012.

Cuny, G., G. Guinot, and S. Enault. *Evolution of Dental Tissues and Palaeobiology in Selachians*. London: ISTE Press, 2017.

Denison, R. *Handbook of Paleoichthyology*, vol. 2, *Placodermi*. München: Verlag Dr. Friedrich Pfeil, 1978.

———. *Handbook of Paleoichthyology*, vol. 5, *Acanthodii*. München: Verlag Dr. Friedrich Pfeil, 1979.

Domeier, M., ed. *Global Perspectives on the Biology and Life History of the White Shark*. Boca Raton, Fla.: CRC Press / Taylor & Francis Group, 2012.

Ebert, D. A., M. Dando, and S. Fowler. *Sharks of the World: A Complete Guide*. Princeton, N.J.: Princeton University Press, 2021.

Everhart, M. J. *Oceans of Kansas: A Natural History of the Western Interior Sea*. Bloomington: Indiana University Press, 2005.

Fox, Rodney. *Sharks, the Sea and Me*. Kent Town, South Australia: Wakefield Press, 2013.

Gayet, M., O. Abi Saad, and O. Gaudant. *The Fossils of Lebanon: Memory of Time*. Editions des Iris, 2003, 2012 (www.adverbum.fr).

Ginter, M., et al. *Handbook of Paleoichthyology*, vol. 3D, *Chondrichthyes: Palaeozoic Elasmobranchii: Teeth*. München: Verlag Dr. Friedrich Pfeil, 2010.

Hall, B. K. *Bones and Cartilage: Developmental and Evolutionary Skeletal Biology*. Cambridge, Mass.: Academic Press, 2005.

Hamlett, W. C., ed. *Sharks, Skates, and Rays: The Biology of Elasmobranch Fishes*. Baltimore: Johns Hopkins University Press, 1999.

Janvier, P. *Early Vertebrates*. Oxford: Clarendon Press, 1996.

Johanson, Z., P. Barrett, M. Richter, and M. Smith, eds. *Arthur Smith Woodward: His Life and Influence on Modern Vertebrate Palaeontology*. London: Geological Society Publication 430, 2016.

Johnson, D. P. *The Geology of Australia*. Port Melbourne: Cambridge University Press, 2009.

Klimley, A. P., and D. G. Ainley. *Great White Sharks: The Biology of Carcharodon carcharias*. San Diego: Academic Press, 1996.

———. *The Biology of Sharks and Rays*. Chicago: University of Chicago Press, 2013.

Koslow, T. *The Silent Deep: The Discovery, Ecology and Conservation of the Deep Sea*. Sydney: UNSW Press, 2007.

Last, P. R. *Rays of the World*. Collingwood, Australia: CSIRO Publishing, 2016.

———, and J. D. Stevens. *Sharks and Rays of Australia*. 2nd ed. Collingwood, Australia: CSIRO Publishing, 2009.

Long, J. A. *Dinosaurs of Australia and New Zealand; Other Animals of the Mesozoic Era*. Cambridge, Mass.: Harvard University Press, 1998.

———. *Mountains of Madness: A Scientist's Odyssey in Antarctica*. Sydney, Australia: Allen & Unwin, 2000 / Washington, D.C.: Joseph Henry Press, 2001.

———. *The Dinosaur Dealers: Mission: To Uncover International Fossil Smuggling*. Sydney Australia: Allen & Unwin, 2002.

―――. *Swimming in Stone: The Amazing Gogo Fossils of the Kimberley*. Fremantle, Australia: Fremantle Arts Centre Press, 2006.

―――. *The Rise of Fishes: 500 Million Years of Evolution*. 2nd ed. Baltimore: Johns Hopkins University Press, 2011.

―――. *The Dawn of the Deed: The Prehistoric Origins of Sex*. Chicago: University of Chicago Press, 2012.

Martill, D. *Fossils of the Oxford Clay*, Palaeontological Association, Field Guide to Fossils, 1991.

Mayor, A. *The First Fossil Hunters: Dinosaurs, Mammoths and Myth in Greek and Roman Times*. Princeton, N.J.: Princeton University Press, 2011.

McGhee, G. R., Jr. *When the Invasion of Land Failed: The Legacy of the Devonian Extinctions*. New York: Columbia University Press, 2013.

McNamara, K. J., and J. A. Long. *The Evolution Revolution: Intelligence Without Design*. Melbourne: Melbourne University Publishing, 2006.

Naish, Darren. *Ancient Sea Reptiles*. Washington, D.C.: Smithsonian Books, 2022.

Pradel, Alan, J.S.S. Denton, and Philippe Janvier, eds. 2021. *Ancient Fishes and Their Living Relatives: A Tribute to John G. Maisey*. München: Verlag Dr. Friedrich Pfeil, 2021.

Reader's Digest. *Sharks: Silent Hunters io the Deep*. Sydney: Readers Digest, 1987.

Shadwick, R. E., A. P. Farrell, and C. J. Brauner, eds. *Physiology of the Elasmobranch Fishes: Structure and Interaction with the Environment*. London: Academic Press, 2016.

Sorbini, L. *La Collezione Baja di Pesci e Piante Fossili di Bolca*. Verona: Museo Civico di Storia Naturale Verona.

Stahl, Barbara B. *Handbook of Paleoichthyology*, vol. 4, *Chondrichthyes III: Holocephali*. München: Verlag Dr. Friedrich Pfeil, 1999.

Westbrook, V., S. Collin, D. Crawford, and M. Nicholls. *Sharks in the Arts: From Feared to Revered*. London and New York: Routledge, 2018.

Wood, S. P. *A Challenging Edinburgh Fossil Site*. Mr. Wood's Fossils, 1992.

Zangerl, R. *Handbook of Paleoichthyology*, vol. 3A, *Chondrichthyes: Palaeozoic Elasmobranchii*. München: Verlag Dr. Friedrich Pfeil, 1981.

## Chapter 1: The Hunt for the Secrets of Sharks

**January 1, 1992, Antarctica.** My book documenting my two early Antarctic expeditions over 1988–89 and 1991–92 provides more details of this entire field trip (published in 2000, it is cited above). For *Portalodus*, see J. Long and G. Young, *Records of the Western Australian Museum*, 1995, 17:287–308). Thomas Rich has been (and still is) the curator of vertebrate paleontology at Museum Victoria since 1974. His wife, Patricia Vickers-Rich, co-supervised my PhD dissertation with Professor Jim Warren, and all have been great mentors to my career. General shark information presented is from D. Ebert et al., *Sharks of the World* (cited above), but chondrichthyan senses are discussed in much more detail throughout this book, with further references cited below.

**The mystery of sharks.** The origin of sharks, and their relationship to the extinct placoderms and modern bony fishes, is still a hotly debated topic. While some would think we are close to resolution, not all agree; for example, see results of a very recent phylogenetic analysis by You-An Zhu et al., *Nature* 2022, 609:955–59; look at the supplementary information for details. Quote from Jean-Luc Godard: quotes.thefamouspeople.com/jean-luc-godard-4309.php.

**A life story in the rocks.** For details on the dating of fossils, see *Nature*'s great guide to the different methods of both radiometric and relative dating using fossils: nature.com/scitable/knowledge/library/dating-rocks-and-fossils-using -geologic-methods-107924044/. For more about sedimentary rocks and how we determine ancient environments using rocks, see BCcampus open textbook chapter: opentextbc.ca/geology/chapter/6-3-depositional-environments-and -sedimentary-basins/.

**Sharks preserved as fossils.** Extinct sharks and their kin (holocephalans and rays) include around three to four thousand known species from the three *Handbooks of Paleoichthyology* cited above (see Cappetta, Ginter, and Zangerl), which provide thousands of extinct species known only from teeth and a very few based on skeletons. Complete preservation occurs in anoxic environments (e.g., Cleveland Shale, Bear Gulch, Solnhofen, Bolca, and others); these fossils' exceptional preservation is explained in chapters 5, 6, 10, and 12.

**The right time to tell this story.** CT scanning of fossil sharks was pioneered by John Maisey of the American Museum of Natural History; his early CT papers started appearing in the 2000s. Only in the past ten years or so has wider access to synchrotrons and micro-CT scanners become common for researchers across the globe. For examples of CT scan techniques used on exceptional fossils, see Mark Sutton, 2008, *Proceedings of the Royal Society B*, rspb.2008.0263, and X. Ni et al., 2012, *Palaeontologia Electronica* 10.26879/288. Neutron beam imaging at ANSTO in Australia: ansto.gov.au/our-facilities/australian-centre-for-neutron -scattering/neutron-scattering-instruments/dingo. Regarding binomial scientific names: All living organisms are classified using a two-part italicized scientific name of which the first part refers to the genus and the second part the species. For example, we humans are classified as *Homo sapiens*—genus *Homo*, species *sapiens*. All other humans species are extinct, like early forms such as *Homo erectus*.

## Chapter 2: The Enigmatic Oldest Shark Fossils

**Fossil hunting in the dead heart of Australia.** This scene was based on a three-week trip I made in 1993, traveling from Perth by car to Alice Springs with two other geologists, sampling rocks for fossils along the way. I published another account of working at Mount Watt; for more details, see chapter 4 in McNamara and Long, *The Evolution Revolution*. For information about *Arandaspis*, see the original paper by A. Ritchie & J. Gilbert-Tomlinson in *Alcheringa* 1977, 1:358–61; revised position of eyes in front of head was first published in my 2011 book *The Rise of Fishes,* p. 33, based on discussions I had with Dr. Alex Ritchie, then of the Australian Museum.

**The Ordovician world.** For Larapintine Sea geology and general paleontology, see David Johnson's 2009 book cited above. For *Arandaspis* anatomy, see the 1977 paper cited above by Ritchie and Gilbert-Tomlinson. About the ancient jawless fish radiation, see chapter 2 of my 2011 book *The Rise of Fishes;* for a more detailed academic treatment, see Philippe Janvier's 1996 book *Early Vertebrates.* On Ordovician seawater temperatures: K. Bergmann et al., *Geochimica et Cosmochimica Acta* 224:18–41. Interview about *Endoceras,* the giant nautiloid of the Larapintine Sea: 2018, abc.net.au/news/2021-07-12/nautilus-outback -australia/100245606. Oldest known fish, *Metaspriggina:* S. C. Morris and J-B. Caron, *Nature* 2014, 512:419–22. Chinese Cambrian fishes, Chengjiang sites: D. Shu et al., *Nature* 1999, 402:42–46. Conodont animals: see summary by the Natural History Museum, London, at https://www.nhm.ac.uk/ natureplus/blogs/micropalaeo/2013/06/12/conodonts—the-most-controversial -of-microfossils.html. *Iowagnathus* as the giant conodont: H. P. Liu et al., *Journal of Paleontology* 2017, 91:493–511. The shark scales named *Tantalepis:* I. Sansom et al., *Palaeontology* 55 2012, 2:243–47.

**Sharks for beginners.** General information about sharks, their cartilage, scales, general anatomy, and much more can be found in D. A. Ebert et al., *Sharks of the World.* Diversity of living chondrichthyans is around 1,225 species, based on the current tally of 536 shark species (Ebert, *Sharks of the World*), 633 rays (Last, *Rays of the World*), and about 56 species of holocephalans (sciencedirect.com/ topics/biochemistry-genetics-and-molecular-biology/holocephali). Shark special tessellated calcified cartilage: J. Maisey et al., *Journal of Fish Biology* 2021, 10.111/jfb.14376, and Hall, *Bones and Cartilage.*

**Shark origins: the mystery years.** Gnathostome (jawed vertebrate) origins is the last big unsolved mystery of vertebrate evolution, in my personal opinion, based on a life working in early vertebrate paleontology. It is one of the few areas of research that hasn't made much progress in the past fifty years or so, as no intermediate "transitional" forms have been found, although we have learned a lot about the sequencing of the anatomical steps leading to gnathostome (jawed vertebrates) characters, especially from various new early placoderm discoveries from the Silurian of China (e.g. *Entelognathus, Qilinyu, Xiushanosteus;* see chapter 3 for all the sources on these forms). General reference to shark placoid scales and skin and their function: Ebert, *Sharks of the World.* Placoid scale functions: see the excellent paper by W. Raschi and C. Tabit, *Marine & Freshwater Research* 1992, 43:123–47. Whale shark placoid eye scales: T. Tomita et al., *PLOS One* 2020, 15:e0235342. Golf ball with dimples going farther: pubs.aip.org/physicstoday/article/73/4/58/1017445/The -speedy-secret-of-shark-skinThe-outward-flaring.

*Tantalepis,* **the first shark?** Main reference to *Tantalepis:* see Sansom cite three notes above. See also discussion on Late Ordovician scales in P. Andreev, *Palae-*

*ontology* 2016, 58:691–704; this paper also gives a good summary of all known Ordovician shark scales. For information on *Shenacanthus:* Y. Zhou et al., *Nature* 2022, 609:954–58. Sharks branching away from jawless fishes: see discussion of shark and thelodont scale structure in P. Andreev et al., *PeerJ* 2016, doi:10.7717/peerj.1850. Oldest known Ordovician thelodont (jawless fish) scales: I. Sansom and D. Elliott, *Journal of Vertebrate Paleontology* 2003, 22:867–70. For information about the scales on the fastest-swimming shark (shortfin mako), see Philip Motta et al., *Journal of Morphology* 2012, 273:1096–1110.

Professor George Lauder of Harvard University is a living guru of functional morphology, or how the mechanics of living creatures make them move more efficiently. He specializes in using biorobotics to study the structure and function of fishes. In 2012 he and Johannes Oeffner published a study that compared shark skins with the new high-performance athletic swimsuits. He showed that as sharks move through the water, it's the small scales on the head of the shark that do most of the work, creating a small layer of turbulent flow that breaks through the "wall of water" as the shark's body moves through it (J. Oeffner and G. Lauder, *Journal of Experimental Biology* 2012, 215:785–95).

**Other early sharks appear.** On the Ordovician sharks of central Australia: G. Young, *Journal of Vertebrate Paleontology* 1997, 17:1–25). Gavin originally named the scales *Areyongia,* but the name was already in use, so it was changed later to *Areyongalepis.* For the Harding sandstone shark scales: P. Andreev et al., *Palaeontology* 2015, 58:691–704. For the first reference to the *Mongolepis* scales from Mongolia: V. Karatjute-Talimaa, *Paleontologicheskii zhurnal* 1990, 1:76–86.

**Ancient shark scale research pays off.** Biography of Wolf-Ernst Reif: R. Schoch et al., "In Memoriam," see https://www.academia.edu/58129720/Obituary _Wolf_Ernst_Reif_27_6_1945_11_6_2009. AeroSHARK announcement by Lufthanasa: cleantechhub.lufthanasagroup.com/en/focus-areas/aircraft-related -hardware/aeroshark.html. Wolf-Ernst Reif's study showing hydrodynamics of scales in fast-swimming sharks: W. Reif and A. Dinkelacker, *Neues Jahrbuch für Geiologie und Paläontologie, Abhandlungen Band* 1982, 164:184–87.

**Sharks endure the first mass extinction.** For details of the end-Ordovician extinction event: D. Harper et al., *Gondwana Research* 2014, 25:1294–1307; also S. Finnegan et al., *Proceedings of the National Academy of Sciences* 2012, 109:6829–34. For depletion of marine faunas, see Sutcliffe et al., *Transactions of the Royal Society of Edinburgh: Earth Sciences* 2001, 92:1–14. For selenium depletion in oceans at the end of the Ordovician, see J. Long et al., *Gondwana Research* 2016, 36:209–18. *Solinalepis* paper: P. Andreev et al., *PeerJ* 2016, peerj.1850. Sinacanthids: M. Zhu, *Palaeontology* 1998, 41:157–71, and recent work: P. Andreev et al., *PLOS One* 2020, 15:e0228589.

All known Ordovician species of mongolepids are from shallow-water estuarine and intertidal settings, and Silurian forms are also from shallow carbonate platforms or near-shore marine settings. The idea that some must have been living in deeper-water habitats is my suggestion, and it can be tested as more deeper-water sediments are sampled for vertebrate microfossils.

**Silurian sharks.** For detailed information about the sinacanthids: P. Andreev et al., *PLOS One* 2020 15:e0228589. For *Shenacanthus:* Y. Zhou et al., *Nature* 2022, 609:954–58; for *Qianodus:* P. Andreev et al., *Nature* 2022, 609:964–68. The story about finding the fossils as the result of a kung fu play fight was told to me by Dr. Lu Jing of the Institute of Vertebrate Paleontology and Paleoanthropology, Beijing. Acanthodians are, strictly speaking, "stem chondrichthyans," but here I use the colloquial term "stem sharks" in a broader sense to keep it simple. For general information about this group, see Robert Denison's *Handbook of Palaeoichthyology*, vol. 5, Philippe Janvier's *Early Vertebrates,* and my *Rise of Fishes* (all cited above). Exquisite Canadian specimens are from the MOTH locality (e.g., G. Hanke et al., *Journal of Vertebrate Paleontology* 2001, 21:740–53), and a well-known UK example is *Climatius* (C. Burrow et al., *Journal of Vertebrate Paleontology* 2015, 35:913421).

Elegestolepids are another widespread group of Silurian chondrichthyans whose scales have a well-developed neck canal, but to keep the narrative focused rather than comprehensive, I left them out of this chapter. For more information about them, start with Plamen Andreev et al.'s excellent 2016 paper (*Journal of Vertebrate Paleontology* e1245664).

**Additional note on the discovery that acanthodians were "stem chondrichthyans" and not related to bony fishes:** The problem of acanthodian relationships was finally resolved by a bright young Canadian student studying at the University of Uppsala, now the center of the newly established Swedish School of Palaeoichthyology. Martin Brazeau, known to his friends as "the Braz," had long, shaggy black hair and the facial features of an emaciated rock star, but his sharp wit and wry sense of humor gave him away as a die-hard intellectual. I remember well the time he stayed with me and my wife, Heather, in 2007 at our inner-city Melbourne apartment when he visited me as a student. We would have quite heated philosophical debates every evening after spending the day examining Devonian fish specimens at Museum Victoria, where I worked at the time. He was then about two-thirds through his doctoral studies in Sweden when he submitted his landmark paper that solved the century-old riddle of the acanthodians. His investigation focused on some very well-preserved remains of an Early Devonian acanthodian called *Ptomacanthus* from the Welsh borderlands of England. The specimens had lain in the collections of the Natural History Museum in London, unstudied, for decades. Martin's new investigation of the same fossil revealed that it had an ossified braincase, overlooked by other researchers, and that this structure was very similar to that in early chondrichthyans. He constructed a new, detailed data matrix of 134 characters—each one representing an anatomical feature seen in the fossils—scored for 47 different fossil species of early jawed fossil fishes. His rigorous analysis demonstrated with a very high probability that *Ptomacanthus,* previously classified without dispute as a "typical acanthodian," was better placed with the sharks. This finding also applied to some twenty other species of acanthodian fishes in his analysis. Martin had proved, using rigorous and testable methods, that acanthodians were no longer a separate group of weird extinct fishes, but that most of them made up a large group that split early on from the evolutionary line leading to modern sharks. We now call them the "stem sharks," as they are on the stem of the branch leading to modern sharks. Martin's landmark sole-authored paper was eventually published in *Nature* on January 15, 2009. It changed forever the way we scientists regard the evolutionary relationships of the two main groups of living jawed fishes. Martin proved that the "acanthodians" were actually basal members of Chondrichthyes that later gave rise to the line leading to modern sharks.

**Chapter 2 biographical notes.** Plamen Andreev: Plamen is a great guy with extraordinary patience who loves to paint in his spare time. Originally from Bulgaria, he completed his PhD at the University of Birmingham with Ivan Sansom, working on the oldest fossil fish remains known from all around the globe, mostly tiny scales and other microfossils. He was based in Qujing, Yunnan, southern China, for some years, where he and his local colleagues from the Institute of Vertebrate Paleontology and Paleoanthropology have been finding a great variety of very old shark microfossils, all of them at present classified as stem chondrichthyan remains, mainly scales, fin spines, dermal armor (*Fangjinshania*), and tooth whorls (*Qianodus*). He loves nothing more than describing tiny shark scales in exhaustingly fine detail. He told me he is convinced that *Qianodus* is more shark-like than anything else because the arrangement of the rows of teeth on the whorl resembles the tiny whorls of teeth seen in some later fossil sharks.

## Chapter 3: Sharks Become Predators

**Searching for the oldest known shark teeth.** My Spain field trip was in conjunction with the International Symposium on Early Lower Vertebrates held in Valencia in June 2022. On Devonian oxygen levels: A. Cannell et al., *Earth-Scence Reviews*, 2022, earscirev.2022.104062. For details of Nogueras Formation near Teruel, see Manuel Perez-Pueyo et al., *Revista de la Sociedad Geológica de España* 2018, 31:89–104. The articulated shark found at the site has not been formally described, but there is an abstract about it: R. Soler-Gijón and O. Hampe, *Ichthyolith Issues Special Publication* 2003, 7:45. For the Santa Cruz de Nogueras fossil museum, see museosantacruzdenogueras.es/museo.htm.

**Sharks' toothy origins**. My paper on the bent *Notorhizodon* tooth: G. Young, J. Long, and A. Ritchie, *Records of the Australasian Museum, Supplement* 1992, 14:1–77. Shark tooth dentine types: see H. Cappetta's 2012 *Handbook of Paleoichthyology* and M. Ginter et al.'s 2010 *Handbook of Paleoichthyology* (both cited above) for a general outline of tissue types found in sharks' teeth. First discoveries of Early Devonian sharks' teeth in Spain, *Leonodus* and *Celtiberina* references: H. Mader *Göttinger Arbeiten Zur Geologie und Paläontologie*, 1986, 28:1–59. It was Oskar Hertwig's original work on teeth and scales published in 1874 that pointed out that the scales of sharks are built a lot like their teeth.

(*Jenaische Zeitschrift für Naturwissenschaft* 8:331).The scientific debate centers on whether teeth evolved from placoid scales on the outside of the body, then moved into the mouth and became functional teeth; or the other way around—that teeth developed and migrated out to cover the body as scales. For more recent discussion on the "outside-in" and other theories of tooth development, see P. Donoghue and M. Rucklin, *Evolution & Development* 2016, 18:19–30, and M. Smith and M. Coates, *European Journal of Oral Sciences* 2014, 106:482–500. For diversity of Early Devonian sharks in the fossil record, see M. Ginter et al. 2010 *Handbook* (cited above). Shark taste buds and teeth development: K. Martin et al., *Proceedings of the National Academy of Sciences* 2016, 113:14769–74. Distinctive tooth shapes for every shark species: see Ebert, *Sharks of the World* (cited above). *Porolepis,* a predatory osteichthyan up to 3 feet long, though not strictly found in the Spanish sites, is known from scales found in Early Devonian sites around the Northern Hemisphere, so was most likely present in these seas.

**Sharks' first superpower: replacing teeth.** Héctor Botella's papers on the dental lamina of *Leonodus: Journal of Vertebrate Paleontology,* 2006, 26:1002–1003; on *Leonodus* tooth replacement rates, Botella et al., *Lethaia,* 2009, 42:365–72. Sand tiger shark replacement teeth at 1.06 teeth per day: Jana Wilmers et al., *Nanomaterials,* 2021, 11:969.

**Welcome to the Early Devonian.** For general information on Devonian floras, see Patricia Gensel and Diane Edwards, eds., *Plants Invade the Land* (New York: Columbia University Press, 2001). For *Ptomacanthus* and acanthodian phylogenetics, see Martin Brazeau's excellent 2009 paper in *Nature* 457:305–308. The first "definite fossil sharks' teeth" are forms that resemble those of modern sharks in overall structure and appearance, yet many are still officially classified as "stem chondrichthyans" (e.g., *Doliodus*) as there is no consensus yet on when the first selachian sharks appeared and how their teeth positively identify them. See M. Ginter, *Spanish Journal of Palaeontology* 2023, 38:47–56; see also M. Ginter et al. 2010 *Handbook* (cited above).

**The first nearly complete fossil shark.** *Protodus* and early *Doliodus* material: original study by Arthur Smith Woodward (*Geological Magazine* 1892, new series, 9:1–6; see also 9 for his bio); the specimens were then studied by R. Traquair, who changed the name *Diplodus problematicus* to *Doliodus problematicus* (1893, *Geological Magazine,* new series, 10:145–49); recent work revising the two sharks from this site based on their teeth: Sue Turner, *Acta Geologica Polonica* 2008, 58:133–45. The expedition to find *Doliodus* is recounted in Sue Turner and Randall Miller's 2005 magazine article at americanscientist.org/article/new-ideas-about-old-sharks. First publication of articulated *Doliodus* fossil: R. Miller et al., *Nature* 2003, 425:501–504; for subsequent papers on *Doliodus* concerning its braincase: J. Maisey et al., *Acta Zoologica* 2009, 90:109–22; its dentition: J. Maisey et al., *Journal of Morphology* 275:586–96; and its fin spines: J. Maisey et al., *American Museum Novitates,* 2017, 3875:1–15. Richard Cloutier told me his story about the discovery of the *Doliodus* specimen in two parts. In the long-snouted placoderm *Brindabellaspis,* we see that the olfactory capsule is open into the eye orbit, so it's unlikely that placoderms like this moved their head from side to side to home in on prey (G. Young, *Paleontographica* A, 1980, 167:10–76). The largest group of placoderms, the arthrodires, still had relatively small olfactory capsules by comparison with similarly sized chondrichthyans.

**Sharks' second superpower: smell.** Sharks' heightened sense of smell cited as "keen sense of olfaction" in Horst Bleckmann and Michael Hofman's chapter in W. C. Hamlett, ed., *Sharks, Skates, and Rays* (cited above), plus work by Jayne Gardiner showing that blacktip reef sharks home in on specific locations from a distance of five miles using mainly their sense of smell (J. Gardiner, *Integrative & Comparative Biology* 2015, 55:495–506). The much-quoted 2010 study suggesting that teleost osteichthyans had a similar degree of olfactory sensitivity to sharks was by Tricia Meredith and Stephen Kajiura, *Journal of Experimental Biology* 2010, 213:3449–56; the new paper on olfactory receptors is by Adnen Syed et al., *Molecular Biology and Evolution* 2023, 40:msad076. White shark homing in on flow of chum near Dangerous Reef: W. Strong, Jr., et al., pp. 229–40 in A. Klimley et al., eds., *Great White Sharks,* cited above. Sharks using olfactory head swinging to home in on prey was first studied by Albert Tester, *Pacific Science* 1963, 17:145–70. Sharks' timing of olfaction, and the bonnethead sharks smelling faint odors in currents: J. Gardiner and J. Artema, *Current Biology* 2010, 20:1187–91.

**Sharks' evolutionary experiments:** For *Tassiliodus:* C. Dereck and D. Goujet, *Geodiversitas* 2011, 33:109–226; for *Protodus,* see Sue Turner's work cited above. For Bolivian *Pucapampella* background and references, see Philippe Janvier's works *Revista Tecnica de YPFB,* 2003, 21:25–35, and P. Janvier and M. Suarez-Riglos, *Bulletin de l'Institut français d'études andines,* Lima, 1986, 15:73–114. For South African material: J. Maisey and E. Anderson, *Journal of Vertebrate Palaeontology* 2001, 21:702–13, and our paper illustrating the braincase (M. Anderson et al., *Journal of African Earth Sciences* 1999, 29:179–94). For *Ramirosuarezia:* A. Pradel et al., *Acta Zoologica Stockholm* 90 (Suppl. 1), 2009:123–33. Philippe Janvier and John Maisey's paper about CT scanning is in D. Elliot et al., eds., *Morphology, Phylogeny and Biogeography of Fossil Fishes* (München: Verlag Dr. Friedrich Pfeil, 2010). Overview of phylogenetic position of *Doliodus* and pucapampellids within the Chondrichthyes: see John Maisey et al., chapter 5 in Johanson, *Arthur Smith Woodward.* On *Zamnopterygion:* Janvier and Suarez-Riglos, 1986, cited above. The story of "The Rock" and its travels is reconstructed from the 2009 paper by A. Pradel et al. cited above. Philippe and Alan's paper has a revealing line that tells us something intrinsically honest about the nature of doing modern science. After all this super-high-tech analytical work, using two different powerful synchrotrons, they laconically state: "neither technique yielded critical information." However, they did reveal some small bits of cartilage still inside the rock that were not clearly seen by the casting of the cavities, so it was worth the effort. New data on increased Middle Devonian oxygen levels: A. Cannell et al., *Earth-Science Reviews* 2022, 231:104062, and Ruliang He et al., *Geochemica et Cosmochimica Acta* 2020, 287:328–40; Middle Devonian plants with extensive root systems and seed plants: W. Stein et al., *Current Biology* 2020, 30:421–31; Middle Devonian land plants raising oxygen levels: M. Elrick et al., *Earth and Planetary Science Letters* 2022, 581:117410. Origins of air breathing in vertebrates in the Middle Devonian: J. Long et al., *Nature* 2006, 444:199–202; A. Clement and J. Long, *Biology Letters* 2010, 6:509–12; J. Graham et al., *Nature Communications* 2014, 5:3022, doi10.1038/n.comms4022. Oldest tetrapod tracks in Zachelmie, Poland: G. Niedzwiedzki et al., *Nature* 2010, 463:43–48; Devonian tetrapods in Ireland, Scotland, and Australia: J. Clack, *Palaeogeography, Palaeoclimatology, Palaeoecology* 1997, 130:227–50. Exciting new data about these tetrapods we have discovered should be published soon after this book comes out, by Per Ahlberg and team.

**Cold killers: ancient sharks of Antarctica:** My book *Mountains of Madness* documenting my two early Antarctic expeditions in 1988–89 and 1991–92 was published in 2000 (and is cited above). Accounts of early New Zealand collecting expeditions from Gavin Young (*Palaeontographica* 1989, A202:1–89), Alex Ritchie (*Australian Natural History* 1971, 65–71), and Errol White ("Trans-Antarctic Expedition 1955–1958," Scientific Reports, *Geology* 1968, 16:1–26). This last paper is also the first description of the *McMurdodus* tooth. *Antarctilamna* paper: Gavin Young, *Palaeontology* 1982, 25:817–43. New paper redescribing *McMurdodus* and creating new genus *Maiseyodus:* J. Long et al. in A. Pradel et al., *Ancient Fishes,* cited above. Earlier papers on *"McMurdodus"* now reassigned to *Maiseyodus* from Australia: S. Turner et al., *Alcheringa* 1987, 11:233–44, and C. Burrows et al., *Acta Geologica Polonica* 2008, 58:151–59. My Antarctic shark paper with Gavin Young describing *Portalodus, Aztecodus,* and *Anareodus:* J. Long and G. Young, *Records of the Western Australian Museum* 1995, 17:287–308; *Anareodus* suggested as being within *Aztecodus:* Ginter et al., 2010 *Handbook* (cited above).

**Big shifts in shark evolution:** Shark tooth size for the Devonian: see Ginter et al., 2010 *Handbook* (cited above). *Carcharhinus* and rivers, plus shark osmoregulation: see Ebert, *Sharks of the World* (cited above). Shark's rectal glands: S. Holmgren and S. Nilsson in W. C. Hamlett, ed., *Sharks, Skates, and Rays.* For *Phoebodus* species and information about *Protacrodus,* see Ginter et al., 2010 *Handbook* (cited above). *Gladbachus* as a shark: U. Heidke, *Pollichia Kurrier* 2009, 25:24–26; revised description and phylogeny showing it as a "stem chondrichthyan" by Michael Coates et al.: *Proceedings of the Royal Society of London* 2017, B285:20172418.

**Biographical Note: Philippe Janvier:** Philippe Janvier's personal stories were relayed directly, but here is a little more background on him. Everyone in this field of research admires and respects Philippe, and not just for his many amazing fossil discoveries. His landmark 1996 book *Early Vertebrates* is as close as we fish paleontologists have to a bible. He has worked in the far-flung reaches of the world, often in remote regions infested with bandits, or in steamy jungles ravaged by war, including Bolivia, Afghanistan, Iran, Turkey, and Vietnam. While working in the hills of Colombia he was approached by a police helicopter to warn him away from the site because a band of militant troops of the revolutionary army were climbing toward him on the other side of the hill. Philippe is heroically driven, yet he remains a very nice, honest, and extremely generous human being, serving as a mentor and inspiration for many of us who study fossil fishes, myself included.

## Chapter 4: The First Rise of Sharks

**Meanwhile, in the Late Devonian:** Late Devonian atmosphere: T. Dahl et al., *Chemica et Geologica* 2020, 547:119665; giant tree lycopsids over 100 feet tall: L. Liu et al., *Communications Biology* 2022, 5:966; *Prototaxites* as a giant fungus, status uncertain but probable: M. Nielsen and K. Boyce, *International Journal of Plant Sciences* 2022, 183, doi.org/10.1086/720688; Late Devonian plankton in the seas: C. Whalen and D. Briggs, *Proceedings of the Royal Society of London* 2018, B285, doi.org/10.1098/rspb.2018.0883; Late Devonian biotic crises and extinctions: McGhee et al., *Palaeogeography, Palaeoclimatology, Palaeoecology* 2004, 211:289–97. Kellwasser and Hangenberg extinction events: D. Boyer et al., *Palaeogeography, Palaeoclimatology, Palaeoecology* 2021, 566:110226.

**Finding a new Devonian shark.** An account of this discovery is also given in my book *Swimming in Stone* (cited above). *Gogoselachus* work: J. Long et al., *PLOS One* 2015, 10:e0126066. Report of acellular bone in an early shark: M. Coates et al., *Nature* 1998, 396:729–30. Most recent phylogeny showing *Gogoselachus* as the sister taxon to crown chondrichthyans (the group including all modern sharks, rays, and chimaerids and extinct lines bracketed by them) is in Christian Klug et al., *Swiss Journal of Palaeontology* 2023, 13358-023-00266-6; see supplementary information for precise phylogenetic details.

**The mountaineer and his magnificent Moroccan sharks.** Christian Klug's discovery of the oldest fossil crab: R. Förster, *Mitteilungen der Bayerischen Staatssammlung für Paläontologie und historische Geologie* 1986, 26:25–31. Papers on Moroccan Devonian sharks: on *Phoebodus:* L. Frey et al., *Proceedings of the Royal Society B: Biological Sciences* 2019, 286:20191336; on *Ferromirum:* L. Frey et al., *Communications Biology* 2020, 3:681; on snaggletooth shark feeding mechanism: A. Chappell and B. Séret, *Journal of Anatomy* 2020, joa.13313; on *Maghriboselache:* C. Klug et al., *Swiss Journal of Palaeontology* 2023, 13358-023-00266-6. Christian's giant ctenacanth is still under investigation, but news of it was presented at the talk he gave at ISELV in Valencia, Spain, in June 2022; he has given me his permission to mention it here.

**Fossil sharks of the golden triangle.** On Shanthai paleogeography: J. Long and C. Burrett, *Geology* 1989, 17:811–13; on *Siamodus* and *"Phoebodus" australiensis:* J. Long, *Journal of Vertebrate Paleontology* 1990, 10:59–71. First *Phoebodus* paper: O. St. John and A. H. Worthen, *Geological Survey of Illinois, Palaeontology* 1875, 6:245–488. Susan Turner's landmark Australian fossil shark paper: S. Turner,

*Journal of Vertebrate Paleontology* 1982, 2:117–31. Michal Ginter's paper naming *Jalodus:* M. Ginter, *Abhandlungen und Berichte für Naturkunde* 1999, 21:25–47. His paper erecting the family Jalodontidae: M. Ginter et al., *Acta Geologica Polonica*, 2002, 52:169–215). Paper erecting order Jalodontiformes (A. Ivanov et al., *Journal of Vertebrate Paleontology* 2021, 41:e1931259). *Jalodus* biofacies as an environmental indicator: *Journal of Vertebrate Paleontology* (M. Ginter, 2001, 21:714–29).

**Sharks scale down their diet.** Michal's first paper on *Phoebodus:* M. Ginter, *Acta Geologica Polonica* 1990, 40:69–81. *Diademodus* and other filter-feeding sharks: M. Ginter, *Acta Geologica Polonica* 2008, 58:147–53; for Cleveland Shale *Diademodus,* see J. Harris, *Proceedings of the Zoological Society of London* 1951, 120:683–97. On *Orodus,* see R. Zangerl, 1981 *Handbook* (cited above). For *Deihim* from Iran, see M. Ginter et al., *Acta Geologica Polonica* 2002, 52:169–215. Other fossil sharks from Iran include forms with strange crushing shapes like *Dalmehodus.* Here is a short account of my trip working on fossil sharks in Iran: In 1998 I was invited to attend a geological conference and field trip in Iran. I became a mentor for two Iranian students, Artemian Artabaz and Vachik Harapetian. I spent time with each of them at the University of Isfahan working on their Devonian fish fossils. Among the collection was a rounded, elongated tooth with central domed cusp, belonging to the well-known shark *Orodus.* While this form is common in younger Carboniferous rocks, it is quite rare in the Devonian. As such it was an important new record of another shark from Iran. Vachik and I named a new fossil shark he had found in the Garigoon Range, near the village of Dalmeh, so we called it *Dalmehodus,* the "tooth from Dalmeh," from the Late Famennian stage (c.363 MYA).

**Sharks reinvent themselves, for the first time.** Holy Cross Mountain's holocephalan tooth: M. Ginter, *Acta Geologica Polonica* 2004, 49:409–15. For *Protacrodus,* see M. Ginter et al., 2010 *Handbook* (cited above). Holocephalan teeth, histology, and cranial anatomy: B. Stahl, 1999 *Handbook* (cited above), and also for Devonian records of stem holocephalans such as *Cochliodus, Synthetodus, Sandalodus,* and others. These are thought to be of unreliable dating, so not proven as definitely Devonian age, per discussion with Michal Ginter.

**Additional note on another possible Late Devonian holocephalan.** About twenty-five years ago, the first fossils of sharks that lived in the extreme polar regions of Gondwana were found. The enigmatic *Plesioselachus* was found by Rob Gess at the famous Waterloo Farm locality near Grahamstown, in South Africa. Rob graduated from being a student when the site was first discovered, to later being appointed an academic at the Bernard Price Institute. He's an affable fellow who hasn't changed a bit since I first met him. He still has the same long, flowing hair and thin, smiling face. The fossil site at Waterloo Farm was discovered in 1985 when a road was found to be too narrow. As it was being widened, Rob arranged to have the many tons of fossil-bearing shale moved to his farm so he could split the rock at his leisure to find fossils. It has taken him decades of painstaking work, but the effort really paid off. The Waterloo Farm locality is one of the most important Devonian sites in the world. It contains a huge diversity of exceptional fishes, plants, and early tetrapods representing a community of organisms that once thrived in a shallow water lagoon close to the South Pole of Gondwana. Two sharks have been found, as well as a late surviving species of stem shark (acanthodian) called *Diplacanthus*. The other shark is a late surviving species of *Antarctilamna*, suggesting that the fauna represents a kind of last refuge; they may have died out earlier in other areas, but in the cold, high latitude of the Gondwana region, some old species just kept hanging on. Back in 1999, Rob and I and others described *Plesioselachus* ("primitive shark") based on a nearly complete but rather strange shark about a foot long from the Waterloo Farm site. We thought at the time that it represented the oldest known complete holocephalan. Although we didn't have the tooth plates, its body shape with whip-like tail and large first dorsal fin spine showed it to be quite like some of the well-known holocephalans from the Carboniferous period. However, we have grown cold on the idea following recent work by Rob and Mike Coates suggesting that while it could still be a holocephalan, the evidence is not strong enough to confirm it. *Plesioselachus* from South Africa: M. Anderson et al., *Records of the Western Australian Museum* 1999, 57:151–56; R. Gess and M. Coates, *Paläontologisches Zeitscrifter* 2017, doi 10.1007/s12542-014-0221-9.

## Chapter 5: Sharks' Armored Rival

**Ancient American Sea, 359 million years ago.** Depositional environment of the Cleveland Shale sharks: Aaron Martinez et al., *Geobiology* 2019, 17:27–42. *Cladoselache* details from R. Zangerl, 1981 *Handbook* (cited above); *Kentuckia* from Dorothy Rainer, *Transactions of the Royal Society of Edinburgh* 1952, 62:53–83. Revised smaller body size of *Dunkleosteus:* R. Engelman, *Diversity* 2023, 15:d15030318.

**The fabulous fossil sharks of Ohio.** Deposition of Cleveland Shale: A. Martinez et al. (as cited above). John Strong Newberry: see his biogeographical memoir published by the National Academy of Sciences, nasonline.org/publications/biographical-memoirs/memoir-pdfs/newberry-j-s.pdf. J. S. Newberry's work on *Cladoselache: Monographs of the U.S. Geological Survey* 1889; for general information on *Cladoselache,* see R. Zangerl, 1981 *Handbook* (cited above). Mike Williams's detailed work on feeding behavior in Cleveland Shale sharks: A. Boucot, ed., *Evolutionary Paleobiology of Behavior and Coevolution* (Amsterdam: Elsevier, 1990), 273–87. *Maghriboselache:* C. Klug et al., *Swiss Journal of Palaeontology* 2023, 13358-023-00266-6; giant *Ctenacanthus* lower jaw in Cleveland Museum, catalog #5238; photograph of jaw specimen and labels courtesy of Caitlin Colleary of that museum. Red Hill fossil site: E. Daeschler and W. Cressler III, "Field Guide," 2011, digitalcommons.wcupa.edu/geol_facpub; Red Hill fossil sharks: J. Downes and E. Daeschler, *Journal of Vertebrate Paleontology* 2001, 21:811–14. Note that *Ageleodus* teeth have also been interpreted as possible fin scale denticles of *Tristychius* (J. Dick, *Earth and Environmental Science Transactions of the Royal Society of Edinburgh* 1978, 70:63–108), or as teeth related to petalodonts (M. Ginter et al., 2010 *Handbook,* cited above, 149–50).

**Meet deadly Dunk.** *Dunkleosteus* estimated size at 28.8 feet: H. Ferrón et al., *PeerJ* 2017, 5:e4081, with a recently revised much smaller body size: R. Engelman, *Diversity* 2023, 15:d15030318). For *Dunkleosteus* to swim with a tuna-like body shape would require it to swim in a thunniform style, which means it would need strong ossifications in the base of the caudal fin. No placoderms have any bones strengthening the caudal fin. Using a typical *Coccosteus*-style tail without reinforced caudal ossification would imply that *Dunkleosteus* swam in typical arthrodire style, with a carangiform swimming motion, using the whole tail in undulations, thus requiring a much longer tail. For this reason I feel that *Dunkleosteus* might have been a bit larger than the 14 feet proposed by Russell Engelman, perhaps more around 20 feet maximum length. For a detailed review of placoderms, see G. Young, *Annual Reviews of Earth and Planetary Sciences* 2010, 38:523–50. *Coccosteus* body shape: R. Miles and T. Westoll, *Transactions of the Royal Society of Edinburgh* 1968, 67:373–476. Another recent placoderm with complete body shape is *Amazichthys:* M. Jobbins et al., *Frontiers in Ecology and Evolution* 2022, 10:fevo.2022.969158. Newberry's work on placoderms used here: J. S. Newberry, *Proceedings of the American Association for the Advancement of Science, 17th Meeting* 1868: 146–47; more detailed descriptions of giant placoderms: J.H.S. Newberry, *Transactions of the New York Academy of Sciences* 1885, 5:25–28, and J. S. Newberry, *Monographs of the U.S. Geological Survey* 1889, 16. Newberry thought some placoderms were more closely elated to lungfishes: J. S. Newberry, *Geological Survey of Ohio,* vol. 2, part 2, *Palaeontology,* 1875. Placoderm armored eyeball: see J. Long and G. Young, *Memoirs of the Australasian Association of Palaeontologists* 1988, 7:65–80. Example of the parasphenoid in placoderms: G. Young, *Zoological Journal of the Linnean Society* 1979, 66:309–52. Placoderm teeth, early work confirming they were real teeth: Smith and Johanson, *Science* 2003, 299:1235–36; Valéria's work on acanthothoracid placoderm teeth and jaw bones: V. Vaškaninová et al., *Science* 2020, 369:211–16. Diversity of placoderms: see my book *The Rise of Fishes* (cited above).

**Sharks, placoderms, and medieval armor.** The superclass Elasmobranchiomorphi, uniting placoderms with chondrichthyans, comes from Erik Stensiö, *Kungliga Svenska Vetenskapakadamiens Hanlingar* 1963, 9:1–419. On Bashford Dean's life, see the Metropolitan Museum of Art's article at metmuseum.org/blogs/now-at-the-met/2014/life-of-bashford-dean, plus information about him in Philippe Janvier, *Early Vertebrates* (cited above). It's worth a visit to the Met in New York to see the Arms and Armor gallery, where a number of artifacts from his collection are on view. Dean's work on placoderms as a distinct group of fishes: *Memoirs of the American Museum of Natural History* 9 (1909): 209–87.

**Revolutionary Chinese placoderms; Yunnan, China, August 2019.** Account based on my field trip to Qujing with Professor Zhu Min in March 2018. Flat micro-CT scanner developed by the Institute of Vertebrate Paleontology and Paleoanthropology in Beijing. Late Silurian placoderm site at Kuanti: *Guiyu,* oldest osteichthyan (M. Zhu et al., *Nature* 2009, 469–74); *Entelognathus* (M. Zhu et al., *Nature* 2013, 502:188–94); *Qilinyu* and toothed jawbones in placoderms (M. Zhu et al., *Science* 2016, 354:334–36, and my commentary on the paper, J. Long, *Science* 2016, 354:280–81). Earliest known placoderm *Xiushanosteus* occurs with the earliest near-complete chondrichthyan *Shenacanthus:* Y. Zhou et al., *Nature* 2022, 609:954–58.

**How placoderms made sharks sexy.** The remarkable Gogo fossil site: J. Long and K. Trinajstic, *Annual Review of Earth and Planetary Sciences* 2010, 38:255–79; K. Trinajstic et al., *Journal of the Geological Society of London* 2022, 179:12pp, jgs2021-105. Accounts of my various early Gogo expeditions and many of the discoveries have been written up in my book *Swimming in Stone* (cited above). Gogo placoderm soft organs including oldest vertebrate heart preserved in Gogo placoderms: Kate Trinajstic et al., *Science* 2022, 377:1311–14; Gogo fossil preservation: I. Melendez. et al., *Geology* 2013, 41:1230–126. Mother fish with embryo, *Materpiscis attenboroughi:* J. Long et al., *Nature* 2008, 453:650–52; next papers on placoderm reproduction were about arthrodires: J. Long et al., *Nature* 2009, 457:1124–27, and Per Ahlberg et al., *Nature* 2009, 460:888–89. Estonian and Scottish *Microbrachius* claspers and reproduction: J. Long et al., *Nature* 2015, 517:196–99. Our recent review on placoderm reproduction: K. Trinajstic et al., in Z. Johanson et al., eds., *Evolution and Development of Fishes* (Cambridge: Cambridge University Press, 2019), 207–26. General chondrichthyan reproduction and intrauterine cannibalism: R. G. Gilmore et al.'s chapter in Hamlett, ed., *Sharks, Skates, and Rays,* and C. Conrath and J. Musick's chapter in J. Carrier et al., eds., *Biology of Sharks,* both cited above.

**Placoderm "rats of the Devonian."** *Bothriolepis* as a global phenomenon: V. Dupret et al., *PLOS One* 2023, 0280208. Victorian *Bothriolepis* species: J. Long and L. Werdelin, *Alcheringa* 1986, 10:355–99. *Bothriolepis* eating conchostracan *Asmusia:* R. Cloutier et al., *Review of Palaeobotany and Palynology* 2011, 93:191–215.

**Sharks and the Devonian extinction events.** End-Devonian extinctions: see McGhee, *When the Invasion of Land Failed* (cited above), S. I. Kaiser et al., *Geological Society, London, Special Publications* 2016, 423:387–439, and D. Boyer et al., *Palaeogeography, Palaeoclimatology, Palaeoecology* 2021, 566:110226. Viluy Traps cause of extinction (general effects and methylmercury poisoning): see M. Rakocinski et al., *Scientific Reports* 2020, doi.org/10.1038/s41598-020 -64104-2. On selenium depletion and mass extinction events: J. Long et al., *Gondwana Research* 2016, 36:209–18. Reefs shrinking by five thousandfold: McGhee, *When the Invasion of Land Failed* (cited above).

**Finale: Devonian sharks triumph over placoderms.** Giant *Ctenacanthus* is based not just on Christian Klug's new Moroccan find, but also on the isolated right jaw ramus from the Cleveland Shale in the Cleveland Museum mentioned in the previous chapter. As the Moroccan jaw is quite a bit larger, a realistic upper size range for *Ctenacanthus* at 15–20 feet is not unreasonable. Tally of Cleveland Shale sharks versus placoderms: R. K. Carr and G. L. Jackson, Ohio *Geological Survey Guidebook 22*, 17; high shark abundance compared to other vertebrate microremains in the Carnic Alps of northern Italy at the end of the Devonian: C. Randon et al., *Geobios* 40:809–26. End-Devonian Hangenberg extinction changes in fish faunas: L. Sallan and M. I. Coates, *Proceedings of the National Academy of Sciences* 2010, 107:10131–35; Lilliput effect: L. Sallan et al., *Science* 2015, 350:812–15.

**Biographical note, Dr. Gavin Young.** The grand maestro of the placoderms is Gavin Young, from Canberra, Australia. Gavin is an old-school paleontologist whose career is built upon his determination and hard work collecting fossils in the field. After entering the public service in his teenage years, he began working at the Australia Bureau of Mineral Resources; he later trained for his PhD in London, mentored by some of the leading fish paleontologists of the time, including Roger Miles, Colin Patterson, Peter Forey, and Brian Gardiner. Gavin has personally collected nearly all the fossil specimens he has ever published on, including many prime specimens from sites around Australia and Antarctica. His highly detailed anatomical descriptions of the Taemas Early Devonian placoderms like *Buchanosteus* and Gogo ptyctodontids shifted the study of placoderms into a serious new direction. Gavin's work has revealed extraordinary new details about the anatomy of the placoderm skull and brain, and about placoderm distributions globally, from which he developed new theories about their evolutionary trends and migrations around the Devonian world.

## Chapter 6: The First Golden Age of Sharks

**An ancient Montana bay, 318 million years ago.** Monsoonal rains and deposition of the sediments to entrap the creatures in the Bear Gulch embayment: E. D. Grogan and R. Lund, *Geodiversitas* 2022, 24:295–315; on *Falcatus:* R. Lund, *Journal of Vertebrate Paleontology* 1985, 5:1–19; larger sharks 10 to 12 feet long: C. Kurtz, interview with Dick Lund, *Montana Naturalist,* Fall 2006:4–5; on *Thrinacoselache:* E. D. Grogan and R. Lund, *Journal of Vertebrate Paleontology* 2008, 28:970–88; on Bear Gulch squid: C. Klug et al., *Communications Biology* 2019, doi.org/10.1038/s42003-019-0523-2; on *Squatinactis:* R. Lund, *Journal of Vertebrate Paleontology* 1988, 8:340–42; on *Rainerichthys:* E. D. Grogan and R. Lund, *Acta Zoologica, Stockholm* 2009, 90:134–51; ray-fin fish *Discoserra:* R. Lund, *Geodiversitas* 2000, 22:171–206; *Stethacanthus* from Bear Gulch: R. Lund, *Geobios* 1984, 17:281–95; eating shrimp is based on actual gut contents of its close relative, *Akmonistion:* from conversation with Mike Coates; on *Echinochimaera:* R. Lund, *Memoirs Muséum National d'Histoire Naturelle, Paris* (ser. C), 1986, 53:195–205; on *Belantsea:* R. Lund, *Journal of Vertebrate Paleontology* 1989, 9:350–68. *Falcatus* mating: While it seems more probable that the sharks were about to mate when the sediment fall buried them, I here use a little artistic license to re-create the mating act.

**Welcome to the age of coal swamps.** Expansion of the Carboniferous flora and cooling: B. Chen et al., *Earth and Planetary Sciences Letters* 2021, 565:116953; *Arthropleura* a giant millipede: N. S. Davies et al., *Journal of the Geological Society* 2022, 179:jgs2021-115; tetrapods: see R. Carroll, *The Rise of Amphibians* (Baltimore: Johns Hopkins University Press, 2009); high global oxygen: A. Cannell et al., *Earth-Science Reviews* 2022, 231:104062; Carboniferous temperatures: E. L. Grossman and M. Joachimski, *Scientific Reports* 2022, 12:8938.

**Dick Lund, the shark whisperer of Bear Gulch.** I've known Dick Lund since 1983, so I asked him and Eileen about the early days at Bear Gulch to get the background for this section. You can see Dick Lund and Eileen Grogan at the Bear Gulch fossil site here: interactive.wttw.com/prehistoric-road-trip/stops/theres-something-fishy-in-montanas-fossil-deposits. Bear Gulch site richness and diversity: R. Lund et al., *Palaeogeography, Palaeoclimatology, Palaeoecology* 2012, 343:1–6. Dick's 1976 *New York Times* article: nytimes.com/1976/11/13/archives/new-jersey-pages-fossils-in-montana-hint-a-golden-age-of-sharks.html.

**The Bear Gulch shark menagerie.** Sources for *Falcatus* and other sharks, holocephalans, and sea creatures mentioned here are cited in full at the start of this chapter's notes above. *Thrinacoselache* eating *Falcatus:* R. Lund and E. Grogan, personal communication. Suction feeding in iniopterygians: R. Dearden et al., *Proceedings of the National Academy of Sciences* 2023, 120:e2207854119. *Echinochimaera* and possible venomous spines in living chimaerids: D. Didier et al., ch. 4 in J. Carrier et al., eds., *Biology of Sharks* (cited above). For more information on paraselachians and to see some really weird Bear Gulch sharks I couldn't mention here for lack of space, see E. D. Grogan et al., ch. 1 in Carrier, *Biology of Shark*s (cited above).

**Thoughts from a 300-million-year-old brain.** *Sibirhynchus* papers and paper on fossil brain: A. Pradel et al., *Proceedings of the National Academy of Sciences* 2009, 106:5224–28, and A. Pradel, *Geodiversitas* 2010, 32:595–661; *Coccocephalus* brain paper: R. Figueroa et al., *Nature* 2023, 614:486–91.

**Sharks activate their dental superpower.** For a great example of many Carboniferous sharks known only by their teeth, see James William Davis monograph, *Transactions of the Royal Dublin Society* 1883, series 2, 1:327–548, and the Ginter et al. 2010 *Handbook* (cited above); *Cladodus* species reduced: C. Duffin and M. Ginter, *Journal of Vertebrate Paleontology* 2006, 26:253–66.

**Mr. Wood's incredible Scottish fossil sharks.** Sources on Stan's life story: T. R. Smithson and W.D.I. Rolfe, *Earth and Environmental Science Transactions of the Royal Society of Edinburgh* 2018, 108:7–17, plus stories related to me by Michael Coates. Stan's 1992 book about fossil hunting at Wardie (cited above). *Diplodoselache woodi:* J. F. Dick, *Earth and Environmental Science Transactions of the Royal Society of Edinburgh* 1981, 72:99–113; Scottish *Rhizodus:* S. M. Andrews, *Earth and Environmental Science Transactions of the Royal Society of Edinburgh* 1985, 76:67–95. *Onychoselache:* J. F. Dick and J. G. Maisey, *Palaeontology* 1980, 23:363–74; *Tristychius:* M. I. Coates et al., *Science Advances* 2019, 5:eaax2742, and see neat movies of this shark's suction power as supplementary information to the paper; on early days with Mike Coates: We met at the September 1982 Cambridge University meeting on Comparative Anatomy and Vertebrate Palaeontology.

**Akmonistion, the Bearsden Shark.** Bearsden Shark poem from *The Herald,* December 7, 2017, heraldscotland.com/life_style/arts_ents/15701468.poem-day -bearsden-shark-edwin-morgan/; *Akmonistion* paper: M. I. Coates and S.E.K. Sequeira, *Journal of Vertebrate Paleontology* 2001, 21:438–59; Rainer Zangerl's idea about erectile dorsal fins and biomimicry: R. Zangerl, *Journal of Vertebrate Paleontology* 1984, 4:372–78; sexual size differences in *Stethacanthus* spine brushes is cited on 73–74 in R. Zangerl's 1981 *Handbook* (cited above). For *Edestus* and *Carcharopsis* teeth, see M. Ginter et al.'s 2010 *Handbook* (cited above).

**Giant predatory sharks rule the seas.** Information about John-Paul Hodnett's life is from my interview with him. Details of the Kentucky Mammoth cave *Saivodus* specimen: B. Katz, February 4, 2020, smithsonianmag.com/smart-news/ jaw-330-million-year-old-shark-found-kentucky-cave-180974115/; *Saivodus* and *Cladodus striatus:* C. Duffin and M. Ginter, *Journal of Vertebrate Paleontology* 2006, 26:253–66, and J. P. Hodnett et al., *Journal of Vertebrate Paleontology* 2024, e2292599, and in M. Ginter et al.'s 2010 *Handbook* (cited above); the Finis Shale supersharks: J. G. Maisey et al., *Journal of Vertebrate Paleontology* 2017, 37:e1325369. For *Dracopristis,* see J.-P. Hodnett et al., *New Mexico Museum of Natural History and Science Bulletin* 2021, 84:391–424; largest *Glikmanius* teeth are nearly an inch wide, so a body size close to 16 feet is not unreasonable: M. Ginter et al., 2010 *Handbook,* for tooth size, cited above; and D. Elliott et al., *Journal of Vertebrate Paleontology* 2004, 24:268–80, citing 1 inch (c.3 cm) wide *"Cladodus" occidentalis,* which is later changed to *Glikmanius.* On *Carcharopsis,* see Allison Bronson et al., *Papers in Palaeontology* 2018:1–14, and Wayne Itano's paper, *Palaeoworld* 2014, 23:258–62.

**Carboniferous shark wrap-up:** summarized from references cited above.

## Chapter 7: Swamp Sharks

**An Illinois delta, 310 million years ago.** For *Bandringa* information, see L. Sallan and M. Coates, *Journal of Vertebrate Paleontology* 2014, 34:22–33, and first paper by Rainer Zangerl, *Fieldiana Geology* 1969, 12:157–69. Thomas Clements et al. paper on Mazon Creek ecosystem and depositional environment and fauna: *Journal of the Geological Society* 2019, 176:1–11; this paper cites that the crustaceans from Mazon Creek include fourteen species of malacostracans, which includes the crabs, shrimps, and lobsters, even though none would look like modern crabs as they evolved in the Jurassic. Tully Monster, *Tullimonstrum:* L. Sallan et al., *Palaeontology* 2017, 60:149–57, and V. McCoy et al., *Geobiology* 2022, doi:10.1111/gbi.12397; lamprey *Hardistiella:* P. Janvier and R. Lund, *Journal of Vertebrate Paleontology* 1983, 2:407–13; shark *Dabasacanthus:* R. Zangerl 1981 *Handbook* (cited above). About the Serpukhovian crisis:

Iris Feichtinger from the Natural History Museum in Vienna recently documented how sixty-seven different genera of fossil sharks fared through the devastating tectonic events that resulted in severe climatic fluctuations during the middle and end of the Carboniferous period. Around 325 million years ago a major decline in shark diversity occurred across the globe, most pronounced around the equatorial-tropical latitudes. Iris became hooked on fossil sharks when collecting fossil teeth on a geological field trip near where she was raised in upper Austria. She was fascinated that a landlocked mountainous country like Austria could once have been a marine diversity hot spot for ancient sharks. Her paper is I. Feichtinger et al., *Journal of Vertebrate Paleontology* 2021, e1925902. Ranking of Serpukhovian crisis as extreme: see McGhee et al., *Palaeogeography, Palaeoclimatology, Palaeoecology* 2004, 211:289–97.

**The amazing Mazon Creek fossils.** Early history of Mazon Creek fossil site collecting and fossil preservation and fauna: see Thomas Clements et al. 2021 article cited in note above; cyclothems basics: digitalcommons.unl.edu/ conservationsurvey/617/; Mazon Creek blobs as "sea anemones": Roy Plotnick, *Papers in Palaeontology* 2023, 9:e1479; snout pieces of *Bandringa:* Donald Baird, *American Museum Novitates* 1978, 1641:1–22.

**Sharks' next superpower: electroreception.** For a good overview on chondrichthyan electroreception in sharks, see chapter 9 in Peter Klimley's 2013 book *Great White Sharks* (cited above). For an overview of the early discovery of shark electroreception, see A. Kalmijn, *Journal of Experimental Biology* 1971, 55:371–83; Ben King's work on origins of electroreception: see B. King et al., *Palaeontology* 2018, 1–34; saw sharks' and sawfishes' use of rostrum to immobilize prey: see B. Wueringer et al., *Reviews in Fish Biology and Fisheries* 2009, 19:445–64.

**Sharks take over the rivers.** Notes and references about *Portalodus* and Antarctic freshwater sharks and fishes are cited above in the section on chapter 2. General information on xenacanths: R. Zangerl 1981 *Handbook* and M. Ginter et al. 2010 *Handbook,* both cited above; on *Barbclaborina:* G. Johnson, *Mitteilungen aus dem Museum Naturkunde Berlin, Geowissen* 2003, 6:125–46; *Bransonella* and oldest xenacanths: M. Ginter et al. 2010 *Handbook* (cited above).

**Antonín Frič and the Bohemian xenacanths.** On his life story: translated from lazne-belohrad.cz/mesto/osobnosti-mesta/antonin-fric/; about him and brother Václav and the Blaschka glass: H. Reiling, *Journal of the History of Collections* 2005, 17:23–43; Antonín Frič's monographs featuring xenacanth sharks: *Fauna dert Gaskohle und der Kalksteine der Permformations Böhmens* 1879, vol. 1(1):1–92; 1889, vol. 2(4):93–114; 1890, vol. 3(1):1–48; see also Oliver Hampe's big paper on the xenacanths of the British Isles, *Transactions of the Royal Society of Edinburgh* 2004; large sizes for female *Lebachacanthus:* K. Beck et al., *Acta Palaeontolgica Polonica* 2016, 61:97–117.

**How xenacanths ruled the rivers:** For data on living sharks here, see D. Ebert et al., *Sharks of the World* (cited above); *Barbclaborina* as filter feeder: M. Ginter et al., 2010 *Handbook* (cited above); see also Gary Johnson's 2003 main reference on this shark cited two notes above. Jan Fischer's isotopic studies on sharks' teeth for determining their environment: *Chemical Geology* 2013, 342:44–62, and *Historical Biology* 2014, 26:710–27.

**The voracious xenacanths of Spain.** Rodrigo Soler-Gijón's life by personal interview. Hannah Byrne's paper on Carboniferous coprolites: *Palaeogeography, Palaeoclimatology, Palaeoecology* 2022, 605:111215; Puertollano xenacanths feeding on each other: R. Soler-Gijón, *Geobios* 1995, 19:151–56; *Orthacanthus* cannibalism: A. O'Gogáin et al., *Palaeontology* 2016, 59:689–724.

**Biographical note on Rainer Zangerl.** Rainer did incredible work on mid- to Late Paleozoic sharks, and he's mentioned in many places in this book, so here is a short bio: He hailed from Switzerland, where he completed his PhD on fossil marine reptiles in 1935. There were no jobs in his field in Switzerland at the time, so he moved to the United States in 1939, aided by the renowned American vertebrate paleontologist Alfred Romer, who secured him a one-year research post at Harvard University. Rainer did fantastic work, so he was able to continue with other science jobs. After a few years moving around to various colleges, teaching courses in veterinary sciences and comparative anatomy, he landed his dream job as the curator of fossil reptiles at the Field Museum in Chicago in 1945. He was eventually promoted to chair of the Geology Department in 1962, which position he held until he retired in 1974. Rainer had a very distinguished career, being a founding member and president of the Society of Vertebrate Paleontology, the world's largest paleontology organization. In 2003 he received its most distinguished award for a lifetime of research excellence, the Romer-Simpson Medal, fittingly named after Alfred Romer, who kick-started his career. While Rainer continued researching fossil reptiles throughout his early years at the Field, he soon realized that Illinois was also ground zero for finding well-preserved fossil sharks, so he spent the rest of his life pursuing this interest. Rainer's most significant discovery was identifying the first iniopterygians. We have also heard about his creative thinking around *Stethacanthus*—the idea that the dorsal fin spine brush was erectile and could be used for mimicry. However, his core work included a raft of detailed papers describing many other kinds of sharks that take our story in a new direction. These include the beautifully preserved symmoriiform sharks he'd collected and described in detail with his star doctoral student, Mike Williams, whom we met earlier when discussing the Cleveland Shale sharks. Rainer's life story above is taken from his obituaries by Hans-Peter Schultze: *The Compleat Mesoangler* 2005, 11:1–3, and H.-P. Schultze and W. Turnbull, *Society of Vertebrate Paleontology News Bulletin* 2005, 188:57–60.

## Chapter 8: Rise of the Buzz-Saw Sharks

**Pangaean sea, 270 million years ago.** The Permian was named by Roderick Murchison in 1841: *Proceedings of the Geological Association* 2010, 121:313–18; Permian seas and the chert event: B. Beauchamp and A. Baud, *Palaeogeography,*

*Palaeoclimatology, Palaeoecology* 2002, 184:37–63. *Helicoprion* body shape is based on other eugeniodontids like *Fadenia:* R. Zangerl, 1981 *Handbook* (cited above); maximum body size: interview with Leif Tapanila; *Helicoprion* feeding on cephalopods: L. Tapanila et al., *Biology Letters* 2013, 9:201300, and L. Tapanila et al., *Anatomical Record* 2020, 303:363–76; *Helicoprion* male battle is not based on fossils, but on assumed territorial male mating aggression.

**Early discoveries and the enigma of *Helicoprion*.** Early finds of *Helicoprion* in Western Australia: H. Woodward, *Geological Magazine* 1886, new series, decade 3, 3:1–7, and C. Teichert, *Journal of Paleontology* 1940, 14:140–49; life and family of Henry Woodward: R. Rastall, ed., *Geological Magazine* 1921, 481–84. Life of Alexander Karpinsky: biography by the Russian Academy of Sciences (translated from Russian), ras.ru/presidents/d87f4048-5927-414c-bcf8-8ed52dc 51550.aspx. Karpinsky's main papers on *Helicoprion: Zapiski Imperatorskoy Akademii Nauk* 1899a, 7:1–67; *Verhandlungen Russisches Kaiserliches Mineralogisches Gesellschaft* 1899b, 36:361–475; *Izvestiya Imperatorskoy Akademii Nauk* 1911, 5:1105–22; *Zapiski Ural'skogo Obshchestva Lyubiteley Estestvoznaniya* 1915, 35:117–45; *Zapiski Ural'skogo Obshchestva Lyubiteley Estestvoznaniya* 1924, 34:1–10. Arthur Smith Woodward's paper: *Bulletin de la Société Belge de Géologie* 1903, 30:230–33. Position of *Helicoprion* tooth whorls: see figure one in Tapanila et al. 2013 (cited above). Svend Bendix-Almgreen paper: *Biologiske Skrifter udgivet af Det Kongelige Danske Videnskabernes Selskab* 1966, 14:1–54.

***Helicoprion*'s secrets finally revealed.** Leif Tapanila's story by interview. Placing the buzz-saw sharks: new classification of eugeniodontids: R. Zangerl's 1981 *Handbook* (cited above); Oleg Lebedev's paper on new *Helicoprion* specimens from Russia: *Acta Zoologica, Stockholm* 2009, 90 (suppl. 1):171–82; the articulated *Helicoprion* head: Ramsay et al., *Journal of Morphology* 2014, 1–18; plus other papers by Tapanila et al. cited at top of chapter notes.

**Other significant buzz-saw sharks.** *Edestus* feeding style: see Wayne Itano's papers *Transactions of the Kansas Academy of Sciences* 2015, 118:1–9, and *Palaeontologia Electronica* 2019, 22.2.39A:1–16. Background on Wayne's life: bvuuf.org/ our-stories-wayne-itano/, and Lebdev's 2009 paper (cited in note above). *Helicoprion* body shape is based on other eugeniodontids like *Fadenia:* R. Zangerl, 1981 *Handbook* (cited above). Papers placing *Helicoprion* and kin as holocephalans: R. Zangerl, 1981 *Handbook* (cited above). *Karpinskiprion:* O. Lebedev et al., *Earth and Environmental Science Transactions of the Royal Society of Edinburgh* 2023, doi.org/10.1017/S1755691022000251; phylogenetic position of the buzz-saw sharks: see *Dwykaselachus* paper, M. Coates et al., *Nature* 2017, 541:208–11; see also how *Helicoprion* dentition fits into holocephalans: Z. Johanson et al., *Integrative and Comparative Biology* 2020, 60:630–43.

***Helicoprion* goes missing.** The full story of the stolen *Helicoprion bessanovi* holotype specimen and other illegal fossil cases (like the *Archaeoraptor* story) was the subject of one of my earlier books, *The Dinosaur Dealers* (cited above), with advice and input from David Ward.

*Helicoprion*'s competition. Sharks living alongside *Helicoprion* including *Tanao-dus* and *Orodus:* A. Ivanov et al., *Acta Palaeontologica Polonica* 2017, 62:289–98; *Wodnika:* R. Zangerl, 1981 *Handbook* (cited above); Kaibab sharks USA: J-P. Hodnett et al., *Historical Biology* 2012, 24:381–94.

A Permian turducken. Jürgen Kriwet's paper: *Proceedings of the Royal Society* 2008, B275:181–86.

The Paleozoic shark wrap-up. This summary is adopted from many works cited in the notes for chapters 2–8 above.

Biographical note: Jürgen Kriwet. Jürgen grew up in southern Germany wanting to become a veterinarian, his mother's profession. Both he and his brother were raised around animals. Young Jürgen became fascinated with sharks after his mother bought him a book titled *The Super Shark* when he was nine years old, but as he grew older, he was steered back toward a more reliable career in medicine. Admitting that he was a bit lazy in school, he told me he flunked the medical school entrance exams, so he joined the German army to study medicine, receiving strong paramedical training. It was while doing patrols in the Alps of southern Germany that he regained his lost interest in fossils and geology. He decided to leave the army and move with his partner to Berlin to study. At that time the Wall still existed, dividing Berlin into two political states. In Berlin he studied geology and paleontology and attended numerous medical and biology classes. Eventually he wound his way back to his childhood love of sharks and grew more interested in their fossil record and evolutionary trends. He carried out detailed geological mapping on a Greek island and completed a paleontology thesis on Cretaceous fossil sharks from northern Spain. While he loved all this fieldwork in interesting countries, he really wanted to research fossil sharks closer to home. Eventually the time arrived for him to choose a doctoral project. His supervisor, the renowned German paleontologist and director of the Natural History Museum in Berlin, Professor Hans-Peter Schultze, shocked him when he said he would not allow him to study fossil sharks because that was his clear passion. Instead, he insisted that Jürgen broaden his knowledge about other groups of fossil fishes and made him study an extinct group of ray-finned fishes for his doctorate. On completing his dissertation, Jürgen resumed following his passion for fossil sharks. He next worked in Bristol with Michael Benton on trends in shark evolution in the Mesozoic era and completed many more detailed studies of sharks and rays of the Mesozoic before being offered the chair in paleobiology at the University of Vienna in 2010. Jürgen's research group has contributed many major studies of Permian and Mesozoic sharks of Europe and documented many new species of fossil sharks from their teeth and fin spines. These include sharks across the Permian-Triassic boundary and well-preserved Jurassic hybodonts from Solnhofen in southern Germany. In fact, many of the studies I discuss throughout the next few chapters were conducted by Jürgen or one of his students.

# Chapter 9: Sharks and the Great Dying

**The Great Dying.** About the Permian extinctions: S. Burgess et al., *Nature Communications* 2017, 8:doi.org/10.1038/s41467-017-00083-9); J. Penn et al., *Science* 2018, 362:doi.10.1126/science.aat1327; V. Vadja et al., *Earth and Planetary Science Letters* 2019, 529:115875; A. Bercovici et al., *Journal of Asian Earth Sciences* 2015, 98:225–46; S. Grasby et al., *Geological Magazine* 2016, 153:285–97. About the end-Permian shark extinctions: P. Vázquez and M. Clapham, *Geology* 2017, 45:L395–98; also Ginter et al., 2010 *Handbook* (cited above). Recovery time after extinctions: W. Foster et al., *PLOS One* 2017, 0172321; Early Triassic atmosphere as low as 12 percent oxygen: R. Huey and P. Ward, *Science* 2005, 308:doi.10.1126/science.1108019; as high as 18.5 percent: M. Wan et al., *Palaeoworld* 2021, 30:593–601; floral changes across the boundary: H. Nowak et al., *Frontiers in Earth Sciences* 2020, 8:fdoi.org/10.3389/eart.2020.613350.

**Sharks make it through the crisis.** On late surviving ctenacanth and other sharks: Guillame Guinot et al., *Nature Communications*, 2013, 4:doi.org10.1038/ncomms3669; Wapiti Lake buzz-saw sharks (eugeniodontids), and mention of China and Spitsbergen sharks: R. Mutter and A. Neuman, *Geological Society of London, Special. Publications* 2008, 295:9–41; for *Waipitodus,* see Raoul Mutter et al., *Zoological Journal of the Linnean Society* 2007, 149:309–37; Lilliput effect: P. Harris and P. Knorr, *Palaeogeography, Palaeoclimatology, Palaeoecology* 2009, 284:4–10.

**The hybodont invasion.** Mostly from my own notes on the Natural History Museum, London, collection, which I examined in late 2022, plus notes from H. Cappetta *Handbook*s 1987, 2012 (cited above); papers on *Hybodus* and *Egertonodus* by John Maisey: both in *American Museum Novitates* 1983, 2878:1–64, and 1987, 2878:1–39; see also T. Scheyer et al., *PLOS One* 2014, 9:e88987. German hybodont from the Posidonia Shale with two hundred belemnites in its belly: R. Hoffman and K. Stevens, *Biological Reviews* 2019, brv. 12557. On hybodont tooth structure and enameloids, see Giles Cuny et al., *Evolution of Dental Tissues* (cited above); on hybodont low-frequency hearing, see J. Maisey and J. Lane, *Comptes Rendus Palevol* 2010, 9:289–309.

**Arthur Smith Woodward and his fossil sharks.** Information on Woodward's life and works is drawn from several chapters in the book about him published in 2016 by Z. Johansen et al., eds., *Arthur Smith Woodward: His Life and Influence* (cited above), especially those by K. Schlinder and M. Smith, C. Duffin, and C. Underwood et al. Arthur Smith Woodward's *Catalogues of the Fossil Fishes in the British Museum* are in four parts, published in 1889, 1891, 1895, and 1901. Woodward's full bibliography is published in the Johansen book cited above. About the twenty original species of *Hybodus,* see John Maisey, *American Museum Novitates* 1987, 2878:1–39.

**Challenges facing sharks of the new era.** Marine reptiles of the Triassic: see Darren Naish's 2022 book *Ancient Sea Reptiles* (cited above). Ichthyosaur rapid evolution in the Triassic: B. Moon and T. Stubbs, *Communications Biology*

2020, 3:article68. For *Shastasaurus, Shonisaurus, Cymbospondylus, Cartorhyn-chus,* and Triassic plesiosaurs, etc., see D. Naish, *Ancient Sea Reptiles* (cited above). Late surviving jalodontids *Kueperodus:* A. Ivanov et al., *Journal of Vertebrate Paleontology* 2021, 41:e1931259.

**Rise of the neoselachians.** Louis Agassiz's work on *Hybodus:* see J. Maisey 1987 cited in hybodont section, three notes above; Chris Duffin's work on *Rhomphaiodon: Belgian Geological Survey Professional Papers 1993,* 264:7–32; neoselachian tooth structure, see Giles Cuny et al. 2017 book *Evolution of Dental Tissues* and H. Cappetta, 2012 *Handbook* (both cited above). Neoselachians like *Synechodus* from Oman: Martha Koot et al., *Journal of Systematic Palaeontology* 2015, 13:891–917; *Pseudodalatias:* J. Sykes, *Mercian Geologist* 1974, 5:197–239; about the cookiecutter shark, see D. Ebert, *Sharks of the World* (cited above). *Mooreodontus:* see M. Ginter et al., 2010 *Handbook* (cited above); *Paracyclotosaurus:* see my 1998 book *Dinosaurs of Australia and New Zealand* (cited above).

**Another mass extinction shakes up sharks.** End-Triassic extinction event papers: L. Mander et al., *Journal of the Geological Society* 2008, 165:319–32; P. Wignall and J. Atkinson, *Earth-Science Reviews* 2020, 193282; C. Belcher et al., *Nature Geoscience* 2010, doi.org/10.1038/NGEO871; L. Opazo and K. Page, *Proceedings of the Geologists' Association* 2021, 132:726–42.

**Biographical Note: John Maisey.** John is mentioned many times throughout this chapter, and indeed throughout this book, as he has prolifically worked on sharks of all ages from the Devonian through to living species. John, born in London in 1949, found his first fossil as he was walking home from school when he was eleven years old—an impression of an ancient sea urchin that had fallen from a truck. John was hooked on fossils from that moment on. He began eagerly collecting fossils along the south coast of England whenever he could persuade his parents to take him on trips. By the age of sixteen he had amassed a huge collection of fossils from all over Britain. He went to the University of Exeter and graduated with honors in 1970. The previous year he worked as intern with the great fish paleontologist Roger Miles at the Natural History Museum, where he learned how to prepare Gogo placoderm fossils out of the rock using weak acetic acid, a technique he would apply later on. He completed his doctoral dissertation on fossil shark fin spines at University College London working with two greats of the fossil fish world, Colin Patterson and Brian Gardiner, with input from a leading fossil mammal expert, Kenneth Kermack. John obtained the position of assistant curator at the American Museum of Natural History in New York in 1978. He became a full curator in 1989, a position he held until he retired in 2018, after nearly forty years of service there. His career saw many landmark studies of fossil sharks completed, including monographic revisions of the Jurassic hybodonts of Britain, X-ray scan studies of fossil and living shark braincases (including Cretaceous hybodonts), and finely executed accounts of Brazilian giant coelacanths and other kinds of bony fishes, as well as histological studies of shark fin spines using thin sectioning techniques. He has also mentored several stars in the paleontology world, including Marcelo de Carvalho, Alan Pradel, Jennifer Lane, Amy Balanoff, Allison Bronson, John Denton, and several others. I once asked John what he is most proud of in his body of paleontology research. He replied that it was the pioneering work he did with CT scanning that enabled him to describe in fine detail every surface feature and foramen of the braincases in fossil sharks to piece together their soft tissue anatomy. He has used his anatomical findings in detailed analyses of shark evolution and gone beyond just fossils in applying his sophisticated CT techniques to elucidate the braincase anatomy of several living sharks. More than anyone else, John showed early on that hybodonts were the closest group to the modern shark radiation, arguing from the evidence of his anatomical findings that all modern sharks (elasmobranchs) must have evolved from them. In later years, he worked on problematic stem sharks from the Devonian of Bolivia, South Africa, and Canada with Philippe Janvier and others. His work at the American Museum saw him involved in much exhibition work, too, including the making of the now-hanging set of megalodon jaws back in 1996, as well as creating the museum's excellent fossil fish displays, all of which show the evolutionary positions of the specimens in little diagrams on the labels, which are still there today (I saw them in late 2022; he also used several of my drawings on the labels). Over the past thirty-five years, I've spent a lot of enjoyable time with John and his paleontologist wife, Gloria, at paleontological meetings, discussing new fossil finds or just shooting the breeze over a pint in some strange *Star Wars*–style bar in a foreign place. He has a typical British sense of humor, dry, witty, and unexpected. Details above are from M. de Carvalho in A. Pradel et al., eds., *Ancient Fishes* (cited above).

## Chapter 10: The Jurassic Rise of Modern Sharks

**Solnhofen Archipelago, 150 million years ago.** Solnhofen background, environmental setting, and fossil preservation plus details of sharks featured in this scene: see E. Villalobos et al., *Diversity* 2023, 15, doi.org/10.3390/d15030386), with behaviors adopted from living sharks from D. Ebert et al., *Sharks of the World* (cited above). *Dakosaurus:* see D. Naish, *Ancient Sea Reptiles* (cited above); other fish, *Pterodactylus, Compsognathus* from Solnhofen paleobiota summary, en.wikipedia.org/wiki/Paleobiota_of_the_Solnhofen_Formation, with scientific peer-reviewed references listed; *Asteracanthus* from Solnhofen: S. Stumpf et al., *Papers in Palaeontology* 2021, 3:1479–1505; hybodonts eating belemnites: R. Hoffman and K. Stevens, *Biological Reviews* 2019, brv. 12557.

**The incredible sharks of Solnhofen.** This scene is based largely on information in E. Villalobos et al. cited in above note; for *Astercanthus,* see Stumpf reference cited in above note; marine reptile scavenged by *Asteracanthus:* see D. Martill's 1991 book (cited above); for *Sphenodus* and its relationships, see J. Maisey, *American Museum Novitates* 1985, 2804:1–28, and S. Klug and J. Kriwet, *Naturwissenschaften* 2008, 95:443–48; further information on *Palaeocarcharias:* C. Duffin, *Zoological Journal of the Linnean Society* 2008, 94:271–86. All other shark information in this passage is from E. Villalobos et al. cited in above note.

**Hybodonts at the top of their game.** Mary Anning background: see A. Berta and S. Turner, *Rebels, Scholars, Explorers* (cited above); Mary Anning's *Hybodus* paper: *Magazine of Natural History* 1839, 12(3):605; see also Edward Charlesworth paper about the *Hybodus* from Lyme Regis found by Mary Anning and Edmund Higgins of Cheltenham: E. Charlesworth, *Magazine of Natural History* 1839, 2(3): 242–48; hybodont diversity in Middle Jurassic: H. Cappetta 2012 *Handbook* (cited above) and comments in J. Rees and C. Underwood, *Palaeontology* 2008, 51:117–47.

**Jurassic sharks and the rise of marine reptiles.** Marine reptile diversification in the Jurassic: see Darren Naish's excellent book *Ancient Sea Reptiles* (cited above). Marine reptiles from the Sundance Shale, Wyoming, eating belemnites: J. Massare et al., *Geological Magazine* 2014, 151:167–82; plesiosaur with hybodont remains in its gut: W. Wahl, *Paludicola* 2005, 15–19.

**Neoselachians pushing the limits.** *Ostenoselache:* Chris Duffin, *Paleontologia Lombarda* 1998, 9:1–27; living hexanchiform sharks: D. Ebert et al., *Sharks of the World* (cited above); for *Hexanchus arzoensis,* see H. Capetta 2012 *Handbook* (cited above).

**A Jurassic crisis changes the game for sharks.** About the Toarcian crisis, see Patricia Rita et al., *Royal Society Open Science* 2019, 6:doi.org/10.1098/rsos .190494; see also T. Scheyer et al., *PLOS One* 2014, 9:e88987; oldest bullhead sharks: see H. Cappetta 2012 *Handbook* (cited above); Toarcian Anoxic Ocean Event: see Rita et al. 2019 article (cited above). Charlie Underwood interview at the Natural History Museum in November 2022. Oldest rays: C. Underwood, *Paleobiology* 2006, 32:215–35.

**Dawn's early rays.** The tame smooth stingrays at Hamelin Bay: hamelinbayholiday park.com.au/things-to-do/see-the-sting-rays, and a paper about tourism and planning at the site (A. Lewis and D. Newsome, *International Journal of Tourism Research* 2003, 5:331–46); Stumpy's demise: perthnow.com.au/news/wa/ popular-hamelin-bay-stingray-slaughtered-in-front-of-screaming-kids-ng -9a16711871399ebb1b8cabe69cd9a3e6. Rays' early fossil dentition: see H. Cappetta 2012 *Handbook* (cited above); the rays from Solnhofen: see E. Villalobos 2023 paper cited in note on Solnhofen Archipelago above. About Steve Irwin's death: web.archive.org/web/20060921002229/http://www.cnn.com/2006/ WORLD/asiapcf/09/04/australia.irwin.stingray.reut/index.html. Electric rays' electric organs: J. Walsh, *Philosophical Transactions of the Royal Society* 1773, 461– 77; giant *Leedsichthys* fish: J. Liston, *Oryctos* 2010, 9:1–36.

**Biographical note: Chris Duffin.** Chris's work appears in many sections of this book from Devonian through to recent species, and his detailed history of science work is featured in the megalodon chapter. Chris Duffin was born in the Wiltshire countryside in England, and as a three-year-old he loved arranging the bones of slaughtered chickens into different groups on his parents' farm. He began searching local outcrops of the Jurassic Cornbrash for fossils, which eventually led him to earn a university degree in geology. He studied the rich bonebeds of the Triassic and Jurassic for his doctorate, becoming hooked on fossil shark teeth as he kept finding loads of them in his samples. He quickly discovered that most of them had not been studied seriously for more than 150 years, so he began investigating many of the key species. His work later switched to the English Triassic fissure fills that held a rich fauna of shark teeth. This led him to work on revising *Cladodus* and led to the discovery of *Saivodus* with Michal Ginter. Chris was not able to land a job in paleontology, so he began working as a teacher, gradually moving up through the ranks to a senior position. His interest in fossil sharks continued throughout his life, resulting in many valuable scientific contributions. His expertise in extinct sharks goes beyond science—he has written many papers on the history of fossil sharks in medieval jewelry and medicine, which are discussed in the chapter on megalodon.

## Chapter 11: Sharks Go Large

**Cretaceous seaway, Italy, 90 MYA.** Concerning the maximum size of *Cretodus* and *Cretoxyrhina:* While Amalfitano et al. 2022a (cited below) propose a possible maximum attainable length of 30–36 feet for *Cretodus,* based on modeling the growth trajectory, current maximum size based on vertebrae and teeth suggests a maximum size of 23 to 26 feet for *Cretodus* and around 26 feet for *Cretoxyrhina*. The largest *Cretoxyrhina* were about 26 feet long, from the Turonian of Europe, with slightly smaller sizes for U.S. Western Interior Seaway specimens. The largest known vertebral disc of any Cretaceous shark is around 5.5 inches diameter for the cardabiodontid from the Tooloebuc Limestone of Australia (data above from Mikael Siversson, personal communication, May 3, 2023). *Cretodus* with turtle inside it from the lastame deposits: J. Amalfitano et al., *Palaeogeography, Palaeoclimatology, Palaeoecology* 2017, 469:104–21, and J.Amalfitano et al., *Journal of Paleontology* 2022a, doi.org/10.1017/jpa.2022.23; on giant *Cretoxyrhina* specimens from same locality: J. Amalfitano et al., *Cretaceous Research* 2022b, 98:250–75; on the mosasaur *Romeosaurus:* A. Palci et al., *Journal of Vertebrate Paleontology* 2013, 33:599–612; on lastame *Onchosaurus:* J. Amalfitano et al., *Cretaceous Research* 2017, 69:124–35; on *Ptychodus* from same site: M. Amadori et al., *PeerJ* 2020, 8:e10167; giant *Inoceramus* clams: nps .gov/articles/inoceramus-shells.htm; protostegid turtles: see Darren Naish, *Ancient Sea Reptiles* (cited above).

**The amazing fossil sharks of the *lastame,* Verona.** This account is based on our car trip to the site in November 2022 and talking with Dr. Cesare Papazzoni from the University of Modena and Reggio Emilia, who kindly drove us there. Shark and marine reptile references are all cited above. Neoselachian diversity rising in the Cretaceous: C. Underwood, *Paleobiology* 2006, 32:215–35.

**Living the Cretaceous life.** Early Cretaceous world, Weissert event: L. Cavalheiro et al., *Nature Communications* 2021, 12:5411; Western Interior Seaway: see

Mike Everhart's 2005 book *Oceans of Kansas* (cited above); marine reptiles and protostegid turtles: see Darren Naish, *Ancient Sea Reptiles* (cited above); new small sharks appearing in the Early Cretaceous based on oldest records: H. Cappetta 2012 *Handbook* (cited above).

**New Cretaceous sharks appear.** Antarctic food chains today: learnz.org.nz/ scienceonice144/bg-standard-f/antarctic-food-webs; polar ice caps in the Early Cretaceous: M. Vickers et al., *Geological Society of America Bulletin* 2019, 131:1979–94. Cretaceous shark records are all from H. Cappetta 2012 *Handbook* (cited above). Oldest sawshark: Mikael Siversson correspondence. Modern shark information is from D. Ebert, *Sharks of the World* (cited above). Epaulette sharks walking: M. Porter et al., *Integrative and Comparative Biology* 2022, 62:1710–24. *Cantoscyllium* from Kent: Arthur Smith Woodward, *Catalogue of Fossil Fishes in the British Museum,* vol. 1; from Spain: J. Kriwet et al., *Zoological Journal of the Linnean Society* 2009, 155:316–45.

**The smallest shark of all time.** Las Hoyas site tiny hybodonts: see R. Soler-Gijón et al., "Chondrichthyes," in F. Poyato-Arriza et al., eds., *Las Hoyas: A Cretaceous Wetland* (München: Verlag Dr. Friedrich Pfeil, 2016); *Etmopterus* body sizes: see D. Ebert et al., *Sharks of the World* (cited above); *Lissodus* tooth plates: see H. Cappetta 2012 *Handbook* (cited above).

**Rise of the deadly lamniforms.** On lamniform tooth histology, see L. Schnetz et al., *Journal of Morphology* 2016, jmor.20597; on goblin and sand tiger sharks, see D. Ebert et al., *Sharks of the World* (cited above). All other records here are from H. Cappetta 2012 *Handbook* (cited above). *Johnlongia* paper: M. Siversson, *Palaeontology* 1996, 39:813–49, and updated range from H. Cappetta's 2012 *Handbook* (cited above); on *Cretalamna:* M. Siversson et al., *Acta Palaeontologica Polonica* 2015, 60:339–84; *Futabasaurus* and *Cretalamana* association: K. Shimada et al., *Journal of Vertebrate Paleontology* 2010, 30:592–97; for anacoracids and information on *Squalicorax:* M. Siversson, *Palaeontology* 2007, 50:939–9509, and H. Cappetta's 2012 *Handbook* (cited above) for various species information; for complete Canadian *Squalicorax:* cbc.ca/news/canada/manitoba/ crow-shark-fossil-believed-to-be-world-s-largest-found-in-manitoba-1.2725575; for complete *Scapanorhynchus* body fossils, see F. Pfeil chapter in A. Pradel et al., eds., *Ancient Fishes* (cited above).

**Mikael Siversson, the fossil shark Terminator.** *Cardabiodon ricki:* M. Siversson, *Transactions of the Royal Society of Edinburgh, Earth Sciences* 1999, 90:49–66, and its vertebral disc study: M. Newberry et al., *Acta Palaeontologica Polonica* 2015, 60:877–97.

**A tale of Ginsu sharks.** Background on the Kansas *Cretoxyrhina* specimens: see M. Everhart, *Oceans of Kansas* (cited above), and K. Shimada, *Journal of Paleontology* 1997, 71:926–33; *Cretoxyrhina* range and diversity: see H. Cappetta, 2012 *Handbook* (cited above); *Cretoxyrhina* swimming speed: H. Ferrón et al., *PLOS One* 2017, e0185185, although the maximum speed calculations are critiqued by J. Semmens et al., *Marine Biology* 2019, 66:doi.10.1007/s00227 -019-3542-0; information on Charles Eastman's life is drawn from his obituary by Bashford Dean, *Science* 1919, 49:139–41, and his obituary in *The New York Times*, September 30, 1918; the Eastman murder case: *Boston Sunday Post*, May 12, 1901, front page article entitled "Eastman Case from Shooting to Verdict."

**Ptychodus, the crushing monster.** See Shawn Hamm's excellent paper (*New Mexico Museum of Natural Science Bulletin* 2020, 81:1–96) for maximum size estimate and paleoecology; also work by K. Shimada et al., *Cretaceous Research* 2010, 31:249–54; Mario Carnavari's detailed work: *Palaeontographica Italica* 1916, 22:35–102; Kensu's life by interview and from *Society of Vertebrate Paleontology News Bulletin* 1999, 176:28–29. *Edaphodon* and giant chimaerids: B. Stahl, 1999 *Handbook,* cited above, and S. Gouric-Cavalli et al., *Journal of Vertebrate Paleontology* 2015, 35:e981128.

**Spectacular fossil sharks from Lebanon and Mexico.** Henri Cappetta's legacy works on Lebanese fossil sharks: *Palaeontographica Abteilung* 1980, A168(1–4):69–148, and rays: *Palaeontographica Abteilung* 1980, A168(5–6):149–229. For an overview of the Lebanese fossil sites, history, geological setting, and preservation, see M. Gayet et al., *Fossils of Lebanon* (cited above). For a recent review of the Lebanese fossil shark fauna, see the chapter by Fritz Pfeil in A. Pradel et al., eds., *Ancient Fishes* (cited above). Mexican *Aquilolamna:* see Romain Vullo et al., *Science* 2021, 371:1253–56.

**Deadly showdowns in Late Cretaceous seas:** Mosasaurs: see Darren Naish, *Ancient Sea Reptiles* (cited above). Mosasaur bones with *Cretoxyrhina* wounds, mosasaurs eating sharks: see M. Everhart, *Oceans of Kansas* (cited above), and B. Rothschild et al., *Netherlands Journal of Geosciences* 2005, 84:335–40. End of the Cretaceous giant hexanchid *Xampylodon:* R. dos Santos et al., *New Zealand Journal of Geology and Geophysics* 2022, doi: 10.1080/00288306.2022.2143382.

**Biographical note: Henri Cappetta.** Henri's former student Sylvain Adnet of the University of Montpellier kindly obtained the biographical information about Henri's life on which my account is based, combined with his extensive publication record.

Henri was born in 1948 at Sète, close to Montpellier, France, where he still lives today. He was introduced to paleontology during the 1960s as a master's student teaching at the University of Montpellier and at the Natural History Museum in Paris. He was influenced by Professor Jean-Pierre Lehman, a fossil fish expert and then director of the museum in Paris. He completed his first doctoral dissertation on Miocene fossil bony fishes and sharks from the Montpellier region in 1969, then the following year became one of the first researchers to work under the direction of Professor Louis Thaler, the director of the newly established paleontological lab called the Institute of Evolutionary Sciences of Montpellier.

When Henri started work on the Miocene faunas of the Montpellier area, he was supported by a great community of French geologists at a time when paleontologists were in demand to participate in the national geological survey and mapping. At the time of his doctorate work, big developments were taking place in excavation techniques and collecting by the new generation of vertebrate paleontologists of the Montpellier laboratory, in particular the use of sieving for fossil shark teeth. This allowed him to add many new layers of information to the work done by the earlier European paleontologists. Henri's great advance at this time was to uncover the microfauna of sharks at many historical French and European localities. He correctly suspected that we must use small species for determining the age of rock units, as he did successfully with small mammals in his first paper back in 1966.

After this, Henri quickly begins to take an interest in many fossiliferous deposits around the globe, especially those in the United States, England, Italy, Niger, and Morocco. At this time he became friends with many fossil shark collectors including David Ward in the UK and Gerard Case in the United States. Under the guidance of Professor Arambourg, who had collected superb fossils from the Cretaceous and Cenozoic deposits of Lebanon, Henri completed his second doctoral dissertation on these superb fossil sharks, which was published in 1980 (in France this old tradition allowed one to teach and supervise PhD students; today it is no longer required). He was then recruited as a researcher at the National Scientific Research Center (CNRS) and became based at Montpellier University. Working with the Belgian shark expert Jacques Herman, they revealed the incredible diversity in microscopic teeth of living and fossil sharks through many detailed studies. Along the way, they amassed a huge working collection of shark jaws and fossil elasmobranchs, which are now housed at the University of Montpellier.

Henri turned his attention to northern Africa during the 1980s, on the advice of several French engineers who worked in the phosphate industry. They were also passionate about collecting fossil shark teeth, and from their work in the mines they could direct Henri to important new sites in Egypt, Israel, Jordan, Tunisia, and especially Morocco. Many of them remained as his close friends. He was awarded a national honorary distinction for his work, the bronze medal of the CNRS, in 1988. Henri's field trips to Morocco spanned almost thirty years. His student Sylvain Adnet told me that Henri really enjoyed this period of field work in exotic locations. Sylvain participated in some of these trips and felt that Henri was happy to work in North Africa. He became director emeritus of the French research institute CNRS in 2010, finalizing the work on his monumental *Handbook* before retiring in 2021.

Another legacy of Henri's work remains at the University of Montpellier—an incredible collection of more than three thousand dried shark jaws and many more thousands of fossil sharks' teeth and other remains from around the world. He trained three PhDs. during his career: the Moroccan researcher Abdel Noubhani, who became a professor at Cadi Ayyad, Morocco, and the French researchers Sylvain Adnet and Guillaume Guinot. These last two continue the great work Henri started on fossil sharks as associate professors at the University of Montpellier. Henri passed away on January 6, 2024.

## Chapter 12: Sharks After the Impact

**Attack of the Killer Asteroid.** Much has been written on the end-Cretaceous extinctions; for recent coverage, see R. DePalma et al., *Proceedings of the National Academy of Sciences* 2019, pnas1817407116; KT event could also have been triggered by Deccan Traps volcanism with a meteorite impact: A. Petersen et al., *Nature Communications* 2016, ncomms12079; vaporization of gypsum at event causing cooling: S. Gupta et al., *Earth and Planetary Science Letters* 2001, 188:399–412; extinction of life: P. Wilf and K. Johnson, *Paleobiology* 2004, 30:347–68, and D. Raup and D. Jablonski, *Science* 1993, 260:971–73. Sharks across KT boundary: G. Guinot and F. Condamine, *Science* 2023, 379:science .abn2080; also J. Kriwet and M. Benton, *Palaeogeography, Palaeoclimatology, Palaeoecology* 2004, 214:181–94, and J. Adolfssen and D. Ward, *Acta Palaeontologica Polonica* 2013, 60:313–38; teleost (Acanthomorpha) explosion after KT event: A. Ghezelayagh et al., *Nature Ecology & Evolution* 2022, 6:1211–20; modern shark diversity: see D. Ebert et al., *Sharks of the World* (cited above).

**Old and new sharks of the new era.** New shark and ray species appearing in the Paleocene: see H. Cappetta 2012 *Handbook* (cited above). Mohamad Bazzi information is from direct correspondence. Tooth shape and shark extinctions and evolutionary trends across KT boundary: M. Bazzi et al., *Current Biology* 2018, 28:2607–15; M. Bazzi et al., *Current Biology* 2021, 31:1–11; and M. Bazzi et al., *PLOS Biology* 2021, 19:e3001108. High-latitude Cretaceous lamniform sharks: M. Bazzi et al., *Gondwana Research* 2022, 103:362–70. Information on living squaliforms and other sharks: see D. Ebert et al., *Sharks of the World* (cited above).

**A thermal crisis hits the oceans.** PETM as two-phase event: S. Jehle et al., *Paleoecology* 2019, 525:1–13; caused by volcanism rather than clathrate release: M. Tremblin et al., *Global and Planetary Change* 2022, 212:103794; foraminiferan extinctions: L. Alegret et al., *Global and Planetary Change* 2021, 196:103372.

**Sharks really get warmed up.** Early Eocene Thermal Maximum temperatures: G. Inglis et al., *Climates of the Past* 2020, 16:1953–68; Ellesmere Island alligators and flora: J. Eberle and D. Greenwood, *Geological Society of America Bulletin* 2012, 124:3–23; Eocene Oregon forests similar to central America: britannica .com/science/Eocene-Epoch; oldest records of sharks cited here: see H. Cappetta 2012 *Handbook* (cited above); comments on living sharks and their ranges: see D. Ebert et al., *Sharks of the World* (cited above).

**The awesome sharks of Bolca.** Lorenzo Sorbini's book on Bolca fish and plant fossils is cited above; for early history, see also M. Friedman and G. Carnevale, *Journal of the Geological Society* 2017, 175:569–79. On *Eogaleus*: G. Marrama et al., *Comptes Rendus Palevol* 2018, 17:443–59; on *Galeorhinus*: F. Fanti et al., *Zoological Letters* 2016, 2:9; on *Brachycarcharias*: G.Marrama et al., *Historical Biology* 2019, 31:102–16; on Bolca rays: G. Marrama et al., *Journal of Vertebrate Palaeontology* 2020, 40:e1803339, and G. Marrama et al., *Zoological Letters* 2019, 5:13; on electric rays: G. Marrama et al., *Journal of Systematic Palaeontology* 2018,

16:1189–1219. *Pakicetus* and the origin of whales: see Philip Gingerich et al., *Science* 1983, 220:403–406.

**Serious predatory sharks appear in our seas.** Tiger sharks and eating: D. Ebert et al., *Sharks of the World* (cited above); shark arm case: 1986 Readers Digest Sharks book (cited above). Tiger shark evolution: J. Türtscher et al., *Paleobiology* 2021, 47:574–90; large fossil tiger shark teeth: fossilera.com/fossils/huge -fossil-tiger-shark-tooth-bone-valley-florida; tiger shark relationships: X. Velez-Zuazo and I. Agnarsson, *Molecular Phylogenetics and Evolution* 2010, 58:207–17. About living species of requiem and hammerhead sharks, see D. Ebert et al., *Sharks of the World* (cited above). Oldest *Carcharhinus:* K. Samonds et al., *PLOS One* 2019, 14:0211789; oldest records of sharks cited here: see H. Cappetta 2012 *Handbook* (cited above).

## Chapter 13: Ascent of the Superpredators

**Krazy Kazakhstan fossils and the mystery of the megatoothed sharks.** The story of this field trip is adapted from my chats with David Ward and notes provided by him on the expeditions. I spent some time staying with David and his wife, Alison, looking at his fossil shark tooth collections to see the incredible transitional forms of *Otodus* as they slowly gained serrations bed by bed. *Otodus aksuaticus:* T. Malyshkina and D. Ward, *Proceedings of the Zoological Institute RAS* 2016, 320:50–65.

**Penguins and whales invade the seas.** Gigantic fossil penguins from New Zealand: D. Ksepka et al., *Journal of Paleontology* 2023, 97:434–53; see also T. Cole et al., *Nature Communications* 2022, 13:3912; overview of early whale evolution: M. Uren, *Annual Reviews of Earth and Planetary Sciences* 2010, 38:189–219; whale and hippopotamus relationship: J. Theodor, *Journal of Paleontology* 2004, 78:39–44; oldest whale, *Mystacodon:* C. de Muizon et al., *Geodiversitas* 2019, 41:401–99; *Llanocetus* and gigantism: E. Fordyce and F. Marx, *Current Biology* 2018, 28:1670–76.e2; ancestral toothed baleen whales and feeding: D. Hocking et al., *Biology Letters* 2017:0348.

**Giant thresher sharks.** For *Alopias palatasi*, see B. Kent and D. Ward, *Smithsonian Contributions to Paleobiology* 2018, 100:157–60; for details of *Alopias grandis*, see H. Cappetta, 2012 *Handbook* (cited above). Early Miocene closure of Indian Ocean and Mediterranean Sea: see O. Bialik et al., *Scientific Reports* 2019, 8842; for *Livyatan*, see O. Lambert et al., *Nature* 2010, 466:105–108.

**Rise of the megatoothed sharks.** I recently visited the Western Australian Museum in September 2023 to see my *Otodus angustidens* tooth and remind myself of the story (it is specimen number WAM 90.12.88; the number on it is written in my hand, as I registered it when I was a curator there in 1990). About chronospecies, see the classic work by Steven Stanley, *Paleobiology* 1978, 4:26–40. Early Cenozoic *Cretalamna* (or *Cretolamna*): see H. Cappetta, 2012 *Handbook* (cited above); *Xiphodolamia:* S. Adnet et al., *Journal of African Earth Sciences* 2009, 55:197–204; *Megalolamna:* K. Shimada et al., *Historical Biology* 2016,

1236795; oldest *Lamna, Isurus:* H. Cappetta, 2012 *Handbook* (cited above). *Cretalamna-Otodus, Otodus subserratus:* D. Ward, 2010, in J. Young et al., eds., *Fossils of the Gault Clay* (The Palaeontological Association, London), 275–99. *Otodus aksuaticus:* T. Malyshkina and D. Ward, *Proceedings of the Zoological Institute RAS* 2016, 320:50–65; *Otodus auriculatus:* D. Ehret and J. Ebersole, *PeerJ* 2014, 2:e625; New Zealand *C. angustidens* specimen: M. Gottfried and R. Fordyce, *Journal of Vertebrate Paleontology* 2001, 21:730–39; transition from *O.chubutensis* to *O.megalodon:* V. Perez et al., *Journal of Vertebrate Paleontology* 2018, 38:1546732; *Otodus* body sizes and trophic levels: E. Karst et al., *Science Advances* 2022, 8:sciadv.abl6529; see also K. Shimada et al., *Historical Biology* 2021, 33:2543–59. New Zealand *Otodus angustidens:* M. Gottfried and R. E. Fordyce, *Journal of Vertebrate Paleontology* 2001, 21:730–39; the Jan Juc, Victoria, specimen of *Otodus angustidens* is unpublished: museumsvictoria.com .au/media-releases/victorian-fossil-find-uncovers-prehistoric-leftovers-of -colossal-shark-feast/.

**Additional note about possible early Miocene shark extinctions.** At around 19 MYA, shark diversity might have declined drastically, as published in E. Sibert and L. Rubin, *Science* 2021, 372:1105–07. Their conclusion is based on deepsea records of shark denticles (placoid scales), and after reading the responses against the hypothesis (I. Feichtinger et al., *Science* 2021, 374:abko632, and G. Naylor et al., *Science* 2021, 374:abj8723) and further responses to these by Sibert & Rubin (*Science,* 374:abj9522 and *Science,* 374:abk1733), I decided that the conclusions reached by the authors were still too contentious to discuss in this book.

## Chapter 14: Megalodon

**Seas off Peru, 6 million years ago.** Environmental setting in the oceans off Peru 6 million years ago: B. Leonard et al., *Journal of South American Earth Sciences* 2011, 31:414–25; *Piscobalena:* G. Pilleri and H. J. Siber, in G. Pilleri, ed., *Beitrage zur Palaontologie der Cetaceen Perus* (Bern: Hirnanatomisches Institut, 1989), 109–15; *Livyatan:* O. Lambert et al., *Nature* 466:105–108; *Atocetus:* C. De Muizon, *Éditions Recherche sur les Civilisations* 1988, 78:1–244. A smaller extinct mako-like species about 25 feet long is *Carcharodon hastalis:* D. Ehret et al., *Palaeontology* 2013, 55:1139–53; megalodon bite force: S. Wroe et al., *Journal of Zoology* 2008, 276:336–42. Fossil whale bones with megalodon bite marks: A. Collaretta et al., *Palaeogeography, Palaeoclimatology, Palaeoecology* 2017, 469:84–91, and S. Godfrey et al., *Acta Palaeontologica Polonica* 2018, 63:463–68.

**Megalodon: Size matters.** Megalodon bite force from Steve Wroe et al., *Journal of Zoology* 2008, 276:336–42; megalodon age ranges: see Catalina Pimiento et al., *Journal of Biogeography* 2016, 43:1645–1655; megalodon size estimates: V. Perez et al., *Palaeontologia Electronica* 2021, 24:1–28; J. Cooper et al., *Scientific Reports* 10: 14596:; K. Shimada, *Historical Biology* 2019, 1666840; megalodon body shape: Philip Sternes et al., *Palaeontologia Electronica* 2024, 27(1):a7; body shape and lifestyle calculated from its scales and its gigantism attributed to endothermy and digestion: K. Shimada et al., *Historical Biology* 2023, doi:10.10 80/08912963.2023.2211597; Catalina Pimiento on megalodon average size: C. Pimiento and M. Balk, *Paleobiology* 2015, 41:479–90; her work in Panama on megalodon nurseries: C. Pimiento et al., *Journal of Paleontology* 2013, 87:755–74, and C. Pimiento et al., *PLOS One* 2010, 5:e10552.

**Megalodon: the ultimate predator.** The Saitama megalodon and other specimens known from more than teeth are all listed in Tyler Greenfield's paper on skeletal materials of megatoothed sharks, *Paleoichthys* 2022, 4:1–9; megalodon bite marks on *Piscobalaena* skeletal remains: A. Collareta et al., *Palaeogeography, Palaeoclimatology, Palaeoecology* 2017, 469:84–91; megalodon attacking sperm whales: A. Benites-Palomino et al., *Proceedings of the Royal Society B* 2022:rspb.2022.0774; on trophic levels using isotopes: J. McCormack et al., *Nature Communications* 2022, 13:2980. New paper on warm-blooded megalodon: L. Griffiths et al., *Proceedings of the National Academy of Sciences* 2023: 120, e2218153120; Endothermic physiology of extinct megatooth sharks: H. Ferrón et al., *PLOS One* 2017, 12:e0185185; intrauterine cannabalism theory: K. Shimada et al., *Historical Biology* 2021, 33:2543–59.

**Megalodon: end of the good life.** Pliocene cooling: A. Fedorov et al., *Nature* 2013, 496:43–49; Great American Biotic Interchange: L. Domingo et al., *Scientific Reports* 2020, 10:1608; megalodon extinction: see C. Pimiento et al., *Journal of Biogeography* 2016, 43:1645–55; on Pliocene marine mass extinction event: C. Pimiento et al., *Nature Ecology & Evolution* 2017, 1:1100–1106; also on extinction of megalodon: R. Boessnecker et al., *PeerJ* 2019, 7:e6088.

**Megalodons in archaeology and history.** From Calvert Cliffs, Hopewell Farm, Ohio, North America: D. Lowery et al., *Archaeology of Eastern North America* 2011, 39:93–108; P. Vogt et al., *Smithsonian Contributions to Paleobiology*, 2018, 100: 3-44; Mayan use of megalodon teeth: S. Newman, *Antiquity* 2016, 90:1522–36. On Pliny the Elder's reference to sharks' teeth and medieval uses of fossil shark teeth, see Chris Duffin's 2017 paper "Fossil Sharks' Teeth as Alexopharmics" (antidotes to poisons), *Toxicology in the Middle Ages and Renaissance*, 2017 (http:I/dx.&ri.org/lo,ltll6ftl9784-I2-i0955&r.00011-3), also his 2019 paper "Snakes' Tongues, Serpents' Eyes and Sealed Earths: Geology and Medicine in Malta," Occasional paper of the St. John Historical Society, Clerkenwall, London, 16pp. Chris Duffin's paper on the *Natterzungen: Jewellery History Today* 14, Spring 2012: 3–5. Background on Steno's life: see Alan Cutler's book *The Seashell on the Mountaintop* (New York: Penguin/Dutton, 2003); Steno and the shark jaws, and Verstegen: J. Davidson, *Proceedings of the Academy of Natural Sciences of Philadelphia* 2000, 150:329–44. I also visited the Stensen Institute in Florence, Italy, to look at some of his works in 2010.

**Hunting for megalodons:** Megalodon hunting in South Carolina: rockseeker .com/where-to-find-megalodon-teeth-in-south-carolina/; see also Rodney Fox's account of hunting for megalodons with Vito in his excellent book *Sharks, the*

*Sea and Me* (cited above). About Vito and his megalodon jaws: theworldslargest sharksjaw.com. **Megalodon fossil remains:** Royal Belgian Institute of Natural Sciences megalodon skeleton: J. Cooper et al., *Science Advances* 2022, 8:eabm9424; recent work by Jack Cooper on megalodon body length: P. Sternes et al. (as cited above); megalodon coprolites: A. Hunt et al., *New Mexico Museum of Natural History and Sciences, Bulletin,* 2012, 57:301; megalodon scales: H. Nishimoto et al., *Mizunami-shi Kaseki Hakubutsukan kenkyu hokoku* 1992, 19:269–72; largest megalodon teeth reliably measured at 16.8 cm, 6.61 inches height: M. Gottfried et al., in A. Klimley and D. Ainley, eds., *Great White Sharks* (cited above); although 7-inch specimens measured along the diagonal have been reported for sale on sellers' websites: fossilera.com/ fossils-for-sale/fossil-megalodon-teeth. Associated megalodon jaws and other fossil remains from Chile and the United States: V. Perez et al., *Palaeontologia Electronica* 2021, 24.1.a09, and from Japan and elsewhere: T. Greenfield, *Paleoichthyes* 2022, 4:109.

**Hunting for megalodons.** I made several collecting trips to Hamilton but met Lionel Elmore only once, around 1970 or 1971, on a trip with my cousin and his parents. My visits to Sharktooth Hill near Bakersfield all took place between 2010 and 2012. Vito's story is from the NBC News site: nbcnews .com/id/wbna43110913; Calvert Museum with a smaller version of Vito's jaws: calvertmarinemuseum.com/DocumentCenter/View/3220/Ecphora-September -2018. For the experience of diving for megalodon teeth and more about Vito, see also Rodney Fox's excellent account in his 2013 book *Sharks, the Sea and Me* (cited above).

**Could megalodon still be out there?** 2013 mock documentary: imdb.com/title/ tt3100780/; The Meg: imdb.com/title/tt4779682/?ref_=fn_al_tt_1); shark week poll: entertainment.time.com/2013/08/07/discovery-channel-provokes -outrage-with-fake-shark-week-documentary/. See also L. Whiteneck et al., *PLOS One* 2022, 17:e0256842. Exploration of the Mariana Trench is recounted in T. Koslow's 2007 book *The Silent Deep* (cited above); see also my 2016 article at theconversation.com/giant-monster-megalodon-sharks-lurking-in-our-oceans -be-serious-53164. Richard Owen, *Odontography* (London: Hippolyte Baillière, 1840–45); Albert Gunther's *Catalogues* were published by the British Museum of Natural History, London.

**After megalodon.** End-Pliocene extinctions of sharks and whales: C. Pimiento et al., *Nature Ecology & Evolution* 2017, 1:1100–1106.

**Biographical note: Louis Agassiz.** Louis Agassiz created the megatoothed shark species *"Carcharodon" megalodon* in 1843, although his earlier published 1835 work referred to this fossil species as *"Carchias megalotis";* see Arnaud Brignon, *Rivista Italiana di Paleontologia e Stratigrafia* 2021, 127:595–625. The first official naming of the megalodon as a species that contains enough information to conform to zoological nomenclature rules goes back to this publication of the same year: J.-L.-R. Agassiz, *Recherches sur les Poissons Fossiles,* 5ème livraison (Neuchâtel: Petitpierre & Prince (text) and H. Nicolet (plates), 1835).

Much has been written, invoking both praise and derision, about the life of Jean-Louis-Rodolphe Agassiz. He founded the science of paleoichthyology—the study of fossil fishes—between 1833 and 1843, when he published *Recherches sur les Poissons Fossiles,* a five-volume set of tomes (four books of descriptions, one of illustrations). His scientific achievements were highly acclaimed for his detailed and beautifully illustrated accounts of not just fossil sharks and fishes, but also jellyfishes, echinoids, turtles, and various other creatures, as well as geological studies on the alpine glaciers and other topics.

Louis Agassiz was born in 1807 at Motier on Lake Morat in Switzerland. He spent his first ten years under the sole tutelage of his father, also named Louis, a clergyman, and his mother, Rose. In 1817 he was sent to a strict public school in Bienne, and at fifteen he attended the College of Lausanne. Here his nascent love of nature flourished when he gained access to the natural history collections of the cantonal museum. His uncle, a physician in Lausanne, encouraged him to study medicine, so he was sent to the medical school of Zurich. Under the mentorship of Professor Heinrich Rudolf Schinz, he received further encouragement to delve into natural history. Agassiz went on to the University of Heidelberg, where he received formal lectures in zoology, botany, and paleontology. In 1827 he continued his studies at the University of Munich. Agassiz became fascinated with fishes and studied a collection of Brazilian freshwater fishes for a research project, later published as a book in 1829.

His interest in fishes then turned to the study of fossil fishes. Agassiz was clearly gaining attention; at age twenty-five, he was appointed to the position of professor of natural history at the University of Neuchâtel in 1832. Agassiz's set of books came out between 1833 and 1843. The success of the first book no doubt contributed to his winning the prestigious Wollaston Medal (£1,000) from the Geological Society of London in 1834. Agassiz's monumental series was the result of his intimate study of the great collections of fossil fishes held in museums and private collections throughout Britain and Europe. This beautiful publication established the names and provided the first descriptions for many of the fossil sharks discussed throughout this book. They include *Acanthodes, Acrodus, Asteracanthus, Chomotodus, Climatius, Ctenacanthus, Ctenoptychius, Hybodus, Orodus, Pleuracanthus, Ptychodus, Synechodus, Sphenodus,* and *Tristychius,* to name just a few. Agassiz also named a huge number of species that were later referred to other genera. However, even as new and more complete fossils were found, nearly all the species Agassiz named are still valid. These include species of living and fossil sharks including *Cladoselache, Hexanchus, Isurus, Notorynchus, Odontaspis, Orthacanthus, Otodus, Symmorium,* and many, many others. After publishing his set of books on fossil fishes, Agassiz produced yet another large volume in 1846, *The Fossil Fishes of the Old Red Sandstone,* turning his attention solely to the Silurian and Devonian fishes. Today, all these books are quite rare and highly sought after by collectors.

# Chapter 15: White Shark

**January 11, 2023, North Neptune Islands, Australia.** I was on board the MV *Rodney Fox* from January 10 to 14, 2023. More information about their shark diving expeditions can be found at rodneyfox.com.au. The birth of the shark cage diving industry in South Australia is also discussed in Rodney Fox's 2013 book *Sharks, the Sea and Me* (cited above).

**A white shark primer.** Carl Linneaus, *Systema Naturae*, 1758, 10th edition, vol. 1: 1–824; Andrew Smith and naming of white shark: L. Anderson, *Scottish Journal of Geology* 2009, 45:59–68; also D. Jordan, *Copeia* 1928, 166:104; juvenile white shark feeding: R. Grainger et al., *Frontiers in Marine Sciences* 2020:10.3389/fmars.2020.00422; off Mexico: E. Tamburin et al., *Marine Biology* 2020, 167:55; tolerance to heavy metals: L. Merly et al., *Marine Pollution Bulletin* 2019, 142:85–92; bite strength: S. Wroe et al., *Journal of Zoology* 2008, 276:336–42, and S. Whitmarsh et al., *PLOS One* 2019, 14, e0224432; metabolic rates in white sharks: F. Carey et al., *Copeia* 1982, 254–60; metabolism when hunting seals and bony fishes: J. Semmens et al., *Scientific Reports* 2013: 10.1038/srep01471; see also white shark energetics: Y. Watanabe et al., *Journal of Experimental Biology* 2022, 4:jeb185603.

**White sharks on the hunt.** White sharks attacking humpback whale: S. Dines and E. Gennari, *Marine and Freshwater Research* 2019, 71:1205–10; specialized hunting with sun behind them: C. Huveneers et al., *American Naturalist* 2015, 185:680010; hunting like serial killers: M. Martin et al., *Journal of Zoology* 2009, 279:111–18; mistaken identity theory: L. Ryan et al., *Journal of the Royal Society Interface* 2019, 18:0533.

**Mating and growth in white sharks.** White shark age and growth: L. Hamady et al., *PLOS One* 2014, 9:e84006; also A. Andrews and L. Kerr, *Environmental Biology and Fisheries* 2014, 98:971–78; white shark nurseries in the United States: T. Curtis et al., *Scientific Reports* 2018, 8:10794; juvenile white shark movements off Australia: B. Bruce et al., *Marine Ecology Progress Series* 2019, 619:1–15; navigating using sense of smell: A. Nossal et al., *PLOS One* 2016, 11:e0143758; tagging white sharks and their ocean migrations: R. Bonfil et al., *Science* 2005, 310:110–103; Simon and Jekyll white sharks: smithsonianmag .com/smart-news/how-two-great-white-shark-buddies-could-change-perceptions -of-the-species-180982752/; Japanese white shark growth rates: S. Tanaka et al., *Marine and Freshwater Research* 2011, 62:548–56; age of sexual maturity: L. Natason and G. Skomal, *Marine and Freshwater Research* 2015, 66:387–98; white shark café: G. Le Crozier et al., *Environmental Science and Technology* 2020, 54:18872–15882, and article by Sarah Emerson, *Vice*, September 18, 2018.

**The origin of white sharks.** End-Pliocene extinctions: see Catalina Pimiento, *Nature Ecology & Evolution* 2017, 1:1100–1106; white shark genome: N. Marra et al., *Proceedings of the National Academy of Sciences* 2019, 116:4446–55; pathology survey of 1,546 sharks: M. Garner, *Veterinary Pathology* 2013, 50:030098581348; Gordon Hubbell's story is drawn from correspondence with

Gordon, also themarinediaries.com/tmd-blog/dr-gordon-hubbell-bitten-by -the-shark-tooth-bug; *Carcharodon hubbelli* and its age: D. Ehret et al., *Paleontology* 2012, 1139–53; mathematical similarity of *C. hastalis* and *C. carcharias* teeth: K. Nyberg et al., *Journal of Vertebrate Paleontology* 2005, 26:806–14; *Carcharomodus escheri:* J. Kriwet et al., *Acta Palaeontologica Polonica* 2014, 60:857–75; see also D. Ehret et al. 2012 paper cited above; DNA analysis of makos: M. Mehlrose et al., *Mitochondrial DNA B* 2022, 7:652–54; ancient Argentine white shark teeth: A.Cione et al., *Geobios* 2012, 45:167–72; trophic relationships between white sharks and megalodon: J. McCormack et al., *Nature Communications* 2022, 13:2980.

**White Sharks are so hot right now.** White shark deepest dives: S. Andrzejacek et al., *Science Advances* 2022, 8:sciadv.abo1754; white shark rete mirabile: see D. Ebert, *Sharks of the World,* and P. Klimley and D. Ainley, eds., *Great White Sharks,* both cited above; for the rete mirabile in humans: I. Fuwa, *Stroke* 1994, 25:1268–70.

**What white sharks fear.** Orcas attacking white sharks: P. Pyle et al., *Marine Mammal Science* 1999, 15:563–68; South Africa: see A. Towner et al., *Ecology* 2023, ecy.3875; and see *National Geographic* documentary *Killer Shark vs. Killer Whale,* 2021; orcas attacking white sharks off South Australia: abc.net.au/news/2015 -02-09/sharks-evacuate-neptune-islands-after-killer-whales-attack/6080788; white shark copepod parasites: G. Benz et al., *Pacific Science* 2003, 57:39–43; M. Palombi et al., *Frontiers in Marine Sciences* 2021, 8; sharks scraping parasites off at Cape Cod: abc.net.au/news/2015-02-09/sharks-evacuate-neptune-islands -after-killer-whales-attack/6080788; tapeworms in white sharks: H. Randhawa, *Journal of Parasitology* 2011, 275–80, and M. Daily and W. Vogelbein, *Journal of the Heminthological Society of Washington* 1990, 57:108–12; post-*Jaws* killing of sharks: https://www.theinertia.com/environment/steven-spielberg-regrets -how-jaws-shaped-the-publics-view-of-sharks/; white shark decline in North Atlantic post-*Jaws:* G. Burgess et al., *Fisheries* 2005, 30 (10):19–26; also J. Baum et al., *Science* 2003, 299:389–92.

**White sharks in our hearts and minds.** Hardwicke Rawnsley, "McDermott's Deed": see V. Westbrook et al., *Sharks in the Arts* (cited above); John Singleton Copley's and Winslow Homer's shark paintings: Westbrook, *Sharks in the Arts;* Andrea Everhart: martinandmacarthur.com/collections/andrea-everhart/ products/bronze-sculpture-ocean-beauty-by-andrea-everhart. Birpai people shark totem: mnclibrary.org.au/research/birpai-studies/totems-of-the-birpai/.

**The largest white sharks and their future.** Average size of white sharks: M. Parrish, "How Big are Great White Sharks?" Smithsonian National Museum of Natural History Ocean Portal, ocean.si.edu/ocean-life/sharks-rays/how-big -are-great-white-sharks, retrieved April 19, 2023; also J. Castro, 2012, pp. 85–90 in M. Domeier, *Global Perspectives* (cited above); Cojimar Monster details: Susan Beegel's excellent essay on the "Monster of Cojimar" details the background, capture, mensuration process, and follow-up research on this large white shark, and dwells on Hemingway's emotional response to its capture, in

*Hemingway Review* 2015, 34:9–35. Deep Blue and Ocean Ramsey: Ocean Ramsey, surfertoday.com/environment/ocean-ramsey-the-shark-whisperer. David Bernvi's estimate of Deep Blue is reported on the wikipedia entry for "Deep Blue (great white shark)"; Bob and Dolly Dyer: reprinted from 1956, vault.si .com/vault/1956/02/20/two-against-the-shark; see also M. McGinness, smh .com.au/national/an-elegant-and-affable-tv-sidekick-20050215-gdkp5k.html; Alf Dean: fishing.net.nz/fishing-advice/general-articles/shark-fishing-legend-alf -dean/; Frank Mundus, Montauk: "Frank Mundus, 82, Dies; Inspired 'Jaws,'" *New York Times* (nytimes.com); Australian and New Zealand white shark populations: R. Hilary et al., *Scientific Reports* 2018, 8:2661; Daniel Botelho: dive photoguide.com/underwater-photography-special-features/article/interview -daniel-botelho-s-out-cage-great-white-shark-photography; Michael Domeier's comments: abc.net.au/news/2019-01-18/conservationist-slammed-for-touching -huge-shark-off-hawaii/10725478; white sharks gaining ground in Canadian Atlantic seas: G. Bastien et al., *Canadian Journal of Fisheries and Aquatic Sciences* 2020, 77:1666–77.

**Future white shark research priorities.** Charlie Huveneers's paper on future priorities: C. Huveneers et al., *Frontiers in Marine Sciences* 2018, 5:00455.

## Chapter 16: Sharks and Humans

**465 million years and counting.** Evidence for the current sixth global mass extinction event taking place right now: Robert Cowie et al., *Biological Reviews* 2022, 97:640–63.

**Sharks as food and goods.** Archaeology of eating sharks, shark tanning leather, etc.: J. Musick, *VIMS Books and Book Chapters,* 2005, 25:243–51, scholarworks .wm.edu/vimsbooks/; shark and ray meat trade value: sharks.panda.org/news -blogs-updates/latest-news/us2-6-billion-global-trade-in-shark-and-ray-meat -revealed-better-rules-and-transparency-needed-to-fight-overexploitation; basking shark fin value: hcas.nova.edu/guy-harvey/forms/news-wildlife-extra -basking-sharks-threatened-by-the-shark-fin-trade.pdf; UK shark fin bill: commonslibrary.parliament.uk/research-briefings/cbp-9591; sharks threatened by overfishing: N. Dulvy et al., *Current Biology* 31:4773–87; Hong Kong shark fins: D. Cardeñosa et al., *Animal Conservation* 2019, 23:203–11; sustainable shark fisheries data: N. Dulvy et al., *Current Biology* 2017, 27:R565–R572; U.S. bycatch report: media.fisheries.noaa.gov/dam-migration/nbr_update_3.pdf; Victoria, Australia, fisheries shark catch: *Victorian Fisheries Authority Commercial Fish Production, Information Bulletin 2021*, Victorian Fisheries Authority, Queenscliff, Victoria, Australia, ISSN 1446-2567.

**Shark "attack" statistics.** Florida Museum shark attack files: floridamuseum.ufl .edu/shark-attacks/. Shark deterrent research: C. Huveneers et al., *PeerJ* 2018, peerj.5554, and A. Gauthier et al., *Scientific Reports* 2020, 10:17869.

**Sharks and human health.** Aaron LeBeau and nurse shark immunity: washington post.com/kidspost/2022/03/08/shark-antibodies-fight-human-diseases and original research by O. Ubah et al., *Nature Communications* 2021, 12:7325; biomimetic study of shark skin flow dynamics: S. Fan et al., *Sustainability* 2022, 14:16662; whale shark eyes: K. Yamaguchi et al., *Proceedings of the National Academy of Sciences* 2023, 120:e2220728180.

**Sharks and climate change.** Epaulette sharks and rising sea temperatures: C. Wheeler et al., *Scientific Reports* 2021, 11:454; for further information on sharks facing climate change, see also G. Osgood et al., *Journal of Animal Ecology* 2021, 90:2547–59, and P. Diaz-Carballido et al., *Frontiers in Marine Sciences* 2021, 8:745501.

**Sharks and ocean pollution.** The great Pacific garbage patch: theoceancleanup .com/great-pacific-garbage-patch; see also R. Hale et al., *JGR Oceans* 2020, 125:e2018JC014719; rates of plastic accumulation: L. Lebreton et al., *Scientific Reports* 2018, 8:4666. All the plastics ever made: R. Geyer et al., *Science Advances* 2017, 3:1700782; shark survey and microplastics: K. Parton et al., *Scientific Reports* 2020, 10:12204; manta rays and microplastics: E. Germanov et al., *Frontiers in Marine Sciences* 2019, 19:00679; ghost nets and sharks: L. Lipej et al., *Plastic Pollution and Marine Conservation* (London: Academic Press, 2022), chapter 5, 153–85.

**The future of sharks.** Overfishing driving extinction of sharks: N. Dulvy et al., *Current Biology* 2021, 31:4773–87.e8; sharks' services on reefs: conservation .reefcause.com/why-are-sharks-important-for-healthy-coral-reefs/; reef sharks endangered: A. Macneil et al., *Nature* 2020, 583:801–806. Great Barrier Reef under threat: washingtonpost.com/climate-environment/2022/11/29/great -barrier-reef-australia-unesco-danger; daggernose shark extinction threat: R. Lessa et al., *Global Ecology and Conservation* 2016, 7:70–81; Ganges shark: A. Haque and S. Das, *Endangered Species Research* 2019, 40:65–73; with extinct species of *Glyphis:* from H. Cappetta's 2012 *Handbook* (cited above). *Glyphis glyphis* is registered as "critically endangered" in all Australian waters since 1999 on the Commonwealth Environment Protection and Australian Biodiversity acts. Turtles, seagrasses in Shark Bay, David Attenborough quote: watoday.com.au/ national/western-australia/sir-david-attenborough-says-wa-tiger-sharks-help -fight-climate-change-20180327-p4z6ie.html.

**Fossil sharks helping to save today's sharks.** The story of Vinnie Valle and the shark tooth necklaces was relayed to me by David Ward, who knows Vinnie well and has made the calculation of how many sharks are saved by the replacement of modern sharks' teeth with fossil ones. For more information on the products, see the V&L Crafts Facebook page.

**Shark tourism and the future.** Neptune Island shark tourism sustainability: Y. Niella et al., *Marine Policy* 2023, 105362; Canadian study: A. Cisneros-Montemayor et al., *Oryx* 2013, 47: 381-88; manta ray tourism and value: M. O'Malley et al., *PLOS One* 2013, 8:e65051; reduction in manta ray hunting at Lamakera: H. Booth et al., *Conservation Science and Practice* 2020, 3:3314; value of basking shark tourism in Scotland and Ireland: C. Gray et al., *Aquatic Conservation, Marine and Freshwater Ecosystems* 2022, 32:537–50; Scottish ma-

rine parks: gov.scot/policies/marine-environment/marine-protected-areas/;
Irish protection of basking sharks: irishmirror.ie/news/irish-news/basking
-sharks-now-protected-irish-28139384.

**A ray of hope.** Lauren Meyer, saveourseas.com/project-leader/lauren-meyer, and
Otlet, otlet.io/about-us.

## Epilogue: The Wisdom of Sharks

**Age of Greenland shark and sexual maturity.** J. Nielsen et al., *Science* 2016,
353:702–704.

# SCIENTIFIC NAMES

angel shark *Squatina* (several species)
basking shark *Cetorhinus maximus*
blacktip reef shark *Carcharhinus melanopterus*
bluntnose sixgill shark *Hexanchus rufogriseus*
bonnethead shark *Sphyrna tiburo*
bramble shark *Echinorhinus brucus*
broadnose sevengill shark *Notorynchus cepedianus*
bronze whaler (copper shark) *Carcharhinus brachyurus*
bull shark *Carcharhinus leucas*
bullhead shark (horn shark) *Heterodontus* (several species)
Caribbean reef shark *Carcharhinus perezi*
catsharks, many kinds; *Scyliorhinus* is a typical form
common thresher *Alopias vulpinus*
cookiecutter shark *Isistius brasiliensis*
daggernose shark *Isogomphodon oxyrhinchus*
deep-sea catsharks, many genera, e.g. *Apristurus, Galeus*
deep-sea lantern shark *Etmopterus* (several species)
deep-sea sand tiger shark *Odontaspis ferox*
dusky shark *Carcharhinus obscurus*
dwarf lantern shark *Etmopterus perryi*
eagle rays *Myliobatis* (several species)
electric ray (numbfish) *Torpedo* (several species)
elephantfish *Callorhinchus milii*
epaulette shark *Hemiscyllium oscellatum*
frilled shark *Chlamydoselachus anguineus*
Galapagos shark *Carcharhinus galapagensis*
Ganges River shark *Glyphis gangeticus*
goblin shark *Mitsukurina owstoni*
gray reef shark *Carcharhinus amblyrhynchos*
great hammerhead *Sphyrna mokarran*
Greenland shark *Somniosus microcephalus*
guitarfish *Rhinobatos, Pseudobatos, Acroteriobatus*

**houndshark,** many kinds, e.g. *Triakis, Mustelus, Galeorhinus*

**kitefin shark** *Dalatias* (several species)

**lemon shark** *Negaprion brevirostris*

**mako shark** *Isurus oxyrinchus* (shortfin mako)

**manta ray** *Mobula* (several species)

**milk shark** and **sharpnose shark** *Rhizoprionodon*

**nurse shark** *Ginglymostoma cirratum*

**piked dogfish** *Squalus acanthias*

**porbeagle shark** *Lamna nasus*

**ratfish** *Hydrolagus* (several species)

**salmon shark** *Lamna ditropsis*

**sand tiger shark (grey nurse)** *Carcharias taurus*

**sawfish** *Pristis* (several species)

**sawshark** *Pristiophorus, Pliotrema*

**sharpnose sevengill shark** *Heptranchias perlo*

**short-tailed ray** *Bathytoshia brevicaudata*

**sixgill shark** *Hexanchus* (several species)

**skates** *Raja* (and several other genera)

**small-spotted catshark** *Scyliorhinus canicular*

**smooth hammerhead** *Sphyrna zygaena*

**smooth ray** *Dasyatis brevicaudata*

**spookfish** *Hariotta* (several species)

**tiger shark** *Galeocerdo cuvier*

**thornback guitarfish** *Platyrhinoidis*

**thresher shark** *Alopias* (several species)

**whale shark** *Rhinodon typicus*

**"whaler" shark, requiem, or ground shark** *Carcharhinus* (several species)

**whiptail stingray** *Dasyatis* (several species)

**white shark (great white shark)** *Carcharodon carcharias*

**whitetip reef shark** *Triaenodon obesus*

**winghead shark** *Eusphyrna blochii*

**wobbegong** *Orectolobus* (several species)

**zebra shark** *Stegostoma tigrinum*

# GLOSSARY OF TERMS

Formal classification terms are given first, with informal names in parentheses. Not all terms in bold in the book are listed here, as some are adequately explained in the sentences following. The terms listed here are in general more complex.

**Acanthodii** (acanthodians). A group of extinct fishes characterized by having fin spines in front of all their fins except the tail fin; some have other spines sticking out of the body or underneath. They are now placed as a group on the stem leading to all other chondrichthyans, so they are nicknamed "stem sharks."

**Ammonita** (ammonites). Group of cephalopod mollusks, squid-like with tentacles, that lived in mostly coiled shells with complex suture patterns where the chamber wall meets the shell.

**Ampullae of Lorenzini**. Special flask-shaped cells full of nerve bundles that detect weak electric currents of other living organsims via a gel-filled canal exposed to the water.

**Anacoracidae** (anacoracids, or crow sharks). Large lamniform sharks of the mid-late Cretaceous period, including many species of *Squalicorax*.

**Antiarchi** (antiarchs). Group of placoderm fishes with long trunk shield, short head shield, and paired bony pectoral appendages. Includes *Bothriolepis, Microbrachius*.

**Arthrodira** (arthrodires). Group of placoderm fishes with jointed necks and generally with two upper pairs of tooth plates. Includes *Dunkleosteus* and common forms like *Coccosteus*.

**Batoidea** (batoids, rays, and skates). Superorder of chondrichthyans within the Elasmobranchii containing all rays and skates and sawfish. They have gill slits on the underside of the head and crushing dentitions made of pavement teeth or very small pointed teeth in some cases.

**Belemnitida** (belemnites). A group of extinct squid-like cephalopod mollusks with a hard guard structure supporting the body; they lived from the Late Triassic to the end of the Cretaceous period.

**Biotic crisis.** A major event where life on Earth decreased in abundance or diversity through rapid environmental changes or extraneous events, but not necessarily on a global scale or as devastating as global mass extinction events. E.g., Kellerwasser event, Paleocene-Eocene Thermal Maximum event.

**Biostratigraphy.** The practice of using relative age ranges of certain key fossil species (often microfossils), by marking their first appearance or disappearance, to determine an approximate or relative age of a rock layer. Such key fossils must be dated using radiometric methods in other sections to tie in their absolute age ranges.

**Brachiopoda** (brachiopods, lamp shells). Hard-shelled bivalved invertebrate animal that filter-feeds using a tentacle-like structure. Their fossils are most abundant in Paleozoic marine rocks.

**Bryozoa** (bryozoan, sea mosses). Fan-like colonial organisms that build a hard exoskeleton housing many tiny chambers where each coral-like creature filter feeds.

**Bundled enameloids.** Tissues of the outer surface of the teeth in sharks (i.e., enameloids) that are composed of irregular bundles of crystallites forming the layer.

**Carcharhiniformes** (carcharhiniforms). Most diverse group of living sharks (c. 291 species in 10 families), including requiem sharks, catsharks, deepsea catsharks, houndsharks, weasel sharks, tiger sharks, hammerheads, and some other groups. They have their largest teeth on the side of the mouth, not up at the front.

**Carchariidae** (carchariid). Sand tiger sharks, a group of lamniforms represented by one living form, *Carcharias taurus*. Many extinct forms.

**Cardabiodontidae** (cardabiodontids). Extinct group of large mid-late Cretaceous lamniform sharks typified by forms like *Cardabiodon*, reaching up to 26 feet in length.

**Ceratitidae** (ceratitids). Family of ammonites (see above) that had highly ornamented shells and thrived in the late Triassic times.

**Chimaeridae** (chimaerids). Family of short-nosed holocephalans in two genera, *Hydrolagus* and *Chimaera*.

**Chimaeriformes** (chimaeriforms). A group within the Holocephali containing chimaeras, ratfishes, spookfishes, and the elephantfish, plus extinct members that are on the stem to this group (e.g., *Echinochimaera*).

**Chondrichthyes** (chondrichthyans). Class of cartilaginous fishes that includes all jawed fishes having a cartilaginous skeleton made of prismatic cartilage: sharks, holocephalans, rays, and many extinct groups that also shared the same type of cartilage.

**Chronospecies.** Evolutionary series of successive species that seemingly grade into one another over time; each species separated mainly by time and slight morphological changes.

**Cladodont.** A type of shark tooth made up of several tusk-like cusps on a rounded base (root), often with smaller cusps between them, after *Cladodus*.

**Cladogram.** Diagram showing inferred relationships of taxa (usually species or genera) based on calculations of similarity of character traits, scored according to basal (generalized, primitive) or derived (apomorphic) conditions, and generated using one of many phylogenetic software programs (e.g., PAUP*, MrBayes, Mesquite, BEAST, and others).

**Cladoselachida** (cladoselachians). A group containing the Devonian sharks *Cladoselache* and *Maghriboselache*, basal sharks with fin spines and cladodont-style teeth, now thought to be on the stem leading to holocephalans.

**Clumped isotope paleothermometry.** Clumped isotopes are heavy isotopes of certain elements like Carbon and Oxygen that bond together with other heavy isotopes. Measuring the temperatures at which these bonds form in living species, in dentine or enamel formation of the teeth, can give a prediction of past temperatures for the biological metabolism of extinct organisms.

**Cochliodontida** (cochliodonts). A group of extinct late Paleozoic holocephalans with peculiar flat teeth.

**Conchostracans.** "Clam shrimps"—small bivaled freshwater-to-brackish crustaceans.

**Ctenacanthidae** (ctenacanths). Group of moderate to large Paleozoic sharks, on the main shark line to modern sharks, with robust cladodont teeth (Devonian-Permian), e.g., *Ctenacanthus, Saivodus, Dracoporistis, Glikmanius.*

**Dasyatidae** (dasyatids). Whiptail stingrays with venomous tail stingers. Includes the common stingray *Dasyatis* with over 40 living and fossil species.

**Deep time.** When time is measured in parcels of millions to billions of years, as for geological and astronomical events.

**Dental lamina.** The epithelial organ below the toothline in animals that gives rise to a cascade of events that makes the teeth. In sharks it produces new teeth like from a factory line.

**Echinorhinidae** (echinorhinids). Bramble and prickly-skin sharks characterized by large, rough placoid scales.

**Elasmobranchii.** Group containing all modern sharks and rays. See John Maisey's paper (*Journal of Fish Biology* 2012, 03245) for a detailed appraisal of the term and its historical uses.

**Enameloid.** Outer shiny layer of tooth in sharks and fishes; not true enamel (as in reptiles and mammals) as the crystallites forming it are not perfectly parallel but irregular, often bundled or clumped. Also called durodentine or vitrodentine.

**Galeomorphii** (galeomorphs). Superorder of sharks including Heterodontiformes, Orectolobiformes, Lamniformes, and Carcharhiniformes.

**Ginglymostomatidae** (ginglymostomatids). Family containing the nurse sharks like *Ginglymostoma*; it includes three living genera and four species, with many extinct forms.

**Gymnotiformes** (gymnotiforms). Teleostean fishes also called knife fishes, which have long bodies and swim using their anal fin. They can produce electric pulses to detect prey, or, as in the electric eel, can use electricity to stun prey.

**Hemiscyllidae** (hemiscyllids). Family of orectolobiform sharks called the long-tailed carpet sharks, including many small forms like *Chilloscyllium* that dwell in intertidal pools and shallow sea environments.

**Heterodont.** Having jaws where the upper and lower teeth differ in shape, size, or both.

**Heterodontiformes** (heterodontiforms). The order of sharks containing the bullhead sharks, today represented by *Heterodontus* with nine species, but including many extinct forms dating back to the Jurassic period.

**Hexanchidae** (hexanchids). The family containing the five species of living sixgill and sevengill sharks, such as *Hexanchus* and *Notorhynchus*. Regarded as the most primitive group of living sharks.

**Hexanchiformes** (hexanchiforms). Primitive living order of sharks containing the hexanchids plus *Chlamydoselachus*, the frilled shark.

**Holocephali** (holocephalans). Group of chondrichthyans with opercular flaps covering gills, crushing tooth plates, head claspers (tentaculum) in males, and upper jaws fused to the braincase (e.g., chimaeras, spookfishes, ratfishes, elephantfishes).

**Homebox genes** (hox genes). Blueprint genes that develop certain patterns in our skeletons and organs through transcription and morphogenesis functions. The same genes can develop consistent patterns throughout evolution, e.g., being responsible for limb development in fishes, sharks, and land animals.

**Hybodontidae** (hybodonts). A family of extinct sharks characterized by their mostly small bodies, two robust dorsal fin spines, broad teeth, and conspicuous head spines above the eyes in most forms, e.g., *Hybodus, Asteracanthus, Strophodus, Lonchidion.*

**Hybodontiformes** (hybodontiforms). Order of extinct shark characterized by their variable teeth (from robust pointed forms to flat pavement teeth). Includes earliest forms like *Tristychius* and *Diplodoselache* from Scotland.

**Ichthyodectidae** (ichthyodectids). Group of medium to large teleostean fishes that became the largest bony fish predators of the Cretaceous period, with some forms reaching up to 20 feet long.

**Ichthyosauria** (ichthyosaurs). Marine reptiles of the Mesozoic era that mostly developed dolphin-like bodies with long tooth-lined snouts for eating fish and squid. Some giant forms developed as toothless filter feeders.

**Iniopterygia** (iniopterygiforms, iniopterygian). Group of small, extinct bizarre-shaped stem holocephalans of the Carboniferous period having large wing-like pectoral fins emerging from behind their necks.

*Lagerstätte* (plural *lagerstätten*). Meaning "storage place," fossil sites showing exceptionally good preservation of organisms, including whole complete individuals in complete articulation, sometimes with soft tissues or color markings preserved intact, e.g., Solenhofen, Germany.

**Lamniformes** (lamniforms). Order of sharks including 15 living species, and many extinct forms, such as mackeral sharks, white shark, basking shark, and others, characterized by their barrel-shaped bodies, conical heads, large mouths extending behind the eyes, and osteodentine tooth structure.

**Lycophyta** (lycophytes). Group of vascular plants including the horsetails and club mosses, today known by some 1,300 mostly small species, but forming large forests of *Lepidodendron* and related forms in the Late Devonian and Carboniferous.

**Manuport.** Archaeological term for a natural object relocated from its source of origin without further information.

**Meckel's cartilage.** The cartilage that forms the entire lower jaw in sharks and becomes the core for lower jaw ossification in bony fishes and land animals.

**Morphometrics.** The study of biological features using the variation of shapes of structures, like bones, teeth, or organs, by applying rigorous mathematical methods.

**Mosasauria** (mosasaur). Extinct group of Late Cretaceous medium to very large sea-going reptiles related to the living monitor lizards, e.g., *Mosasaurus, Tylosaurus.*

**Myliobatidae** (myliobatids). Family of eagle rays including common forms like *Myliobatis,* with characteristic elongated replacement teeth.

**Mysticeti** (mysticetes). Group of baleen or filter-feeding whales, often very large, like the blue whale or humpback whale, that lack teeth; some early fossil forms still had teeth, e.g., *Janjucetus, Balaenoptera.*

**Neoselachii** (neoselachians). A group that evolved in the Triassic that includes all modern sharks and rays plus some early fossil species that may not be directly related to the modern sharks. They are defined by having three distinct layers of enameloid on the teeth (see G. Cuny et al., 2017, in Notes on Sources). John Maisey (see reference above under "Elasmobranchii") regards the term as synonomous with Elasmobranchii.

**Nudibranchs.** Sea butterflies, naked slug-like mollusks that swim through the water, often brightly colored.

**Odontoceti** (odontocetes). Group of extinct toothed whales and dolphins, including forms like the sperm whale, orca, and Pacific dolphin.

**Odontodes.** Tiny tooth-like units made of dentine around a pulp cavity. They form the scales or adorning fin-spines, or as denticles lining the inside of the mouth in sharks and other fishes. Not replaced in the same way as true teeth.

**Orectolobiformes** (orectolobiforms). A major order of sharks containing some 45 living species, including the carpet sharks, blind sharks, whale sharks, wobbegongs, nurse, and zebra sharks, plus many extinct kinds. The nostrils have barbels coming off them.

**Orthodentine.** Dentine tissue forming most of the shark's tooth, having conspicuous vascular canals from the pulp cavity running through it.

**Osteichthyes** (osteichthyans). Bony fishes (those jawed fishes with skeletons made of bone). Includes two major groups, the Actinopterygii (ray-finned fishes such as salmon and marlin) and the Sarcopterygii (lobe-finned fishes such as coelacanths and lungfishes).

**Osteoblasts.** Bone cells, seen in fossils as well-defined spaces with radiating tendrils.

**Osteodentine.** Dentine tissue forming most of the shark's tooth that is dense, lacking visible large vascular canals, or with extra dentine infilling pulp cavity regions.

**Otodontidae** (otodontids). The family containing *Otodus* and closely related extinct lamniform sharks, e.g., *Otodus megalodon*.

**Paraselachii.** An extinct group containing holocephalans and stem holocephalan groups like the petalodont, cochliodont, and menaspid groups.

**Parasphenoid.** Median toothed or smooth bone of the palate in many groups of fishes and land animals; it is absent in sharks.

**Petalodontida** (petalodonts). A group of extinct Late Paleozoic holocephalans with peculiar flat teeth and long roots. Each individual had very few teeth in each jaw (three to five).

**Phylogenetic position.** The relative position in the evolutionary tree of life where the degree of similarity is determined between one species and another species (or genus, family, etc.).

**Placodermi** (placoderms). An extinct group of armored shark-like jawed fishes with bony overlapping plates covering the head and trunk area. Reproduced by internal fertilization.

**Placoid scales.** Type of very small (generally under 1 mm) scale found on sharks, with a thin, flat base or root and a crown with neck canals. In some living forms the base can be secondarily lost.

**Platyrhinoidea** (platyrhinoids). Group of rays called the guitar fishes.

**Plesiosauridae** (plesiosaurs). Extinct small to large marine reptiles with long necks, small heads, and long tails, primarily fish and squid eaters, e.g., *Plesiosaurus, Elasmosaurus*.

**Pliosauridae** (pliosaur). Extinct medium to very large predatory marine reptiles with long crocodile like-heads and short tails, incuding the largest marine predators of the Jurassic period, e.g., *Liopleurodon*.

**Pristidae** (pristids). The sawfishes, a group of medium to very large rays with long snouts lined with pointed teeth along the sides; the gill slits are underneath the head, e.g., *Pristis*.

**Pristiophoridae** (pristiophorids). Small saw sharks, having long snouts with side teeth and gill slits exposed on the sides of the head.

**Psilophytopstida** (psilophytes). Primitive low-growing vascular plants lacking true roots and leaves, common in the Silurian-Devonian.

**Ptyctodontida** (ptyctodontids). A group of long-bodied, viviparous placoderms with large eyes and robust crushing toothplates, e.g., *Materpiscis*.

**Pycnocline.** Geological term for a stable density gradient of water bodies that sit on top of each other in layers, created by differences in salinity levels, temperature variations, and other factors.

**Quadrate.** The bone or cartilage that forms the upper jaw hinge joint in sharks, fishes, and lower vertebrates. In mammals it is reduced to become the malleus bone in the inner ear.

**Rajidae** (rajids). Rays of the skate family, including some 150 species of living forms and many extinct species dating back to the Late Cretaceous period.

**Regional endothermy.** The ability to keep a higher body temperature than the surrounding environment due to a higher metabolic rate; equates to warm-bloodedness in some very large sharks, like white sharks and the extinct megalodon.

**Rete mirabile.** Meaning "wonderful net," a complex of closely placed arteries and veins for exchanging heat between vessels, maintaining warmer body temperatures; found in white sharks, threshers, and makos.

**Rhinobatidae** (rhinobatids). The family of rays known as guitarfish or shovelnose rays, having an elongated body and small ray-like pectoral fins, e.g., *Rhinobatos*.

**Rosetta stone specimen.** Named after the famous Rosetta Stone, which helped decipher ancient languages, such fossils help resolve multiple scientific names of fossils that are really all attributed to just one species.

**Scapulocoracoid.** The cartilage element forming the shoulder girdle in all chondrichthyans, or bone supporting the pectoral fin or limb in osteichthyans and most tetrapods.

**Sclerorhynchoidei** (sclerorhynchids). An extinct group of medium to large Late Cretaceous rajiform rays with long snouts (rostra) armed with robust side teeth, similar to sawfishes.

**Selachii** (selachians). A group within the elasmobranchs containing all typical sharks (excludes rays).

**Sinacanthidae** (sinacanthids). Extinct group of stem-chondrichthyans commonly found in the Silurian period, featuring broad, short fin spines with many fine ridges adorning them.

**Sox2 gene.** A homeobox gene (see above) that develops many kinds of tissues from embryonic stem cells. It is turned on when sharks make their teeth and taste buds, suggesting teeth might have first formed from ancestral taste buds.

**Squaliformes** (squaliforms). Order of sharks containing about 135 living species in six families, such as sleeper sharks, dogfish, gulper sharks, kitefin and lantern sharks, and cookiecutter sharks, as well as many extinct forms.

**Squalodontidae** (squalodonts). Extinct group of predatory shark-toothed dolphins and small whales (up to 20 feet long) bearing multicuspid teeth with coarse serrated edges, e.g., *Squalodon, Eosqualodon*.

**Squalomorphii** (squalomorphs). Superorder of sharks including Hexanchiformes, Squaliformes, Squatiniformes, and Pristiophoriformes.

**Stem.** A "stem taxon," or group of taxa that are on the stem or branch leading to the main branch of a living group of organisms (called the **crown group**). We regard the extinct acanthodians as containing taxa with some on the stem to other chondrichthyans (hence "stem chondricthyans," or stem sharks) with other taxa not on the stem.

**Stromatoporoidea** (stromatoporoids). An extinct group of marine sponges built of hard calcite layers; they formed huge reefs in the Devonian period.

**Symmoriiformes** (symmoriiforms). An order of extinct Paleozoic sharks on the stem leading to holocephalans, characterized by large eyes, whip-like extensions on tails, and dorsal fin spines, which can be developed as overhanging "swords" (e.g., *Falcatus*) or robust "anvils" in some cases (e.g., in stethacanthids like *Akmonistion*).

**Synechodontiformes** (synechodontiforms). Common group of extinct sharks that lived in the Mesozoic era with long bodies and terminal mouths, defined by features of their tooth and root histology, e.g., *Synechodus, Palaeospinax*.

**Teleostei** (teleosteans, teleosts). Bony ray-finned fishes that have a robust tail skeleton and protruding jaws. They are the most common group of fishes alive today, e.g., trout, salmon, swordfish, cod, groupers, eels, halibut.

**Temnospondyls.** An extinct group of large-headed, salamander-like amphibians.

**Theropoda** (theropod). Large group of mostly predatory dinosaurs, including all forms with three-toed feet and hollow bones, e.g., *Tyrannosaurus*.

**Thylacocephalans.** A group of large predatory arthropods, possibly crustaceans, with large bulbous eyes and long pairs of hunting appendages.

**Trophic ecology.** Research that determines the hierarchy of the food chain in an ecosystem, from the apex predators at the top of the food chain to base-level organisms like plankton.

**Viviparity** (viviparous). When creatures give birth to live young rather than laying eggs.

**Xenacanthidae** (xenacanthid, xenacanth). Family of extinct, mostly freshwater sharks with prominent neck spines and pronged teeth. See next two entries for higher-classification-level terms.

**Xenacanthiformes** (xenacanthiforms). Much as for Xenacanthimorpha, except they have three cusps on each tooth with the middle one the smallest, and lacking inverted V-nesting ornament on teeth.

**Xenacanthimorpha** (xenacanthimorphs). Group of extinct eel-shaped selachian sharks living mostly in freshwater or estuarine habitats. The group is defined by their teeth having a pronged crown with two or three main cusps, and a root with a coronal button and a basal tubercle under the root. Most had large neck spines.

# INDEX

Page numbers in *italics* refer to illustrations and photographs

hunting for, 7, *7*, 323–26
origin of, 290–95, 306–7
reconstruction, 312–13, *313*
seas off Peru, 308–10
size of, 310–15, *312*, 326–28, *327*
teeth, 7, *7*, 290–91, *306*, 306–7,
   315–17, *316*, 319, 320–21, *323*
as ultimate predator, 315–18
urban myths, 326–28
*Otodus obliquus*, 274, 291–95, 301–3,
   *302*, 350–51
*Otodus sokolovi*, 303–4
Oukherbouch, Said, 75
Owen, Richard, 326–28


paddlefish, 155, 162
Padua University, 241
*Pakicetus*, 289, 296
*Palaeeudyptes*, 295
*Palaeocarcharias*, 218, *220*, 223
*Palaeocarcharodon*, 302
*Palaeorhincodon*, 276–77
Paleocene-Eocene Thermal Maximum
   event, 274–76
Paleocene period, *13*, 271–79
   Bolca sharks, 279–84
   global warming, 276–79
paleontology, overview, 12–15
Paleo-Tethys Ocean, 150–51, 166–67,
   *199*, 230
Paleozoic era. *See also specific periods*
   shark evolutionary tree, *13*, 190–93,
   *192*, 365
*Paleozoic Fishes of North America, The*
   (Newberry), 97
Panama, 314–15, 319
Pangaea, 200, 201, 243
Pangaea sea, 166–69
Panthalassa Sea, 87, 166, 178, 243
Papazzoni, Cesare, 279
*Paracestracion*, 230
*Paracyclotosaurus*, 214
*Paraorthacodus*, 221

*Pararhincodon*, 246
parascyllids, 246
parasphenoid lining, 97
Paris Natural History Museum, 280
Patterson, Colin, 421*n*
Pauling, Linus, 180
pectoral fins, 26, *26*, 52, 54
*pejus* temperature, 374
pelvic fins, 26, 29, 96, 108–11, 123,
   128, 220, 247
penguins, 295–96, *296*
Pennsylvania
   Red Hill site, 94–95
Pennsylvanian epoch, 125
Permian Chert Event, 167
Permian period, 159, 166–91
   food chains, 188–90
   *Helicoprion. See Helicoprion*
   shark evolutionary tree, *13*, *192*,
   192–93
Permian-Triassic extinction event,
   192–93, 197–203, 390
Perm Mountains, 166, 167, 171–73
Peru, 303–4, 308–10
   Atacama Desert, 348–49
*Perucetus*, 303–4
petalodonts, 131–32, 187
Philadelphia Academy of Sciences,
   94
*Phoebodus*, 66, *75*, 75–76, 78–82, 142,
   200
*Phoebodus australiensis*, 80
*Phoebodus saidselachus*, 75
*Phorcynis*, 223
Phosphoria Formation, 174–78
photic zone, 81–82
phylogenetic position, 76
Piccard, Jacques, 328
piked dogfish sharks, 263, 273, *273*
Pimiento, Catalina, 313–15, *314*, 319
*Piscobalaena*, *267*, 308–9, 316, 349
Pisco Formation, 348–49
Placoderm Kama Sutra, 110
placoderms, 32, 42, 87–112

Sharktooth Hill, 325

shark tourism, 334–35, 384–86, 387–88

Shark Week (Discovery Channel), 328,
    385

sharpnose sevengill sharks, 248, 287

*Shenacanthus, 1,* 31, 41–41, *42, 43, 43,*
    104

Shimada, Kenshu, 259–61, 313, 318

*Shonisaurus,* 209–10, 215

short-tailed rays, 233–34

shoulder girdles, 71–72, 139

Shubin, Neil, 94

Siberian Traps, 198

Silurian period, *13,* 31, 40–43, 98, 101,
    104
    shark evolutionary tree, *43,* 116,
        190–93, 192, *192*
    *Shenacanthus, 1,* 31, 41, *42, 43,*
        104

Simon Fraser University, 368

sinacanthids, 41–42

Siversson, Mikael, 245, *251,* 251–55, *253*

sixgill sharks, 47, 48, 96, 221, 224,
    229, *229,* 263, 305, *305*

skates, 229, 248, 264

Skelton Névé, 63

slave trade, 391

"slow cooking period," 31

small-spotted catsharks, 48

smell (olfaction), 9, 30, 55, 56–57, 88

Smith, Andrew, 338–39

Smith, James, 285–86

Smithsonian National Museum of
    Natural History, 312, *313,* 315

smooth hammerheads, 288

smooth rays, 232, *233*

Soler-Gijón, Rodrigo, 162–63, *163*

*Solinalepis,* 40

Solnhofen site, 216–24, 235

Sorbini, Lorenzo, 279

South Africa, 59, 64, 338–39, 341, 346,
    353, 362

South Carolina, 325–26

Southern Shark Ecology Group, 387

South Pole, 39, 243

Sox2 gene, 48

Spain
    Las Hoyas sharks, 248–50, *249*
    Nogueras Formation, 44–46
    Puertollano, 162–64

*Spathiobatis,* 218, 235, *236*

*Sphenodus,* 217, 221

*Sphyrna,* 154, 288

Spielberg, Steven, 355

spiracles, 61, 238, 353

spookfishes, 134

*Squalicorax,* 252, 272

*Squalicorax pristodontus,* 252

squaliforms, 273

squalodonts, 303

*Squalus, 273*

*Squalus acanthias,* 273

*Squalus carcharias,* 338

*Squatina,* 220, 247–48

*Squatinactis,* 122, 129–30

squatinids, 247

Stahl, Barbara, 130

Stairway Sandstone, 21

Staite, Brian, 5, 64

*Stanwoodia,* 142

*Stegostoma,* 277

stem holocephalans, 125, 127–28, 130,
    131–32

stem sharks. *See* acanthodians

Steno, Nicolas, 322, *323*

Sternberg, Charles H., 256

Sternes, Phillip, 312

stethacanthids, 141, 142, 147

*Stethacanthus,* 115, 122–23, 140, 141

Stevens, John, 342

stingrays. *See* rays

stingrays of Bolca, 283–84

stout fins, *26, 27*

*Striatolamia,* 302

stromatoporoids, 69, 114

Stumpf, Sebastian, 222

suction feeding, 75–76, 77, 130–31,
    139–40, 146–47, 192, 220

supervolcanoes, 8
Swansea University, 311–12
Swedish Natural Sciences Research
    Council, 253
symmoriiforms, 142, 200–201
synchrotrons, 18, 31, 60, 134, 174
synechodontiforms, 212
*Synechodus*, 212

Tallin University of Technology, 109
*Tanaodus*, 187
Tang, Youhong, 372–73
*Tantalepis*, 25, *34*, 34–36, 42–43
Tantalus, 34
Tapanila, Leif, 175–77, 178–79
*Tassiliodus*, 58
taste buds, 47–48
*Tatanectes*, 227–28
Taylor, Ron and Valerie, 333–34
teeth (tooth whorls)
    Carboniferous period, 134–35, 141
    *Doliodus*, 54–55
    *Edestus*, 179–82, *180*
    *Helicoprion*, *171*, 171–79, *175*, *177*, 180,
        181, 183, 184
    hybodonts, 203–7, *204*
    *Karpinskiprion*, 182–83, *183*
    mammals, 203–4
    megalodons, 7, *7*, 311–12, 315–17, *316*,
        319, 320–21, *323*
    neoselachians, 211–13, *213*
    origins of, 46–50
    *Otodus*, *306*, 306–7
    placoderms, 97–98
    replacing, 47, 50–51
    shapes and structures, 48–49,
        221, *229*, 258–59, 272–73, 276,
        349–50
    white sharks, 334, 336, 339,
        349–52
teleosteans, 231, 270–71
*Temnodontosaurus*, 226
temnospondyls, 214

Tethys Ocean, 239, 243, 283
tetrapods, 61–62
Texas
    Finis Shale, 143
    Red Beds, 159, 161
thelodonts, *35*, 35–36
theropods, 14, 234
thornback guitarfishes, 248
thresher sharks, 250, 257, 278, 297–99,
    *299*, 317
*Thrinacodus*, 81–82
*Thrinacoselache*, 122, 129, *130*
*Thrissops*, 217, 219
thylacocephalans, 76–77
tiger sharks, 7, 8–9, 233–34, 244, 252,
    271, 284–87, *285*
*Tiktaalik*, 94
*Titanic*, RMS, 391
*Titanonarcine*, 283
Toarcian oceanic anoxic event, 230–31
"tongue stones," 320–21, 322
*Torpedo*, 274, 275
tourism, 334–35, 384–86, 387–88
Transantarctic Mountains, 4
"transitional" fossils, 10
*trappa*, 198
Triassic-Jurassic extinction event,
    202–3, 213–15
Triassic period, *13*, 197–98, 199–215
    challenges facing sharks, 208–11
    early. *See* Early Triassic period
    end-Triassic extinction event, 202–3,
        213–15
    hybodonts, 201–6
    neoselachians, 204–6, 211–13
    Permian-Triassic extinction event,
        192–93, 197–203, 390
    world map, *199*
*Trieste* (bathyscaphe), 328
trilobites, 25, 40, 58
Trinajstic, Kate, 106, 108–9
*Triodus*, 159, *159*, 163–64, *164*
*Triodus sessilis*, 188–89
*Tristychius*, 138–40, 147, 156, 192, 205

# ABOUT THE AUTHOR

JOHN LONG is the Strategic Professor of Paleontology at Flinders University in Adelaide, one of Australia's largest paleontological research groups. The former vice president of research and collections at the Museum of Natural History of Los Angeles County in California, he has published more than 200 peer-reviewed papers, some 25 books, and 150 popular science articles. He writes regularly for *The Conversation,* including a viral piece on megalodon. A passionate communicator of science, he has given talks at venues ranging from the American Museum of Natural History and the Philadelphia Academy of Sciences to the University of California, Berkeley, and the Hammer Museum. His work has appeared in *Nature, Science,* and *Scientific American.*

## About the Type

This book was set in Galliard, a typeface designed in 1978 by Matthew Carter (b. 1937) for the Mergenthaler Linotype Company. Galliard is based on the sixteenth-century typefaces of Robert Granjon (1513–89).